evolution

The Greatest Deception In Modern History

Scientific Evidence for Divine Creation

Creation vs Evolution

Roger G. Gallop, Ph.D.

Second Edition (2014)

Revised (2016)

Roger G. Gallop, Ph.D.
evolution - The Greatest Deception In Modern History

Library of Congress Control Number: 2014947859

ISBN-13: 978-0-9829975-7-4 (paperback, 2nd ed.)
ISBN-13: 978-0-9829975-8-1 (case laminate, 2nd ed.)
ISBN-13: 978-0-9829975-5-0 (eBook, 2nd ed.)

Published by:

Red Butte Press, Inc.
P. O. Box 711
Ponte Vedra Beach, Florida 32004-0711

Graphics: Roger Gallop
Chris Gallop (www.chrisgallop.com)
Patrick Gerrity
Anton Zakharov

Book Design/Layout: Roger Gallop

Website Design: Rachel Blaisdell, Interactive Media Design (www.raylay.com)

This book was written for educational and ministerial purposes.

Visit www.CreationScienceToday.com

reviews

"...the book is so good that it should be required reading for all Christians preparing to attend college as well as anyone that has watched The Discovery Channel, Disney Channel, National Geographic, and on and on."

Paul D. Laymon, Professional Geologist, President of Dominion, Inc.
A Professional Environmental Geosciences Company

"...this book provides an excellent description...of the science of creation as we know it today...it is also an excellent read...[and] should be on the bookshelf of everyone interested in creation science and the Biblical truth of origins."

Richard Overman, M.S., President, Creation Education Resources, Inc.

"Of all the creation/evolution books I've read, this is by far, the best! It is masterfully organized, well-written, easy to read and well-documented....clearly presenting the many complex scientific facts crucial to understanding evolution and creation. It is fair to say that you have succeeded brilliantly in producing a comprehensive, scholarly review of the key issues. I highly recommend this book for anyone who wants to better understand the evidence in support of Creation and the shocking lack of evidence underlying the fantasy of molecules-to-man evolution."

Don Scanlan, D. Scanlan Associates, LLC, Spine, Orthopedic, Orthobiologics

"...a well illustrated tour de force which strongly refutes the notion that 'science supports' the evolution/long age view."

— as described by Creation Ministries International, Creation.com

dedications

To the memory of my beloved Mom, Iva May Gallop
(1920 – 2009)
and
To the memory of my beloved Dad, Parron G. Gallop
(1903 – 1982)

To my best friend and soulmate
My beautiful wife, Katharine
Kind and loving, and forever patient and giving
1 Corinthians 13: 4-8

contents

preface

This book is a compilation and review of commonly known scientific evidence supporting Divine creation—evidence in the fields of geology, biology, genetics, medicine, paleontology, chemistry, biochemistry, physics, astrophysics, and astronomy, and specific research by esteemed scientists, and from my own personal experience as a consulting geologist and marine scientist.

I was compelled to write this book, having grown weary over the many years of continual evolutionary doctrine promoted "as truth" by public television, national networks and other media, and taught in our public schools. This is a comprehensive, well-illustrated book that I wanted to read in high school and college but could never find. It was written for the high school and college student, parents and the general public, and the professional scientist.

Although the "theory of evolution" is filled with many gaps and failures, it continues to be taught by teachers as "proven fact"—and then later in life, students perpetuate this doctrine without question as teachers, journalists, and parents. What many people today never hear is the fact that evolutionary theory is not based on known scientific laws or evidence but rather, on false assumptions and poor science.

Creation, as described in the Book of Genesis, is perfectly consistent with all known scientific laws and evidence—and the scientific evidence is *overwhelming*. I challenge any skeptical person to give this book an impartial reading before dismissing the creationist viewpoint.

in reading this book

Technical terms are defined in Appendix A (Glossary) and some terms or explanations are found in "text boxes." Appendices (B through I) provide a slightly more in-depth review of various topics. Some words are italicized or bold for emphasis. All Scripture is italicized.

prologue

> *"For the time will come when men will not put up with sound doctrine. Instead, to suit their own desires, they will gather around them a great number of teachers to say what their itching ears want to hear. They will turn their ears away from the truth and turn aside to myths."* —2 Timothy 4:3–4, NIV

nature of mankind

In the beginning God created man (male and female) in His image and gave man dominion over the entire world (Genesis 1:26-28, 5:1-2). God created a perfect world (Genesis 1:31)—no death, struggle, violence, cruelty, or bloodshed—and everything was *"upheld"* by God (Colossians 1:17 and Hebrews 1:3). Because of sin (rebellion) by the first man, Adam, against his Creator (Genesis 3:6, Romans 5:12), creation was *cursed* by God (Genesis 3:14–19)—in effect leading to a contaminated and corrupt world. Creatures under man's rule, though morally innocent, shared in God's judgment. While this may seem implausible to many, in fact, creation, rebellion, and the curse are in perfect alignment with scientific laws and the moral nature of man.

At the time of creation God allowed man to be a free moral being. There are multitudes of verses in the Bible supporting free will—among them, Genesis 2:16-17, 3:5; Joshua 24:15; Psalm 119:30; Isaiah 7:15, 66:3; Ezekiel 33:11; Matthew 16:24; Romans 10:13; 1 Corinthians 7:37; 2 Peter 3:9; and Revelation 22:17. For example, Joshua 24:15 (NIV) proves that man is a free moral agent: *"...choose for yourselves this day whom you will serve."* Free will simply means "the ability of an individual to make his or her personal choice for or against God, and it is an indisputable fact of Scripture."[1]

Why did God give humanity the choice between good and evil as described in Genesis 3:5? The answer can be simply stated: without choice there is no love or fellowship between humans and their Creator—and humans chose to walk out of God's will. Evil—with its ensuing pain and suffering—and freedom of choice remain two of the great mysteries of the Bible.[2]

Since the beginning of human history, humans have always had an inherited or inborn tendency to sin and rebel—that is, oppose the Creator's sovereignty over their lives. In the days of Noah (4004 BC–2385 BC), mankind was rebellious and exceedingly wicked—

> Regarding the issue of "free will," Adam and Eve were placed in a beautiful land and were forbidden by God to eat of *"the tree of the knowledge of good and evil"* or they would die, but they could eat of the *"tree of life."* (Genesis 2:9, 16-17, NIV) This was a test of their "free will." Many Biblical scholars believe that if they had eaten of the "tree of life," the test would have ended and sin would never have entered the world. Instead, they disobeyed God—consequently, sin infiltrated humanity.

> John 3:18 (NIV) states, *"Whoever believes in him is not condemned, but whoever does not believe stands condemned already because he has not believed in the name of God's one and only Son."* In this verse, God gives every human being the right to believe or not believe that Jesus Christ is the Son of God, and to choose good or evil (that is, living a righteous life or sinful life). "Believes" implies choice.

and with this rebellion and wickedness came eventual consequences—the Flood that destroyed humanity.

In Genesis 6:5 (NIV), *"The Lord saw how great man's wickedness on the earth had become, and that every inclination of the thoughts of his heart was only evil all the time. The Lord was grieved that He had made man on the earth, and His heart was filled with pain. So the Lord said, I will wipe mankind, whom I have created, from the face of the earth—men and animals, and creatures that move along the ground, and birds of the air—for I am grieved that I have made them."*

In Matthew 24:37–39 (NIV), Jesus Christ says, *"As it was in the days of Noah, so it will be at the coming of the Son of Man. For in the days before the Flood, people were eating and drinking, marrying, and giving in marriage, up to the day that Noah entered the ark, and they knew nothing about what would happen until the Flood came and took them all away. That is how it will be at the coming of the Son of Man."*

The Flood served as the basis for Apostle Peter's comparison with the final destruction of the world and the Second Coming of Jesus Christ: *"First of all, you must understand that in the last days* ***scoffers will come,*** *scoffing and following their own evil desires. They will say, 'Where is this "coming" He promised? Ever since our fathers died, everything goes on as it has since the beginning of creation.' But* ***they deliberately forget*** *that long ago by God's words the heavens existed and the earth was formed out of water and by water. By these waters also the world of that time was deluged and destroyed. By the same word the present heavens and earth are reserved for fire, being kept for the day of judgment and destruction of ungodly men."* (2 Peter 3:3–7, NIV) [Bold added]

Rebellious and idolatrous history of humanity is graphically portrayed in the Bible and in historical accounts of the Dark Ages, World War I and World War II, and by today's continual worldwide ethnic conflicts. Consider the world today, with its political corruption, lying, slander, public displays of moral depravity, bizarre behavior, violent crimes against humanity, abortion, theft, adultery, drug-taking, drunkenness, gambling, greed of all kinds, prostitution, sexual perversions, wars and rumors of wars, and terrorism. Immorality is tolerated and accepted as the norm by our youth. Take a moment to read Romans 1, verses 18–32 (see epilogue)—it reads like a commentary of today's world.

When the 12 tribes of Israel and Judah ignored the law of Moses (the Ten Commandments as described in Exodus 20:1-17) and forgot their Creator in the times of the Judges (1375–1050 BC), and no one was leading them in obedience and faithfulness to God,*" ...everyone did what was right in his own eyes"* (Judges 21:25, NAS) and chaos reigned. As described in the Old and New Testaments, people often turned their backs on God and worshipped idols and false gods—but eventually they faced harsh punishment for their rebellion with the fall of Jerusalem (Israel and Judah) to Babylon in 586 BC and to Rome in AD 70. People today and throughout history are no different from people in the days of the Old and New Testaments.

Countries such as Great Britain and the United States, where people once honored God, experienced unprecedented security and prosperity for their faithfulness. These same countries today are morally bankrupt and collapsing economically as people turn away from God. *"Righteousness exalts up a nation, but sin is a disgrace to any people."* (Proverbs 14:34, NIV) When nations turn their backs on God and live as if He does not exist, corruption and wickedness abound. Economic woes follow as taxes increase and governments print and borrow money to pay for larger police forces, prisons, and social services to repair the myriad of social maladies. Does this sound familiar?

evolutionary thinking

In today's world, one of the primary reasons for abandoning faith in God is the widespread indoctrination and acceptance of evolutionary thinking—that everything made itself by natural processes and God is unnecessary. Some secular scientists admit there is "design" in the world and universe and exclaim that the designed thing somehow designed itself!

This rationale leads naturally to atheism, materialism, and secular humanism (man can chart his own course without God; essentially, that man is God), which is the primary reason for our social ills today. Such thinking abounds in our universities and governments and is the foundation of the United Nations and European Union.

Evolution is an atheistic doctrine that tries to explain the presence of life on earth without a Divine Creator. Secular scientists assert the earth developed through purely natural processes following a cosmic explosion called the Big Bang—a primeval explosion from nothingness (see chapter 9). They believe that living microbes simply developed spontaneously from nonliving chemicals (dust or rock) following a series of unknown random molecular reactions.

In other words, secular scientists believe that all organisms on earth are related to the same common primordial ancestor, and that when we die, we simply cease to exist. Such thought was first introduced to the world by Charles Darwin with his book, *The Origin of Species,* published in 1859.

This rationale is in stark contrast to the Bible, which teaches that God created the universe and many "kinds" of plants and animals on earth during a time period of six solar days, and God created humans distinct from animals during the same period of time. Also, the Bible tells us that God is holy, righteous, merciful, and offers humans the hope of eternal salvation.

Worshiping "Mother Earth"
Sketch by Anton Zakharov

scientific creation

What many people today never hear and realize is the fact that *scientific creation is based on scientific evidence*—and such evidence is ***overwhelming***. So why do secular scientists continue to adhere to a false evolutionary doctrine? This book provides the reasons and summarizes much of the evidence for scientific creation, including research by esteemed scientists in almost all fields of science.

Consider the following evidence for Creation

- Evolution is contrary to the First and Second Laws of Thermodynamics, the Law of Biogenesis, and the Law of Causality whereas creation is consistent with such laws.[3] These laws have always proved valid wherever they could be tested.
- Evolution has no known biological processes or mechanisms to form higher levels of organization and complexity—gene mutations are overwhelmingly degenerative and none are "uphill" (that is, unequivocally beneficial) in the sense of adding new genetic information to the gene pool.
- The probability of getting an average-size protein of left-handed amino acids (found only in living cells) by random, natural processes is **zero**. And the probability of getting a living cell, synonymous to the most sophisticated supercomputer yet microscopic in size, is likewise **zero**.
- Geologic landforms and sedimentary features throughout the world are completely consistent with a worldwide flood as described in the Book of Genesis. For many esteemed geologists who have researched geologic landforms and catastrophic processes, evidence of a global flood is **indisputable**.
- Enormous limestone formations, huge coal and oil formations, and immense underground salt layers are indicative of a worldwide flood—*not slow and gradual processes over billions of years.* Such features are satisfactorily explained by a global flood and known geophysical and geochemical processes.
- A worldwide flood as described in Genesis 6-8 is within the boundaries of known geophysics—see phase diagram in chapter 4 and **Pangaea Flood Video** at **www.CreationScienceToday.com.**
- There is no credible technique for establishing the age of sedimentary rock—fossil dating used to establish the age of sedimentary rock suffers from circular reasoning and guesswork, all based on the assumption of evolution.
- The standard geologic and fossil column, as depicted in most science textbooks with transitional creatures evolving toward more complex forms, is utterly *fictitious* and *misleading* and does not represent the real world. In reality, it perfectly represents the aftermath of a worldwide flood.
- There are ***no*** transitional fossils or living forms—there is not one single example of evolution! Evolutionists look for "the" missing link—ironically, they are in desperate search for just one! But there should be billions of examples of transitional forms with transitional structures if evolution were true, but there are none.

- Contrary to popular belief, evidence indicates that early man was intelligent and highly skilled with an advanced social structure. There is also evidence suggesting their belief in the existence of an afterlife.
- Soft tissue with traces of red blood cells has been found in dinosaur fossils supposedly 70 to 250 million years old. (Soft tissue and red blood cells have relatively short life spans.)
- Carbon-14 has been found in coal and diamonds and in deep geological strata, all supposedly hundreds of millions of years old. Researchers have been unable to find carbon (stable forms: carbon-12 and -13) without carbon-14 (unstable form). (C-14 has a relatively short life span.)
- Radioisotope dating suffers from broken, unprovable assumptions. The technique is "fatally flawed"—yet scientists contend as fact what they cannot prove.
- Abundant daughter isotopes are indicative of accelerated nuclear decay associated with creation (expansion, stretching out, or acceleration of the universe from an extremely hot, dense phase when matter and energy were concentrated) and a worldwide flood with massive restructuring of the earth's lithosphere—*not slow and gradual processes over billions of years.*
- Powerful evidences of accelerated nuclear decay in igneous rocks found worldwide are helium in zircon crystals, radiohalos and fission tracks, and rapid magnetic field reversals and decay.
- Gravitational time dilation offers a credible explanation for a young earth—that is, it explains how light from the extremities of the universe has the potential of reaching the earth in a relatively short period of time (from earth's perspective).
- Over a hundred geochronometers (techniques to date the earth and universe) indicate a young earth and universe, for example, helium in zircon crystals, rapid magnetic field reversals and decay, carbon-14 found in coal and diamonds, and lack of continental erosion, ocean sediments, and salt in the sea.

Each of these evidences is enough to convince most rational people that evolution is a false doctrine and the earth is, in fact, young!

Do you want to know more? Read this book and review the website CreationScienceToday.com.

Also, visit Creation Ministries International (CMI) at Creation.com and be sure to browse their bookstore, media center, and magazine section. Readers may also find the following websites very informative: ICR.org, AnswersinGenesis.org, and CreationResearch.org.

notes: prologue

1. Hagee, J. (2006). *Jerusalem Countdown: A Warning to the World*. Lake Mary, FL: FrontLine, 145-146.

John 6:44 (NIV) states, *"No one can come to me unless the Father who sent me draws him, and I will raise him up at the last day."* Although this scripture implies election or predestination, an individual must first make a personal choice (according to scripture cited in the second paragraph of the prologue), either to accept or deny the Creator.

God gives every human being the freedom to believe or not believe that Jesus Christ is the Son of God and to choose good or evil (that is, choosing to live a righteous life or sinful life). In Matthew 4:19 (NAS), Jesus said to his disciples, *"Follow Me, and I will make you fishers of men"* and in 2 Peter 3:9 (NAS), *"The Lord is not slow about His promise....not wishing any to perish but for all to come to repentance"*—implying that all men have freedom of choice.

Once a person *"calls on the name of the Lord"* (Romans 10:13, NIV), the Holy Spirit *"draws him...."* (John 6:44, NIV) into a spiritual union that cannot be broken or removed (John 10:27-28). Believing or calling on the name of the Lord is a continual and sincere desire of the heart, and people do not come to Christ strictly on their own initiative, but the Father draws them. God is omnipotent and omnipresent—he knew you in ancient times (John 1:1, Romans 8:29, Ephesians 1:4-5, 1 Corinthians 2:7, Revelation 1:4) and foreknew the choice you will make, although you still have freedom of choice today.

Regarding election or predestination, many were divinely selected in advance (for example, Abraham, Isaac, Jacob, Moses, Joshua, David, the disciples, John the Baptist, Paul, Timothy, and all the Old Testament Prophets to just name a few) for specific divine purposes—God knew them in advance and knew they would believe and dedicate their lives to God of the Holy Bible.

2. The problem of evil and suffering has been addressed by theologians and authors throughout the centuries; more prominently by C. S. Lewis in *The Problem of Pain;* by best-selling author Randy Alcorn in his book, *If God Is Good – Faith in the Midst of Suffering and Evil* (2009, Colorado Springs, CO: Multnomah Books.); by Dr. Carl Wieland in his book, *Beyond the Shadows - Making Sense of Personal Tragedy* (2011, Atlanta, GA: Creation Book Publishers); and by Dr. N. L. Geisler in his book, *If God, Why Evil?* (2011, Minneapolis, MN: Bethany House Publishers).

Also see articles by Mitchell, T. (December 2011). Death and Steve Jobs. *Answers Update*, 18 (12), Hebron, KY: Answers in Genesis, 1-2 (answersingenesis.org), and Johnson, J. (November 2011). Human suffering: Why this isn't the "best of all possible worlds." *Acts & Facts*, 40 (11), Dallas, TX: Institute for Creation Research, 8-10 (ICR.org). Further discussion is found in note 1 of the Epilogue.

In the beginning, God created man in God's image (Genesis 1:27), and He created a world without death and suffering (Genesis 1:31). God also allowed man to be a free moral being—that is, He gave man the ability "to choose what is true, what is right, what is good." (p. 10, Johnson) The first man, Adam, decided to walk out of God's will (Genesis 3)—a moral decision that brought death and corruption to a perfect creation (Genesis 3:14-19; Romans 5:12, 6:23). Wickedness and depravity with its pain and suffering are "man's fault, not God's." People ask why would a loving God create a world full of pain and suffering—but the real question is, "why would a loving God come into the world He created to suffer and die to pay for my sin?" (p. 2, Mitchell)

The Bible tells us about the entry of evil into the universe—how Lucifer and 1/3 of the angels chose to rebel against God the creator by attempting to usurp God's power or authority (Ezekiel 28:11, 14–15; Revelation 12:3-9, 9:1; and Luke 8:30). (p. 48–49, Alcorn) Rebellion by these angelic beings led ultimately to the temptation and fall of mankind. Allowing evil, hence pain and suffering, "was a necessary price to achieve a far greater eternal result." (p. 41–42, Alcorn)

> *"...our present sufferings are not worth comparing with the glory that will be revealed in us."* (Romans 8:18, NIV), and *"For our light and momentary troubles are achieving for us an eternal glory that far outweighs them all."* (2 Corinthians 4:17, NIV)

As Alcorn explains, "From the beginning, God planned that his Son [Jesus Christ] should deal the death blow to Satan, evil, and suffering, to reverse the Curse (see chapters 2 and 3), redeem a fallen humanity, and repair a broken world." (p. 51, Alcorn) *"The reason the Son of God appeared was to destroy the devil's work."* (1 John 3:8, NIV)

"Evil's ultimate origin remains a mystery...[and] God has chosen to remain silent on this question." (p. 50, Alcorn) *"The secret things belong to the Lord our God."* (Deuteronomy 29:29, NIV) The question of evil and suffering will "ultimately make good sense, in the fullness of time." (p. 10, Johnson).

Sin is breaking the moral law (see chapter 3, Human Nature and the Moral Law), or any one of the Ten Commandments, or a departure from goodness. Evil is sin, iniquity, wickedness, immorality, or corruption. "Evil, in its essence, refuses to accept God as God and puts someone or something else [idolatry] in His place... For this reason, the Bible treats idolatry as the ultimate sin, since it worships as God what is not God." (p. 24–25, Alcorn)

3. There is another law just as valid as the laws of chemistry, physics, and biology—the law of Human Nature, or Moral Law, or the law of "right and wrong behavior." Unlike other natural laws, it is a law that humans are free to disobey. See chapter 3, Human Nature and the Moral Law.

Chapter 1

Creation versus Evolution

"In the beginning was the Word, and the Word was with God, and the Word was God. He was in the beginning with God. All things came into being through Him; and apart from Him nothing came into being..."—John 1:1–3, NAS

"For in Him all things were created, both in the heavens and on earth, visible and invisible, whether thrones or dominions or rulers or authorities—all things have been created through Him and for Him."—Colossians 1:16, NAS

Scientific Creationism and Evolutionary Theory

Evolutionists believe the Big Bang created the universe out of nothing 10 to 20 billion years ago; our solar system formed about 5 billion years ago; single-celled organisms formed from nonliving matter 3 to 5 billion years ago; multicellular organisms slowly evolved about 1 billion years ago; humans evolved from higher life forms 185,000 to 2 million years ago; and modern civilization emerged within the last 5,000 to 10,000 years. This doctrine holds that man descended from the apes, all vertebrates descended from fish, all fish descended from invertebrates, and all life descended from single-celled organisms which arose spontaneously from nonliving chemicals (dust or rock).

Theory of evolution is the belief that all living organisms made themselves by their own natural processes, with no supernatural input. It implies increasing organization and complexity in the universe from a single-celled organism which arose spontaneously from nonliving chemicals (dust or rock). It is a doctrine of continual design without a designer. Chaos has become cosmos, all by itself.

Creationists believe the straightforward interpretation of Scripture—the earth and all living things were supernaturally created in six solar days by the God of the Bible about 6,000 years ago as described in Genesis 1—and today's physical and biological world provides overwhelming scientific evidence of Divine creation. According to the Holy Bible, Jesus Christ is the Creator of the heavens and earth and all that exists (John 1:1–3, Colossians 1:16, Hebrews 1:2).

A term often used to describe change within animal (and plant) populations is microevolution—a "natural selection" process. Although microevolution is a mechanism for change, it is not evolution—that is, there is No creation of new genetic information to form a new type of

Macroevolution is another name for evolution. It is theoretical changes in an individual because of new genetic information introduced into the gene pool which, in turn, produces a new kind (or category) of organism. Such changes have ***never been observed*** within living populations.

Scientific creationism is the belief that basic "kinds" or groups of animals (or plants) appeared abruptly without arising from a different kind of animal (or plant). See chapter 3.

animal (or plant). Rather, it is a term commonly used to describe reshuffling of genes within an "existing gene pool" of the original kind of animal (or plant) in response to a specific environment. This is sometimes referred to as "survival of the fittest." Most people are unaware that natural selection (or survival of the fittest, adaptation, or speciation) is a thinning out process that leads to loss of genetic information. This is explained in chapter 3.

Some progressive theologians and secular geologists, who are influenced by the theory of evolution and uniformity theory, have coined certain terms in an attempt to reconcile the Holy Bible with old earth, evolutionary doctrine. Such terms include progressive creation (also called the Day-Age Theory) and theistic evolution. But such ideas have the insurmountable problem of allowing death before sin which is contrary to Romans 6:23, 5:12, 8:20; and Genesis 3.

Progressive creation maintains that the days of Genesis equal geologic ages (millions of years), and each basic category of life was supernaturally created at various times throughout earth's history. **Theistic evolution** maintains that God created life and started the evolutionary process.

Such ideas dishonor the Christian church, especially in light of overwhelming scientific evidence in favor of creation. Further, the Bible states that, *"For since the creation of the world God's invisible qualities—his eternal power and divine nature—have been clearly seen, being understood from what has been made, so that* ***men are without excuse."*** (Romans 1:20, NAS) [Bold added]

Evolutionary model includes:

- naturalistic origins of all things by chance; random favorable mutations,
- transitional forms of animals and plants,
- genetic gain of information (net increase in complexity) over time,
- tendency for things to move from simple to more complex forms, and
- earth was dominated by uniform events with local catastrophic events (separated by many millions of years).

When people put their faith in evolution, they are putting their faith in the ideas and unproven assumptions of morally corrupt mankind, and ultimately they must conclude that evolutionary doctrine is the product of 'blind chance,' that 'life is meaningless,' and there is 'no life after death.'

Creation model includes:

- supernatural origin,
- separate, distinct kinds of animals & plants,
- genetic loss of information (net decrease in complexity) over time,
- tendency to decay or degenerate, and
- earth was impacted by a catastrophic worldwide flood event. See chapter 4.

When people put their faith in our Lord Jesus Christ and the Holy Bible, they are putting their faith in God the Creator of the heavens and earth and all that exists (John 1:1-3, Colossians 1:16, Hebrews 1:2)—a God who is holy, righteous, and merciful, and offers the gift of eternal salvation (John 3:16, 5:24, 14:6; Romans 10:9, 13; and Ephesians 2:8-9) to those willing to accept.

Empirical Science and the Past

Today's science relies on empirical analysis—that is, verification through repeated measurement and testing. It is the basis for what is known as the "scientific method," the common steps that biologists and other scientists use to gather information to solve problems. These steps include observation, hypothesis (prediction), data collection, experimentation to test the hypothesis under controlled conditions, and conclusions. So what does empirical science have to do with the past and origin of life?

> **Science** is an organized body of knowledge in the form of ***testable*** explanations or predictions about the universe; that is, all matter and energy. The definition of 'science' is associated with the scientific method, a systematic way to study the natural world, including physics, chemistry, geology, biology, and paleontology.

Empirical analysis is a wonderful testing tool but its application is *limited to the present*—the way things are and the way they work in the present. Empirical science can build jet airplanes, automobiles, skyscrapers, bridges, supercomputers, and find antibiotics and cures for diseases, but empirical science *cannot* deal directly with the past, as most people believe.

> **Empirical analysis** is verification through repeated measurement and testing. Testing is carried out under controlled conditions to validate a hypothesis or perhaps to verify a known law.

Science that puts men on the moon is based on scientific principles that can be tested and repeated *in the present*. Theories about the origin of man, the earth, and the universe **cannot** be observed in the past or tested, and therefore **cannot** rely directly on "empirical analysis." Nevertheless, observation and empirical analysis are often used to indirectly evaluate the past, e.g., testing to determine the presence of blood cells in fossil dinosaur bones or carbon-14 in diamonds, coal, or ancient fossils. Although results are extrapolated to the past, such testing applications are *limited to the present.* The primary difference between creation and evolution is not about the accuracy of data but rather, the "interpretation" of data.

> The primary difference between scientific creation and evolution is not about the data but rather, the "interpretation" of data—for example, interpretation of observed geological landforms, the fossil record, and radioisotopes.

When geologists or anthropologists have only the end results of an event, a full reconstruction of the one-time ancient event is routinely evaluated based on evolutionary assumptions about the past. For example, the reconstruction of events causing immense rock formations or fossil deposits—what happened in the unobserved past—is simply conjecture or guesswork based on 'slow and gradual' old age assumptions of evolutionary doctrine.

It is important to realize that secular scientists *assume evolution and old age as their foundation or basis for reconstruction or interpretation—evolution and an old earth are assumed to be true.* An individual fact is accepted or rejected as valid only if it fits the old earth, evolutionary model. This is a significant concept to understand. The presumption of evolution "as fact" exists in many sciences including biology, geology, astronomy, paleontology, and anthropology.

The question we are examining in this book is, Is evolution true or is it a great deception? If evolution is false, the foundation of many science disciplines that examine the past will crumble. As you continue reading and, if you set aside preconceived notions, you will soon realize the *preponderance of scientific evidence refutes evolution and overwhelmingly supports the creation model—not evolution.*

Assumption of Evolution and Old Earth

The whole process of dating the earth (specifically rocks) begins with the *presumption* of evolution and an old earth. Data is analyzed and interpretations consistent with evolution are retained—and all contrary evidence and data are rejected or ignored. Contrary evidence to evolution is viewed as an anomaly or simply wrong. An individual fact is accepted or rejected as valid only if it fits the evolutionary model.

The Institute for Creation Research summarized the problem succinctly in its *Acts & Facts* news magazine:[1]

> If evolution was merely a scientific theory that was open to evaluation based on the evidence, then its evidentiary failings would be freely acknowledged and additional theories could be considered as they are warranted. But far from being a free marketplace of ideas where scientists consider themselves at liberty to pursue the evidence where it leads, the modern scientific establishment has bound itself to a **single system of interpretation**, with myriad variations but one bottom line: evolution is a fact, and alternatives must be rejected out of hand. [Bold added]

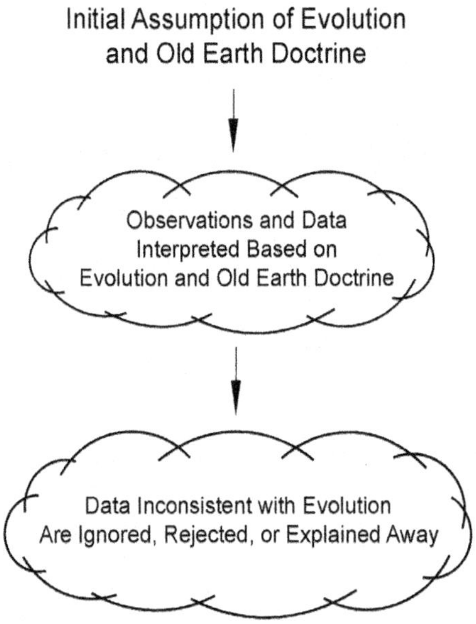

Initial Assumption of Evolution and Old Earth Doctrine

↓

Observations and Data Interpreted Based on Evolution and Old Earth Doctrine

↓

Data Inconsistent with Evolution Are Ignored, Rejected, or Explained Away

Secular scientists maintain that evolution is a 'science' and creation is a 'religion.' But the fact is, evolution is not testable using empirical analysis and, therefore, does not meet the definition of 'science.' Also, evolution does not meet the definition of theory, as in "evolutionary theory." A theory is an explanation of a set of related observations based on hypotheses and verified by independent researchers—but the fact is, evolution (genuine gain in genetic information or net increase in complexity) has never been observed in fossils or living populations. See chapters 3 and 7.

In fact, evolution is not even worthy of the term hypothesis which is an educated guess based upon observation. At best, evolution is an unsubstantiated hypothesis. The bottom line is that evolution has never been observed or proven by empirical science—it is just assumed to be true. "Once again, we emphasize that evolution is not science...it is a philosophical worldview, nothing more."[2]

So what motivates scientists to maintain a tight grip on evolutionary doctrine? The simple answer is that the term "supernatural" or the concept of God or creation is considered to be outside the realm of real science. Theories of how the universe, the earth, and man originated may come and go, but the belief that it happened by chance is an "unshakeable faith" for many today.

Divine Creation is inconceivable to many scientists because the science community is largely atheistic. Life is here on earth, so secular scientists feel they must explain life "naturalistically" —consequently, they believe that evolutionary doctrine and ignoring data contrary to evolution is legitimate. Evolution is a belief system that many, if not most, scientists (that is, biologists, geologists, astronomers, paleontologists, and anthropologists) assume as fact and routinely use to interpret their observations.

Evolution has **never been observed** within fossils or living populations, there are no transitional types, and there are no known biological processes for evolution (mutations are overwhelmingly destructive; see chapter 3).

Other reasons for this belief include: acceptance of evolution without independent, unbiased investigation, pressure by the public school system on the teacher (separation of church and state issues), and never having a chance to consider alternatives. There are also social and cultural pressures, family upbringing, academic peer pressure, and the need to be accepted and recognized by colleagues. To avoid having to recognize God, *"professing to be wise, they became fools."* (Romans 1:22, NAS)

But there is an even more ominous reason for belief in evolutionary doctrine. It is tied to the ancient (spiritual) rebellion of men against their Creator as foretold in Genesis 11:4 and found throughout biblical history. It refers to the fact that humanity, ever since the rebellion of the first man, Adam, has had an inherited tendency to oppose the Creator's rule or sovereignty over their lives (Romans 1:18-32).

While neither the creation model nor the evolutionary model can be proven or disproven, these two views can be compared to see which one fits the data better. When compared, the young earth model (creation and a worldwide flood) fits the data perfectly while the old earth model (evolution and 'slow and gradual' geologic events) has continual flaws—it is essentially *upside down science*.

> Over the last 100 years the scientific community has blindly pursued an evolutionary agenda rather than scientific truth.

Creation scientists maintain the creation model is reasonable and rational and backed up by the weight of overwhelming scientific evidence observable in the present. And Christians who believe in the existence of an all-powerful God and claim to believe in Scripture should never feel obligated to accept the evolutionary, old earth doctrine.

Consider the following quotes:[3]

"Biologists are simply naïve when they talk about experiments designed to test the theory of evolution. It is not testable. They may happen to stumble across facts which would seem to conflict with its predictions. These facts will invariably be ignored and their discoverers will undoubtedly be deprived of continuing research grants." —Professor Whitten (professor of genetics, University of Melbourne, Australia), 1980 Assembly Week address.

Sketch by Anton Zakharov

"The fact of evolution is the backbone of biology, and biology is thus in the peculiar position of being a science founded on an unproved theory—is it then a science or a faith? Belief in the theory of evolution is thus exactly parallel to belief in special creation—both are concepts which believers know to be true but neither, up to the present, has been capable of proof." —L. Harrison Mathews, FRS, *Introduction to Darwin's The Origin of Species*, (London: J.M. Dent & Sons Ltd, 1971), p. 11.

"Question is: Can you tell me anything you know about evolution, any one thing, any one thing that is true? I tried that question on the geology staff at the Field Museum of Natural History and the only answer I got was silence. I tried it on the members of the Evolutionary Morphology Seminar in the University of Chicago, a very prestigious body of evolutionists, and all I got there was silence for a long time and eventually one person said, 'I do know one thing—it ought not to be taught in high school." —Dr. Colin Patterson (senior paleontologist, British Museum of Natural History, London). Keynote address at the American Museum of Natural History, New York City, November 5, 1981.

"Durant concludes that the secular myths of evolution have had 'a damaging effect on scientific research,' leading to 'distortion, to needless controversy, and to the gross misuse of science.'" —Dr. John Durant (University College Swansea, Wales), as quoted in "How evolution became a scientific myth," *New Scientist*, September 11, 1980, p. 765.

"I am convinced, moreover, that Darwinism, in whatever form, is not in fact a scientific theory, but a pseudo-metaphysical hypothesis decked out in scientific garb. In reality the theory derives its support not from empirical data or logical deductions of a scientific kind but from the circumstance that it happens to be the only doctrine of biological origins that can be conceived with the constricted world view to which a majority of scientists no doubt subscribe." —Wolfgang, Smith, "The Universe Is Ultimately to Be Explained in Terms of a Metacosmic Reality" in *Cosmos, Bios, Theos*, Margenau and Varghese (Eds.), p. 113.

"Evolutionism is a ***fairy tale for grown-ups.*** This theory has helped nothing in the progress of science. It is useless." [Italics and bold added] —Professor Louis Bouroune, former president of the Biological Society of Strasbourg and director of the Strasbourg Zoological Museum, later director of research at the French National Centre of Scientific Research, as quoted in *The Advocate*, March 8, 1984. Bouroune's quote taken from: "Evolution: A Fairy Tale for Adults," V. Long, *Homiletic and Pastoral Review*, vol. 78, no. 7, 1978, pp. 27-32.

"Scientists who go about teaching that evolution is a fact of life are ***great con men***, and the story they are telling may be the ***greatest hoax ever.*** In explaining evolution, we do not have one iota of fact." [Italics and bold added] —Dr. T. N. Tahmisian, Atomic Energy Commission, USA, as quoted in: *Evolution and the Emperor's New Clothes* (3D Enterprises Limited, 1983), title page.[4]

Consider these additional quotes:[5]

[The Big Bang] "... is only a myth that attempts to say how the universe came into being ..." —Hannes Alfvén, "The Big Bang Never Happened," *Discover 9,* June 1988, p. 78.

[The Big Bang] "... represents the instantaneous suspension of physical laws, the sudden, abrupt flash of lawlessness that allowed something to come out of nothing. It represents a true miracle—transcending physical principles ..." —Paul Davies, *The Edge of Infinity* (New York: Simon & Schuster, 1981), p. 161.

"In fact, evolution became, in a sense, a scientific religion; almost all scientists have accepted it and many are prepared to *bend* their observations to fit with it. To my mind, the theory does not stand up at all." [Italics added] —H.S. Lipson, "A Physicist Looks at Evolution," *Physics Bulletin*, 1980, vol. 31, p. 138.

***notes:* Chapter 1**

1. Evolution's evangelists (May 2008). *Acts & Facts*, 37 (5), Dallas, TX: Institute for Creation Research, 10. Copyright © 2008 Institute for Creation Research, used by permission.

2. Morris, H.M. The scientific case against evolution. Institute for Creation Research. Retrieved October 2014, from http://www.icr.org/home/resources/resources_tracts_scientificcaseagainstevolution/. (Dr. Henry M. Morris (1918-2006) was founder of the Institute for Creation Research.)

In the article by Dr. Morris, it is interesting to note the candid statements by various university professors concerning the 'belief in modern evolution.'

3. Truth and Science Ministries, In Search of the Truth. Retrieved April 2008, from http://www.truthandscience.net/quotes.htm and http://www.errantskeptics.org/Quotes_Regarding_Creation_Evolution.htm. Also see Morris, H.M. (1997). *That Their Words May Be Used Against Them*. Green Forest, AR: Master Books, 127, 129.

4. T. N. Tahmisian, as quoted in title page: Mitchell, N.J. (July 1983). *Evolution and the Emperor's New Clothes.* United Kingdom: Roydon Publications. Also retrieved April 2008, from http://www.truthandscience.net/quotes.htm and http://www.errantskeptics.org/Quotes_Regarding_Creation_Evolution.htm.

5. Evolutionism: Is evolution a religion? Northwest Creation Network, Mountlake Terrace, WA. Retrieved December 2007, from http://nwcreation.net/evolutionism.html (contact@nwcreation.net). Also see Morris, H.M. (1997). *That Their Words May Be Used Against Them*. Green Forest, AR: Master Books, 72, 112.

Chapter 2

Physical Scientific Evidence for Creation

"Lift up your eyes to the heavens, look at the earth beneath; the heavens will vanish like smoke, the earth will wear out like a garment and its inhabitants die like flies [Second Law]. *But my salvation will last forever, my righteousness will never fail."*
—Isaiah 51:6, NIV

"For this is what the Lord says—he who created the heavens, he is God; he who fashioned and made the earth, he founded it; he did not create it to be empty, but formed it to be inhabited—he says, I am the Lord, and there is no other."
—Isaiah 45:18, NIV

Universal Decay and Conservation of Matter/Energy

In your everyday life, have you noticed that everything tends to fall apart and disintegrate over time? Decaying buildings, bridges, roadways, automobiles, and clothing—everything is subject to deterioration and is in constant need of repair. Material things and all known processes proceed from organization to disorganization—material things are not eternal. Everything deteriorates—eventually, all things wear out and return to dust.[1]

Do you wonder why we get old—why people often refer to life as an "uphill battle" or that "life is hard and then you die?" Age, disease, and death of all living things are tied directly to the Second Law of Thermodynamics (sometimes referred to as the Law of Increasing Entropy), which states that usable energy in the universe available for work is decaying or running down. This, of course, is contrary to the "chaos to cosmos, all by itself" doctrine of evolution.[4]

The First Law of Thermodynamics, on the other hand, simply states that matter/energy cannot be created or destroyed, but can be transferred from one form to another. This law confirms that *creation is no longer occurring*—but it also implies that *creation occurred at sometime in the past!* In today's world, there is no creation of new matter/energy rising to

The **Second Law of Thermodynamics** states that matter/energy in the universe available for work is decaying or running down. **Entropy** is a measure of disorder or unusable energy—it represents energy that is no longer available for doing work. Every energy transformation reduces the amount of usable or free energy of the system and increases the amount of unusable energy. It is essentially a mathematical formula of the useless energy in a system.[2]

Thermodynamics is the study of heat power; or it is the branch of physics that examines kinetic energy, or the efficiency of energy transfer.[3] Simply stated, heat is energy in transit, or energy transfer or transformation from one system to another because of a difference in temperature.

The **First Law of Thermodynamics,** or the Law of Conservation of Matter, states that matter/energy cannot be created or destroyed. Although quantity of matter/energy remains the same, energy can be transferred from one form to another. For example, it can change from solar energy to fossil fuel to mechanical energy to electrical energy to light energy.

higher levels of organization and complexity as evolutionists would have you believe! Again, this is contrary to the so-called theory of evolution.

Although quantity remains unchanged (according to the First Law), the quality of matter/energy has the tendency to decline or deteriorate over time (according to the Second Law). Every energy transformation reduces the amount of usable or free energy of the system and increases the amount of unusable energy. In other words, while usable energy is used for growth and repair, it is "irretrievably lost in the form of unusable energy."[5]

The effects are all about us, touching everything in the world, including ecological degradation. Just look around your physical world. Each year, vast sums of money are spent on maintenance and medical bills to counter the unrelenting effects of decay.

Energy is defined as the ability to do work or to move something. Types of energy include heat or thermal energy, fossil fuel (coal, oil, and natural gas), chemical energy, electrical energy, solar energy, wind energy, hydropower, mechanical energy, ocean energy, and atomic energy.

With every energy transformation, the quantity of energy remains unchanged but the quality of energy diminishes—there is a reduction in the amount of usable energy and increase in the amount of unusable energy. Examples include hot to cold; high pressure to low pressure; solar energy to wind energy to mechanical (turbine) to electrical energy to light energy.

When we eat, our bodies transform energy stored in food (potential energy) into energy to do work (kinetic energy). When we run or walk, or think, read or write, we "burn" food energy in our bodies. Automobiles, airplanes, boats, light bulbs, and machinery also transform energy into work—but, eventually, usable energy is irretrievably lost.

Remains of rural home built in early 20th century in Currituck, North Carolina (old Gallop farmstead) *Photo by Roger Gallop*

Urban decay—everything is subject to decay over time

Worn out, abandoned truck near Dalton, Georgia. *Photo by Caroline Blochlinger*

Laws of Science are facts of nature accepted to be true and absolute (unwavering), and have been subjected to extensive and repeated measurements using empirical analysis. Examples include the law of gravity, laws of motion, and laws of thermodynamics.

Energy Flow and Dispersion

Energy flows from an area of concentration to an area of dispersion—simply, it wants to diffuse or spread out (for example, air compressed in a tire, sound waves, a battery slowly running down, or lightning in the atmosphere). Also, energy tends to flow from an area of high temperature to an area of low temperature.

Examples are all around us. A hot pan cools down when it is taken off the stove, or hot coffee cools down when it is placed on the table, or a car engine cools down when the ignition is turned off. Thermal energy (heat) always flows to cooler surroundings—so, in the summer, warm air outside tries to infiltrate into an air-conditioned house or car, and in the winter, warm air in the house tries to escape to the cold air outside.

All energy will try to find an avenue of dispersion and eventual depletion. The opposite never happens unless there is "preprogrammed machinery" forcing the opposite effect (for instance, refrigeration, or cooling a house during the summer and heating a house during the winter), but inevitably, if left on its own, energy will always disperse or flow from hot to cold. Dispersion—the universal tendency for all matter/energy to run down or move toward a state of unusable energy—is a process described by the Second Law of Thermodynamics.

The subject of open and closed systems should be mentioned at this point. Evolutionists have tried to water down the Second Law by stating that it applies only to closed systems, but this is not the case. The Second Law of Thermodynamics is just as valid for open systems as it is for closed systems. John Ross, Harvard University, states; "... there are no known violations of the Second Law of Thermodynamics. Ordinarily the Second Law is stated for isolated [closed] systems, but the Second Law applies equally well to open systems ... There is somehow associated with the field of far-from-equilibrium phenomena the notion that the Second Law of Thermodynamics fails for such systems. It is important to make sure that this error does not perpetuate itself."[7]

According to G. Mulfinger, chemist and physicist, "It is probably no exaggeration to claim that the laws of thermodynamics represent some of the best science we have today. While the utterances in some fields (such as astronomy) seem to change almost daily, the science of thermodynamics has been noteworthy for its stability. In many decades of careful observations, not a single departure from any of these laws has ever been noted."[8]

No evidence refutes it, say physicists G. N. Hatspoulous and E. P. Gyftopoulos: "There is no recorded experiment in the history of science that contradicts the second law or its corollaries...."[9] Further, the Second Law "yields a general expression for the entropy of a system... applicable to all states...of all systems."

Evolutionist Isaac Asimov confirmed that "we can say: 'Energy can be transferred from one place to another, or transformed from one form to another, but it can be neither created nor destroyed.' Or we can put it another way: 'The total quantity of energy in the universe is constant.'... This law is considered the

> The first and second laws *apply equally as well* to open and closed systems—in all systems, these laws have *never been refuted.*

> **Open and Closed Systems:**[6]
>
> **Closed systems** are able to exchange energy (heat) but not matter with their environment. A greenhouse is an example of a closed system.
>
> **Open systems** are able to exchange energy (heat) and matter with their environment. The earth is considered an open system.

most powerful and most fundamental generalization about the universe that scientists have ever been able to make."

Asimov also said, "Another way of stating the Second Law, then, is, 'The universe is constantly getting more disorderly!' Viewed that way we can see the Second Law all about us. We have to work hard to straighten a room, but left to itself it becomes a mess again very quickly and very easily. Even if we never enter it, it becomes dusty and musty. How difficult to maintain houses, and machinery, and our own bodies in perfect working order; how easy to let them deteriorate. In fact, all we have to do is nothing, and everything deteriorates, collapses, breaks down, wears out, all by itself—and that is what the Second Law is all about."[10]

What's the difference between Laws of Science and Theory? "Scientific laws must be simple, true, universal, and absolute. They represent the cornerstone of scientific discovery, because if a law ever did not apply then all science based upon that law would collapse."[11] The First and Second Laws have always proved valid whenever they could be tested—there are *no exceptions* to these laws. A theory is an explanation of a set of related observations based on hypotheses and verified by independent researchers—but theories are not laws, and they are often disproven and replaced with other theories. A "law" differs from theories, hypotheses, and principles in that a law can be expressed by a single mathematical equation with an empirically determined constant.

Entropy and Heat Death

Evolutionists have attempted to harmonize entropy and evolution, but "this is an impossible task, because really the one is itself the negation of the other. Creation (or what biologists imply by 'evolution') actually has been accomplished by means of creative processes, which are now replaced by the deteriorative processes implicit in the second law."[12] Evidence indicates that our sun is shrinking (0.1% per century) and that every star in the universe consumes trillions of tons of energy every second—thus, usable energy in the universe cannot last forever.

Heat death is a possible final state of the universe, in which it has "run down" to a state of no available energy to sustain motion or life. In physical terms, it is a point of maximum entropy. To put it simply, in the real world the long-term flow is downhill—not uphill as many evolutionists would have you believe.

The Law of Increasing Entropy states that usable energy in the universe available for work is decaying or running down. Although quantity of energy remains unchanged (according to the First Law), the quality of matter/energy has the tendency to decline or deteriorate over time (according to the Second Law). Entropy (disorder or decrease in usable energy) is increasing to a maximum as matter/energy in the universe degrades to an eventual state of inert uniformity, or heat death (nothingness or randomness). Ultimately, when all the energy of the cosmos has been degraded, all molecules will move randomly, and the entire universe will be cold and without order.

Let's take a step back in time. Because the universe is constantly losing usable energy and never gaining, one can reasonably conclude the universe had a beginning—a moment of least entropy—a time of minimal disorder with a minimal amount of unusable energy. This was a time *when the First Law of Thermodynamics did not apply—a time when systems were rising to higher levels of organized complexity* (which is no longer occurring today)—a time of Creation!

Because the universe is winding down, logically it would mean the universe was created with plenty of usable energy at the beginning. The question one might logically ask: "Who wound up the clock?"[13] In the Bible, a moment of least entropy is fully described in Genesis 1. Applied to the whole universe, this is a *fundamental contradiction* to the "chaos to cosmos, all by itself" doctrine of evolution.

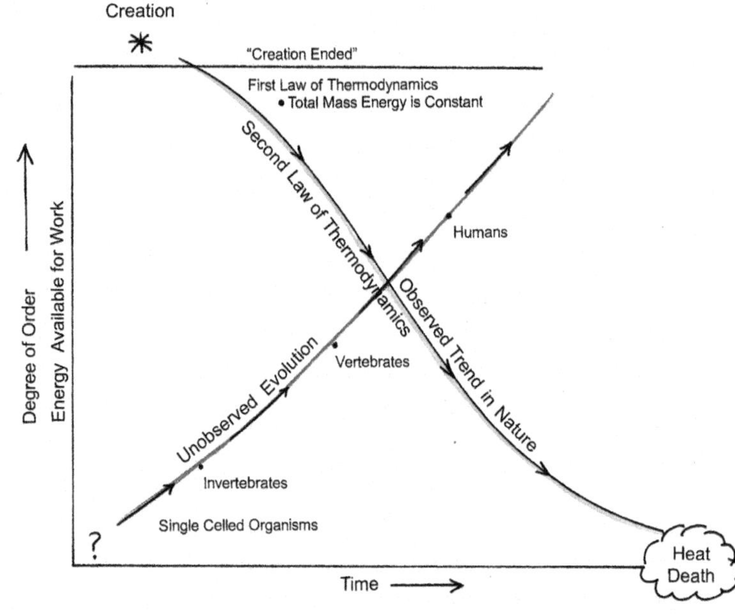

Creation, Entropy and Heat Death
First and Second Laws of Thermodynamics
Support Creation—Not Evolution
Sketch by Roger Gallop

At the start of creation, energy flow (from hot to cold) would be taking place as sun and stars were heated by nuclear fusion reaction, volcanoes spewed lava and dispersed heat into the oceans, and as Adam began to breathe and digest food for the first time. A concept of "zero" entropy at the very moment of creation is debatable—therefore, a moment of "least" entropy is used to describe this instant in time.

De-Evolution

The Second Law is a major problem for evolutionists. Dr. Duane Gish, Ph.D. in biochemistry, comments: "Of all the statements that have been made with respect to theories on the origin of life, the statement that the Second Law of Thermodynamics poses no problem for an evolutionary origin of life is the most absurd. ... the observations on which the Second Law is based do absolutely exclude the possibility of an unaided, spontaneous, naturally occurring, evolutionary origin of life. ... The operation of natural processes on which the Second Law of Thermodynamics is based is alone sufficient, therefore, to preclude the spontaneous evolutionary origin of the immense biological order required for the origin of life."[14]

Evolution maintains that high levels of organized complexity suddenly appeared out of nothingness, and that somehow the cosmos (organized universe) was produced by some primordial explosion. This process would require that "atoms organize themselves into increasingly complex and beneficial, ordered arrangements. Thus, over eons of time, billions of things are supposed to have developed *upward*, becoming *more* orderly and complex."[15]

Evolutionists have it completely upside down. The theory of evolution (so-called) is contradicted by the Second Law, which stipulates that all real systems and processes naturally diminish to lower levels of organization and complexity. This law also reveals the "natural tendency of complex, ordered arrangements and systems is to become simpler and more *disorderly* with time."[16] In summary, there is an irreversible downward trend at work throughout the universe.

Mutations—which are an expression of the universal tendency for decay and disorder—also force us to realize that we, as human beings, are not nearly as intelligent as our early ancestors—and, of course, this is contrary to popular belief. Because of evolutionary teaching, many people are under the *false assumption* that our generation is the most intelligent. *This is simply not the case*. Modern technology such as airplanes, automobiles, computers, and medical diagnostics and vaccines is the result of "accumulation of knowledge. We stand on the shoulders of those who have gone before us."[17]

> Just look around: our brains and bodies have suffered (mainly through genetic mutations and diseases) for 6,000 years, since the days of Adam because of sin and the curse. It is important to make a distinction between evolution and accumulated knowledge—accumulated knowledge and new technology are *NOT* evolution.[18] See chapter 3, section 6,000 Years of Genetic Burden.

The conclusion of many is that evolution is *not* possible because of the Second Law of Thermodynamics—it is a barrier to evolution—it is a law that has always proved valid wherever it could be tested. Harmonizing entropy and evolution is impossible because one is the exact opposite of the other. Many scientists believe that the Second Law is enough to disprove the theory of evolution and is *one of the important reasons* why many esteemed scientists have abandoned evolutionary doctrine in favor of creationism.

As stated by Dr. Emmett Williams, "The second law of thermodynamics is an empirical law, directly observable in nature and in experimentation. This law implies that the direction of all natural processes is toward states of disorder.... All natural systems degenerate when left to themselves.... Huxley states that evolution is an irreversible process which leads to greater variety, to more complex, higher degrees of organization. His assertion contradicts the prediction of the direction of natural processes called for by the second law! Either evolution has occurred in spite of the second law, or evolution has not occurred at all. There is no question about the correctness and universality of the second law..."[19]

At the present time, "it seems as though evolutionary doctrine will never be able to raise the curtain on the mystery of creation. For the scientist who has lived by his faith in the power of reason, the story ends like a bad dream. He has scaled the mountains of ignorance; he is about to conquer the highest peak; as he pulls himself over the final rock, he is greeted by a band of theologians who have been sitting there for centuries."[20]

Preprogrammed Design

Preprogrammed genetic systems are sometimes referred to as "specific complexity"—intelligent design or systems with high informational content. Examples are an encyclopedia, an automobile, an airplane, a supercomputer, sculpture of presidents carved into the granite of Mount Rushmore (Keystone, South Dakota), and the unparalleled complexity of a DNA molecule within the living cell. For biological systems to grow, for example, there must be preprogrammed genetic machinery—cellular parts functioning interdependently from the beginning—to transform incoming energy into useful forms. For plants such systems include photosynthesis, and for animals it includes respiration.

> The genetic code is so enormously complex that it is impossible for such information to have arisen by random chance without a Creator. Genetic systems are discussed in chapter 3 and Appendix B.

> Repetition such as XYZXYZXYZ, space pulsars, or the geometric arrangement of crystals or snowflakes (also known as 'order') is not specific complexity or proof of intelligent design.

Such preprogrammed systems do not violate the Second Law, but they do permit the tendency of running down to be temporarily overcome. Regarding the Second Law and living organisms, E. Williams writes, "Living organisms also tend to reach a state of maximum entropy, but at a relatively slower rate... There is no valid experimental evidence of a violation of the second law of thermodynamics in either animate [living] or inanimate [nonliving] material. This places the evolutionist in the position of passively or actively denying the observable (second law) to believe the unobservable (macroevolution)."[21] In the real world, even preprogrammed machinery eventually wears down because of the Second Law.

Nevertheless, evolutionists believe entropy poses no problem because they maintain uphill drift (order and creation of new information) can occur if abundant energy is available in an open system. But this type of thinking is nothing more than make-believe fantasy. If all it took were the right chemical ingredients, we should routinely see dead organisms springing back to life—but this, of course, does not happen in nature. When a dead plant or animal receives energy from the sun, its genetic organization quickly diminishes—and the sun's heat speeds up the downhill, deterioration process.

Energy cannot create genetic machinery; natural systems must comprise this information at the beginning. Applied to the origin of first life, such machinery cannot arise except from outside intelligent design. In his book, *Is the Big Bang Biblical?,* Dr. John Morris states, "the time has come for evolutionists to drop their weary claim that the sun's raw energy is all that is necessary to produce order from disorder."[22]

Consider the following quotes regarding the sun's energy and complex living systems:[23]

"A source of energy alone is not sufficient, however, to explain the origin or maintenance of living systems. The additional crucial factor is a means of converting this energy into the necessary useful work to build and maintain complex living systems ..." —Charles Thaxton, Walter Bradley, and Roger Olsen, *The Mystery of Life's Origin*, 1992, p. 124. (Thaxton is a Ph.D. chemist, Bradley has a Ph.D. in materials science, and Olsen has a Ph.D. in geochemistry.)

"The presumed 'evolution' of chemical elements in the primeval oceanic 'biotic soup' into single, and then multiple celled organisms, propelled by the sun's powerful rays, also sidesteps the crucial factor of information. Without some inbuilt informational mechanism to translate heat energy into a system-building function, the effect of heat on chemicals in the supposed primeval sea would have been to destroy them." —Douglas F. Kelly, professor of systematic theology, *Creation and Change*, 1997, p. 70.

Another property that is often overlooked is the ingredient that "makes it alive," which cannot be explained by referring to its chemical or genetic properties. This property is the "life ingredient" of an organism—sometimes referred to as "teleonomy" which is a concept of design and purpose. Genesis records that God first formed Adam out of *"the dust of the ground"* and then *"breathed into his nostrils the breath of life, and the man became a living soul."* (Genesis 2:7, KJV) Also see Job 33:4 and 34:14–15. The "life ingredient" is the soul of living creatures. See chapter 3, Uniqueness of Human Beings.

It is important to make a distinction between animal and plant life. The Bible always applies the term "life" to living, moving creatures—but not to vegetation. As stated in Genesis 1:21 (NIV), *"God created...every living creature that moves..."* Life has independent movement—although plants can reproduce after their kind, they are not alive.[24]

Vegetation was designed to be a source of food for all living creatures (Genesis 1:29-30). God announces in Leviticus 17:11 (NIV), *"For the life of a creature is in the blood..."* Blood is the life source of all living things—if a moving creature has blood, then it is alive. Cain's offering of *"the fruit of the soil"* was rejected by God because plants were not living creatures suitable for the covering of sins (Genesis 4:3-5).

In summary, energy cannot cause a dead creature to come alive. It takes more than energy—it takes super-intelligent design at the beginning and the "breath of life." Living things get their information from their parent organisms, but we never see genetic information arise from unprogrammed matter. Nonetheless, even preprogrammed machinery eventually wears down because of the Second Law.

Outside forces (for instance, man's design of automobiles, computers, and other technology through intelligent design and expenditure of large amounts of energy) can increase order for a time, but such reversal cannot last indefinitely. Once the preprogrammed force is released, processes return to their natural direction—*downhill with greater disorder and chaos*. Energy is once again transformed into lower levels of free energy—that is, energy that is less available for further work. The tendency of all natural systems is to become simpler and disorderly with time, which completely contradicts evolutionary doctrine.

See the book, *World Winding Down*, by Carl Wieland (2012, Creation Book Publishers, Powder Springs, GA). Dr. Wieland provides an easy to understand description of the Second Law of Thermodynamics (i.e., the relentless tendency of all things to become more disorderly with time).

The Anthropic Principle

Another aspect of preprogrammed design is the Anthropic Principle, or the realization by scientists that the universe was designed in a very precise manner to support human life. This principle is an attempt to explain the fact that fundamental constants of physics and chemistry are *fine-tuned* to allow the universe and life to exist.

This principle was first suggested by astrophysicist and cosmologist Brandon Carter in 1973 when he concluded that essential structures from the sub-atomic level to the entire universe depend on delicate balances between different physical forces. Such balances could not have arisen by random chance. There is overwhelming evidence that our universe, galaxy, solar system, stars and planets, and the earth were fine-tuned for the existence and well-being of human, animal, and plant life.

There are over fifty different scientific laws relating to physics and chemistry that have been designed to precise thresholds to uphold the existence of human and animal life. Any minute deviation of these mathematical constants would have a catastrophic effect, and life would be impossible. If any one of these forces (constants) was just slightly different, the universe and life could not exist. The following are just a few examples:[25]

- Astronomers are able to accurately measure the speed at which galaxies are expanding, and it is exactly the speed that allows our universe, galaxy and solar system, and human life to exist. If the expansion rate was faster, galaxies, stars, and planets would not have formed, and if it was slower, the force of gravity would have caused the universe to implode at the moment of conception.

- Sir Isaac Newton described universal gravitation and the three laws of motion, laying the groundwork for classical mechanics that dominate the scientific view of the universe and modern engineering. Newton demonstrated that the motions of objects on earth and of celestial bodies (e.g., stars and planets) are governed by the same set of natural laws. Any deviation from these laws would collapse the orbits of the earth, moon, and other celestial bodies, and life would be impossible. Further, if earth's gravity was greater, ammonia and methane gases could not escape the atmosphere and life would be impossible. Likewise, if earth's gravity was weaker, water vapor would escape the atmosphere and, again, life could not exist.

- At the sub-atomic level, solid matter is composed of atoms, which are made up of neutrons and protons (comprising the nucleus) and electrons spinning around the nucleus. If the force responsible for binding protons and neutrons into the atomic nucleus was just slightly weaker or slightly stronger, then atoms (except hydrogen, which has only one proton and no neutrons) could not exist, and our universe and life would be impossible.

Newton's Three Laws of Motion: law of inertia, law of force = mass x acceleration, and law of action/reaction. See Glossary.

The nucleus of an atom consists of protons and neutrons with very small orbiting electrons.

- The rotation of the earth every 24 hours and its axial tilt of 23 degrees were designed to support life. If the rotation was faster, atmospheric winds would have a destructive effect on the earth's surface, and if the rotation was slower, long periods of darkness and light would make it very unlikely for vegetation to flourish. If the tilt of the planet was any more or any less, surface temperatures would be too extreme—in either case, life would be unable to thrive.
- Our atmosphere and ocean basins comprise the right mixture and balance of gases (atmosphere comprises 20.95% oxygen, 78.98% nitrogen and trace gases including 0.93% argon and 0.038% carbon dioxide). If these gases were altered in any way, vegetation and animal life on earth would soon die. For example, if there was more carbon dioxide in the atmosphere, the earth's surface temperature would soar to extreme levels, and if there was less, the earth's surface would be extremely cold—in either case, the earth would remain uninhabitable.
- Unlike other chemical molecules, water (H_2O) is lighter in its solid or frozen form than in its liquid form. Therefore, when water freezes it expands (that is, it becomes less dense) and floats to the surface instead of sinking to the bottom of lakes and oceans. If water did not possess this unique quality, lakes and oceans would freeze from the bottom up and, in the process, kill all fish and benthic life.

There are hundreds of other examples of a purposeful design. The universe is not a random or chance event. As stated in Isaiah 45:18 (NIV), *"For this is what the Lord says—he who created the heavens, he is God; he who fashioned and made the earth, he founded it; he did not create it to be empty, but formed it to be inhabited—he says, I am the Lord, and there is no other."*

Biblical Explanation

Before sin entered the world, the world was good (Genesis 1:31); God "upheld" and *continuously restored* everything in the beginning (Colossians 1:17 and Hebrews 1:3), but when sin entered the world (because of the first man, Adam—Genesis 3:6), God *cursed the world* (Genesis 3:14–19), so the perfect creation began to *degenerate—that is, suffer death and decay* (Romans 5:12, 6:23, and 8:22). The dominion of man (male and female) over the entire world (Genesis 1:26-28, 5:1-2) meant that when Adam sinned, all of creation was cursed as well.

During the six days of creation, matter/energy was created and organized into increasingly complex and energized systems, in exact contradistinction to the universal tendency toward disorganization and de-energization which is experienced today! It was not until the "curse" that creation was replaced with the First Law and the Second Law (*unrestrained* universal decay as a dominating trend in nature). The Second Law recognizes that energy is becoming less and less available to maintain the natural processes (biological and physical) of

> The First Law recognizes that in today's world, there is no creation of new matter/energy rising to higher levels of organized complexity! But the First Law implies that creation existed some time in the past. The Second Law recognizes decay and deterioration as a dominating trend in nature.

the universe. Eventually, the universe will reach a point of maximum entropy (heat death) if the Second Law continues to function.

In Isaiah 51:6, Hebrews 1:10–11, and Romans 8:20–22, God indicated that these creative processes are no longer in operation—facts thoroughly verified by these universal laws.

> *"Lift up your eyes to the heavens, look at the earth beneath; the heavens will vanish like smoke, the earth will wear out like a garment and its inhabitants die like flies* [Second Law]. *But my salvation will last forever, my righteousness will never fail."* (Isaiah 51:6, NIV)

> *"He also says, 'In the beginning O Lord, you laid the foundations of the earth, and the heavens are the work of your hands. They will perish, but you remain; they will all wear out like a garment.'"* [First and Second Laws] (Hebrews 1:10–11, NIV)

> *"For the creation was subjected to frustration, not by its own choice, but by the will of the one who subjected it, in hope that the creation itself will be liberated from its bondage to decay and brought into the glorious freedom of the children of God. We know that the whole creation has been groaning as in the pains of childbirth right up to the present time."* [First and Second Laws] (Romans 8:20–22, NIV)

Genesis 1:31 (God created a perfect world), Genesis 3:3-7 (first sin), and Romans 5:12, 6:23 (consequences of sin) indicate that there was no death in the animal kingdom until Adam sinned. The first death was that of an animal to provide clothing for Adam and Eve (Genesis 3:21). Our experience points to the fact that every living organism eventually dies—and we are reminded that we are dust and to dust we will return.

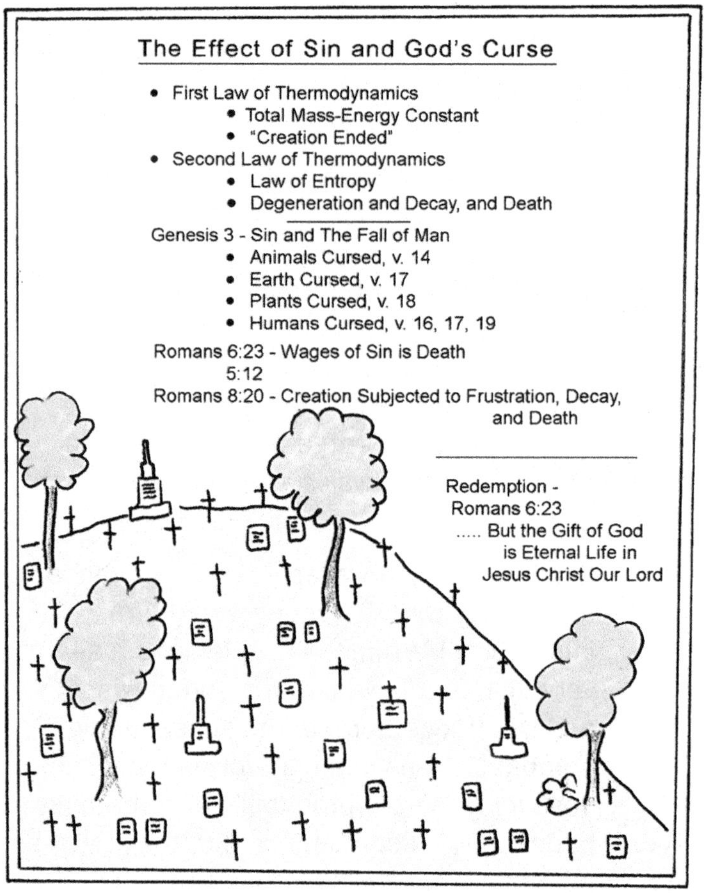

The Effect of Sin and God's Curse

- First Law of Thermodynamics
  - Total Mass-Energy Constant
  - "Creation Ended"
- Second Law of Thermodynamics
  - Law of Entropy
  - Degeneration and Decay, and Death

Genesis 3 - Sin and The Fall of Man
- Animals Cursed, v. 14
- Earth Cursed, v. 17
- Plants Cursed, v. 18
- Humans Cursed, v. 16, 17, 19

Romans 6:23 - Wages of Sin is Death
5:12
Romans 8:20 - Creation Subjected to Frustration, Decay, and Death

Redemption -
Romans 6:23
..... But the Gift of God is Eternal Life in Jesus Christ Our Lord

Aging and Death Are Implicit in the Second Law of Thermodynamics. *Sketch by Roger Gallop*

In Romans 8:22 (as cited above), "*...will be liberated from its bondage to decay...*" implies the universe is not destined for destruction but for renewal, and living things will no longer be subject to death and decay (Revelation. 21:1). Revelation 22:3 (NIV) states: "*No longer will there be any curse*"—that the curse has been lifted!

Beneficial decay processes would have operated in the beginning to naturally facilitate creation. Examples include classic energy dispersion (hot to cold; high pressure to low pressure), animal digestion (breakdown of food molecules), friction that prevents slipping (conversion of mechanical energy to heat energy), and heating of the sun by nuclear fusion reaction. When God cursed the world, creation was replaced with the First Law and the Second Law (*unrestrained* universal decay).

Summary

The First Law of Thermodynamics states that energy and matter cannot be destroyed, which implies that creation existed some time in the past. The Second Law states that disorder increases with time, which forces us to conclude the earth was at one time *more organized and integrated*, and *more beautiful* than it is now. The universal tendency of things to run down and to fall apart shows the universe had to be "wound up" at the start. The universe is winding down—it is not eternal.

The First and Second Laws imply that the origin of first life can only arise from Divine Creation—and these are universal laws of nature, according to observation and empirical analysis. All biological and physical processes operate by these laws. Energy and matter (a form of energy) are not being created at the present time.

> The acceptance of this "undiscovered trend" of chaos to cosmos by secular scientists in support of evolutionary doctrine is the most astounding and perplexing paradoxes in science!

Evolution has been "assumed" as the universal principle of change in nature, despite the fact that it is completely contrary to the laws of thermodynamics—in fact, there is ***no*** experimental evidence supporting the assumption of evolution. In evolutionary doctrine, life's complexity is attributed to an "undiscovered trend" in nature which produces order from disorder, merely by random natural processes, although the observed trend in nature is toward decay, not improvement.

A rational look at the laws of thermodynamics and entropy insists that evolution is not possible and that time does not perform miracles of evolution. Simply stated, the laws of thermodynamics preclude evolution—***time is the enemy of evolution***. Here are a few comments by scientists about thermodynamics and time:[26]

> "The First Law teaches that a natural process cannot bring into existence something from nothing. If the First Law is correct, which seems to be the case, and if the universe had a beginning, which seems to be scientifically accepted, then one conclusion is that something unnatural created the universe. ... The thought that the universe may have originated supernaturally is **unsettling to many people.** Yet, taken at face value, this conclusion is consistent with the total sum of evidence before us." (bold added)—Robert Gange, *Origins and Destiny*, 1986, p. 18. (Gange is a Ph.D. research scientist in the field of cryophysics and information systems, and he is also a Christian.)

This begs the question: **why is it unsettling to many people?!** Evolution is a doctrine of death—when humans die, we cease to exist. It works toward the destruction in the belief in the Resurrection—that ultimate hope. The Bible tells us that God is holy, righteous, merciful, and offers eternal salvation. So why are so many people working so hard to remove Christian faith from the world? Does such motivation make sense to any rational person? Why is there so much hate—a hatred and rebellion described in Romans 1:18-21? (See epilogue, why do people believe so strongly in evolution?)

> Evolutionist's viewpoint: "Time is the hero of the plot. The time with which we have to deal is of the order of two billion years. ... Given so much time the 'impossible' becomes possible, the possible probable, and the probable virtually certain. One has only to wait: time itself performs miracles." —Dr. George Wald, "The Origin of Life," in *The Physics and Chemistry of Life*, 1955, p. 12.

> Reality: "Time, however, does not increase the chance that the penny would turn into a nickel, a dime, a quarter, and a silver dollar which would then sprout wings and fly off together into the sunset doing aerobatic stunts in a tight formation. Time does increase the probability of something happening if it can happen, but the statement 'time itself performs miracles' is false." —Thomas F. Heinze, "How Life Began," Chick Publications, 2002, p. 22.

notes: Chapter 2

1. Taylor, Paul S. (1998–1999). Second law of thermodynamics – Does this basic law of nature prevent evolution? Gilbert, AZ: Eden Communications, Christian Answers Network, 1. Retrieved May 2008, from http://www.christiananswers.net/q-eden/edn-thermodynamics.html (adapted from Taylor, Paul S. [1995]. *The Illustrated Origins Answer Book.* [5th Ed.] Mesa, AZ: Eden Communications), 1 and Emmett L. Williams (Ed.). (June 1981). *Thermodynamics and the Development of Order.* Norcross, GA: Creation Research Society, 18.

2. Entropy (May 2008). In Wikipedia, the free encyclopedia. Retrieved May 2008, from http://en.wikipedia.org/wiki/Entropy.

3. Taylor, op. cit., 1; and King, Allen L. (1962). *Thermophysics.* San Francisco, CA: W. H. Freeman & Company, 5.

4. Thaxton, C.B., Bradley, W.L., and Olsen, R.L. (1984). *The Mystery of Life's Origin*. Dallas, TX: Lewis and Stanley; as cited in Catchpoole, D., Sarfati, J., and Wieland, C. (2008). *The Creation Answers Book.* (D. Batten, Ed.). Atlanta, GA: Creation Book Publishers, 21.

5. Second law of thermodynamics (2008). (M. Houdmann, P. Matthews-Rose, R. Niles, editors). All About Science. Retrieved April 2008, from http://www.allaboutscience.org/second-law-of-thermodynamics.htm (AllAboutScience.org).

6. Thermodynamic system (2009). In Wikipedia, the free encyclopedia. Retrieved December 2009, from http://en.wikipedia.org/wiki/Thermodynamic_system. Also see Taylor, op. cit., 4.

In thermodynamics, a thermodynamic system is defined as that part of the universe that is under consideration. A hypothetical boundary separates the system from the rest of the universe, which is referred to as the environment or surroundings. A useful classification of thermodynamic systems is based on the nature of the boundary and the quantities flowing through it, such as matter, energy, work, heat, and entropy. A system can be anything, for example a piston, a solution in a test tube, a living organism, an electrical circuit, a planet, etc.

In reality, a system can never be absolutely isolated from its environment. In analyzing a system in steady-state, the energy entering into the system is equal to the energy leaving the system.

7. Ross, John (July 7, 1980). Letter in *Chemical and Engineering News*, 58: 40; as cited in Morris, Henry M. (1997). *That Their Words May Be Used Against Them.* Green Forest, AR: Master Books, 74. (Dr. John Ross is a Harvard University scientist and evolutionist.) Also see Taylor, op. cit., 4 and 9.

8. Mulfinger, G. (1981). History of Thermodynamics. In: *Thermodynamics and the Development of Order.* (E. Williams, Editor). Norcross, GA: Creation Research Society, 7-8. Also cited in Taylor, op. cit., 3.

9. Hatsopoulos, G.N. and Gyftopoulos, E.P. (1970). Deductive Quantum Thermodynamics. In: *A Critical Review of Thermodynamics.* (E.B. Stuart, B. Gal-Or, and A.J. Brainard, Editors). MD: Mono Book Corporation, 78. Also cited in Taylor, op. cit., 3.

10. Asimov, Isaac (June 1970). In the game of energy and thermodynamics you can't even break even. *Smithsonian Institution Journal*, 10–11; as cited in Morris, Henry M. (December 1997). *That Their Words May Be Used Against Them.* Green Forest, AR: Master Books, 65; also cited in Taylor, op. cit., 6.

11. Wilson, J. (2007). Scientific laws, hypotheses, and theories. Retrieved April 2008, from http://wilstar.com/theories.htm.

12. Whitcomb, J.C., and Morris, H.M. (1961). *The Genesis Flood.* Phillipsburg, NJ: The Presbyterian and Reformed Publishing Company, 224-225.

13. Second law of thermodynamics (2008), op. cit.

14. Gish, Duane (March 1979). A consistent Christian-scientific view of the origin of life. *Creation Research Society Quarterly*, 15 (4), 199, 186; and Taylor, op. cit., 3.

15. Taylor, op. cit., 2.

16. Ibid., 2. Also see Lindsay, R.B. (1968). Physics – To what extent is it deterministic? *American Scientist*, 56 (2): 100–111; Hatsopoulos and Gyftopoulos, (1970), op. cit., 78.

17. Ham, K., Sarfati, J., and Wieland, C. (2000). *The Revised & Expanded Answers Book.* (D. Batten, Ed.). Green Forest, AR: Master Books, 139.

18. Paradox: As our brains and bodies have suffered because of mutations—which are an expression of the universal tendency for decay and disorder—information has increased because of accumulated knowledge over the 4,000-year period (Noah to present day). In the last days, as described in Daniel 12:4, "many will go back and forth, and knowledge will increase." See Alvin Toffler's book, *Future Shock*, that refers to an exponential or geometric increase in knowledge—but at the same time, our bodies and physical world (including morality) are decaying. Medical technology, for example, is barely keeping pace with *increasing mutational load* or "genetic burden" and disease. The world of the 1940s and 1950s has ceased to exist and since the year 2000 it is already disappearing. More information was generated in 2008 than in the previous 5,000 years, and the amount of new technical information is doubling every two years. By 2012, it is predicted to double every 72 hours—in other words, a human will be unable to keep up with technology. But accumulation of knowledge and new technology is *NOT* evolution—it is simply an "accumulation of knowledge." Hal Lindsey Report, television, August 10, 2008.

19. Williams, E. (1981). Thermodynamics and Evolution: A Creationist View. In: *Thermodynamics and the Development of Order.* (W. Williams, Editor). Norcross, GA: Creation Research Society, 19.

20. Jastrow, Robert (1992). *God and the Astronomers.* New York: W. W. Norton, 107. Also cited in Wikiquote. Retrieved from http://en.wikiquote.org/wiki/Robert Jastrow.

21. Williams, E. (1981). Resistance of Living Organisms to the Second Law of Thermodynamics. In: *Thermodynamics and the Development of Order*. (W. Williams, Editor). Norcross, GA: Creation Research Society, 91.

22. Morris, J.D. (2003). *Is the Big Bang Biblical?* Green Forest, AR: Master Books, 141.

23. Truth and Science Ministries, In Search of the Truth. Retrieved April 2008, from http://www.truthandscience.net/quotes.htm.

24. Morris, Henry (August 2012). It's alive. *Acts & Facts*, 41 (8), 4-5.

25. Jeffrey, G.R. (2000). *Journey Into Eternity*. Toronto, Ontario: Frontier Research Publications Inc., 101-107; and Anthropic principle. Retrieved May 2008, from http://ourworld.compuserve.com/homepages/rossuk/c-anthro.htm.

26. Truth and Science Ministries, In Search of the Truth, op. cit. Also see Morris, H.M. (1997). *That Their Words May Be Used Against Them*. Green Forest, AR: Master Books, 57.

Chapter 3

Biological Scientific Evidence for Creation

"Then the Lord God formed man of dust from the ground, and breathed into his nostrils the breath of life; and man became a living being." —Genesis 2:7, NAS

Law of Biogenesis - Origin of Life

Teachers, scientists, and journalists talk as if evolution were observed fact, but no one has any scientific explanation of how the immensely complex, information-bearing molecules could have arisen from "nonlife" without preprogrammed design and outside intelligence. Also, the property of a cell which "makes it alive" cannot be explained by just referring to its chemical properties.

Law of Biogenesis states that life can only arise from pre-existing life, not from nonliving matter. This is a fundamental law of science which contradicts the idea that organisms spontaneously formed from nonliving chemicals (dust or rock).

Living things get their infomation from their parent organisms and, contrary to evolutionary doctrine, scientists have never observed life arise from raw, unprogrammed matter. As proved by French scientist, Louis Pasteur, in the mid-1800s, life cannot be produced from nonliving matter. Biogenesis describes a process whereby living organisms can only arise from other living organisms—and this biological law has become the cornerstone of biological sciences.

No one has ever observed or demonstrated spontaneous generation of life from nonliving things (a hypothetical term known as abiogenesis) and, for that matter, no one has ever observed any organism give rise to a different type of organism (macroevolution)—it remains purely conjecture. No one has ever found a single unequivocal transitional, in-between type where there should be billions if evolution were true.

Abiogenesis is a term coined by evolutionists to describe how life arose spontaneously from nonlife. No one has ever observed or demonstrated spontaneous generation of life from nonlife.

Variations of species within animal and plant groups (kinds) are *NOT* evolutionary changes but merely reshuffling of genes within the "*existing* gene pool" that was originally present in a certain "kind" of animal or plant population. Animal kinds are represented by mankind, bear kind, horse kind, dog kind, cat kind, and so forth. Species variations within kinds may include dominant color, long-haired versus short-haired animals, mating calls and gestures, and less obvious variations such as insect resistance to an insecticide, and bacterial resistance to antibiotics (discussed later in this chapter). No one has ever observed uphill drift—that is, the addition of new genetic information to the gene pool. Genetic drift has *always* been "downhill," which is consistent with the Second Law.

Major problems for evolutionists are the origin of life from nonlife and a mechanism for an expanding, more complex gene pool. Consider the unimaginable improbabilities involved in just getting the evolutionary process started—in spite of the First and Second Laws, and Law of Biogenesis. Professor Christian de Duve, a Nobel Prize-winner and non-creationist, admits that each step in the evolutionary process is "highly conjectural" and even simple genetic molecules are exceedingly complex and far beyond human understanding.[1] Yet, as a naturalist, he must comply with evolutionary secular pressure that it happened by natural processes and "any hint of teleology [i.e., design] must be avoided." Life is here on earth, so scientists must explain life "within the realm of science" [that is, naturalistically] without the possibility of God and Divine Creation.

Evolution - A Statistical Impossibility

Let's take a look at life's basic building structures—amino acids. An interesting phenomena in nature is that all amino acids that make up proteins in living things—from higher order animals and plants to bacteria, molds and viruses—are 100 percent 'left-handed.' When scientists have attempted to produce amino acids (in the lab) necessary for life by random natural processes, what was produced was a mixture of left- and right-handed amino acids which are incompatible with life.

In such experiments, oxygen was excluded because biological molecules (amino acids) are destroyed in the presence of oxygen—although evidence (e.g., oxidized iron in Precambrian rocks) indicates the earth has always had oxygen in the atmosphere. If the earth's ozone layer (oxygen layer) did not exist, ultraviolet rays would have destroyed biological molecules. Also, secular scientists maintain that life originated in the oceans but the process of hydrolysis (water splitting) would have destroyed amino acids.

The fact of the matter is, scientists do not have the slightest idea why biological proteins use only left-handed amino acids. Statistically, the probability of just getting an average-size protein of left-handed amino acids (300 amino acids) occurring naturally is 1 in 4.9×10^{191}.[2] According to the laws of probability, anything smaller than 1 in 1×10^{50}, the chance of an event occuring is zero (impossible; this is our threshold or marker). To provide yet another perspective, the chance of finding just one unique atom in the entire known universe is 1 in 1×10^{80}. Yet, textbooks continue to promote the idea that life originated by natural processes. The only way to link together left-handed amino acids is through purposeful design.

Also, left-handed amino acids have to be arranged in perfect sequence (in three dimensions) for the protein to work. Imagine the odds against evolution. Here's a practical analogy. If all the parts of a Porsche sports car were thrown into a small pond, how long would it take for the parts to assemble themselves into a car? The answer is NEVER. According to evolutionary doctrine, the only difference between the car and the origin of the first living cell is that parts of the car were created by intelligent beings, whereas the cell with unparalleled complexity and design was supposedly made and assembled by random interaction of lifeless molecules.

Statistically, a protein molecule, consisting of thousands of precisely arranged left-handed amino acids with complex genetic machinery functioning interdependently, could never arise by random chance. How did we get a brain, heart, lungs, stomach, teeth, tongue, eyes, ears, skin, muscles, bones, a nervous and vascular system, and all the other organs working together as male and female? The bottom line is the genetic code—synonymous to a supercomputer software program—is so enormously complex that it is impossible such information could have arisen by random chance without a Creator. See App B for more information.

Astronomers and mathematicians, Sir Fred Hoyle, Ph.D., and Chandra Wickramasinghe, Ph.D., Sc.D., calculated the probability of getting a living cell by natural processes:[3]

> "Precious little in the way of biochemical evolution could have happened on the Earth. It is easy to show that the two thousand or so enzymes that span the whole of life could not have evolved on the Earth. If one counts the number of trial assemblies of [left-handed] amino acids that are needed to give rise to the enzymes, the probability of their discovery by random shuffling turns out to be less than 1 in 10 to the power of 40,000." —Sir Fred Hoyle and Chandra Wickramasinghe, "Where Microbes Boldly Went," *New Scientist*, vol. 91, August 13, 1991, p. 415. (This ratio is unfathomably small and the probability is zero.)

There are about 2,000 types of amino acids, but only 20 (L-handed) are used in living organisms, and the chance of obtaining them all in a random trial is 1 in $(10^{20})^{2000} = 10^{40,000}$. And these amino acids must be arranged in a very specific sequence for each protein.[4]

Purposeful Design

Dr. Michael Denton effectively illustrates the complexity and design of the cell in his book, *Evolution: A Theory in Crisis* (Bethesda, MD: Adler & Adler, 1986), 161-165.

> Viewed down a light microscope at a magnification of several hundred times, such as would have been possible in Darwin's time, a living cell is a relatively disappointing spectacle; appearing only as an ever-changing and apparently disordered pattern of blobs... To grasp the reality of life...we must magnify a cell a thousand million times...until it resembles a giant airship large enough to cover a great city like London or New York. What we would then see...an object of unparalleled complexity and adaptive design. On the surface of the cell we would see millions of openings...opening and closing to allow a continual stream of materials to flow in and out. If we were to enter one of these openings we would find ourselves in a world of supreme technology and bewildering complexity...endless highly organized corridors and conduits branching in every direction...some leading to the central memory bank in the nucleus and others to assembly plants and processing units. The nucleus itself would be a vast spherical chamber more than a kilometer in diameter...all neatly stacked in ordered arrays, the miles of coiled chains of the DNA molecules. A huge range of products and raw materials would shuttle along all the manifold conduits in a highly ordered fashion to and from the assembly plants in the outer regions of the cell.
>
> We would wonder at the level of control...of so many objects down so many seemingly endless conduits, all in perfect unison. We would see all around us..... all sorts of robot-like machines. We would notice that the simplest of the functional components of the cell, the protein molecules, were astonishingly complex pieces of molecular machinery, each one consisting of about three thousand atoms arranged in highly organized 3-D spatial conformation. We would wonder even more as we watched ... these weird molecular machines...the task of designing one such molecular machine—that is, functional protein molecule—would be completely beyond our capacity... Yet life of the cell depends on the integrated activities of thousands...probably hundreds of thousands of different protein molecules.
>
> What we would be witnessing would be an object resembling an immense automated factory, a factory larger than a city and carrying out almost as many unique functions as all the manufacturing activities of man on earth. However, it would be a factory which would have one capacity not equaled in any of our own most advanced machines, for it would be capable of replicating its entire structure within a matter of a few hours. To witness such an act at a magnification of a billion times would be an awe-inspiring spectacle.

The question is, who wrote the genetic software program? Who placed this functioning code inside the nucleus of the cell? A former atheist, leading philosopher, author and debater, considering the complexity and interdependency of biological systems, once stated, "Super-intelligence is the only good explanation for the origin of life and the complexity of nature."[5] See Appendix B for more information including the sports car example.

Such programming and design, including the "breath of life" (Genesis 2:7, NIV) and consciousness (awareness), would be expected from a Divine Creator—not from random, unobserved evolution. Every living thing gives powerful evidence for intelligent design and only *willfully ignoring the facts* would lead someone to assign such complex design to mere chance. In today's world, the science community simply rejects or ignores all evidence for "supernatural creation."

Here is what some scientists have said concerning creating life from nonlife:[6]

> "Scientists have not been able to cause amino acids dissolved in water to join

together to form proteins. The energy-requiring chemical reactions that join amino acids are reversible and do not occur spontaneously in water." —George B. Johnson and Peter H. Raven, *Biology, Principles & Explorations* (Holt, Rinehart and Winston, 1996), p. 235.

"Both right- and left-handed amino acids take part equally well in ordinary chemical reactions, but in living organisms the shape is so important that only proteins made entirely of left-handed amino acids will connect properly, and they never happen in nature outside of living cells." —Thomas F. Heinze, "How Life Began," Chick Publications, 2002, p. 15. [Ordinary amino acids' chemical reactions are both left- and right-handed, whereas life uses only the left-handed variety.]

"This is a very puzzling fact. ... All the proteins that have been investigated, obtained from animals and from plants, from higher organisms and from simple organisms—bacteria, molds, even viruses—are found to have been made of (left-handed) amino acids." —Linus Pauling (Nobel Laureate in chemistry), *General Chemistry*, (3rd Ed), 1970, p. 774.

"Since science has not the vaguest idea how (proteins) originated, it would only be honest to admit this to students, the agencies funding research, and the public." —Hubert P. Yockey, "Self Organization Origin of Life Scenarios and Information Theory," *Journal of Theoretical Biology*, vol. 91, 1981, 13–31.

"The idea of such a 'soup' containing all desired organic molecules in concentrated form in the ocean has been a misleading concept against which objections were raised early." —S. J. Mojzsis, *The RNA World*, ed. 2, 1999, p. 7.

"One has only to contemplate the magnitude of this task to concede that the spontaneous generation of a living organism is impossible. Yet here we are—as a result, I believe, of spontaneous generation."—George Wald, "The Origin of Life," in *The Physics and Chemistry of Life* (Simon & Schuster, 1955), p. 9.

Zenith pocket watch 1900-1910

A metaphor supporting the argument for creation was best stated by William Paley in his book, *Natural Theology*, first published in 1802. Paley trained for the Anglican priesthood, graduated from Christ's College, University of Cambridge, United Kingdom, and was appointed a fellow and tutor of his college. In his book, Paley introduced one of the most famous metaphors in the philosophy of science, the image of the watchmaker.[7]

> ...when we come to inspect the watch, we perceive... that its several parts are framed and put together for a purpose, e.g., that they are so formed and adjusted as to produce motion, and that motion so regulated as to point out the hour of the day; that if the different parts had been differently shaped from what they are, or placed after any other manner or in any other order than that in which they are placed, either no motion at all would have been carried on in the machine, or none which would have answered the use that is now served by it... the inference we think is inevitable, that the watch must have had a maker—that there must have existed, at some time and at some place or other, an artificer or artificers who formed it for the purpose which we find it actually to answer, who comprehended its construction and designed its use.

Paley argued that living organisms are even more complicated than watches. Only an intelligent designer could have created life and the universe, just as only an intelligent watchmaker can make a watch: "The marks of design are too strong.... Design must have had a designer. That designer must have been a person. That person is GOD."

Darwinism - Unnatural Selection

A common belief today is that Charles Darwin discovered "evolution"—assumed by many to be "unquestionable fact." With a college degree in theology, Darwin published his *The Origin of Species* (On The Origin of Species By Means of Natural Selection) in 1859 at the age of 50—and although his "contributions" were considered rather mundane at the time, this book eventually became the basis of today's so-called "theory of evolution." Probably no other book has so influenced Western civilization and the world by serving as the catalyst of reviving paganism and rebellion of man against his Creator.

According to Duane P. Schultz in *A History of Modern Psychology* (1981), "The theory of evolution...freed scholars from constraining traditions and superstitions and ushered in the *era of maturity and respectability for the life sciences*. The theory of evolution was also to have a tremendous impact on contemporary psychology."[8] [Italics added] Although most people today, including scientists and teachers, have never read it, the entire educational and scientific community regards Darwin's book as the academic foundation of evolutionary doctrine.

Most people don't realize that Darwin was ignorant of genetics and the human genome, having died before this field of science was established around the turn of the 20th century. In ignorance, Darwin believed in the "inheritance of acquired characteristics"—that is, if an animal acquired physical characteristics during its lifetime, those characteristics would be passed on to its offspring. For example, Darwin believed that if a man developed large arms by heavy lifting, he will have strong-armed children. It is known fact, however, if a man's phenotype (body) changes, his genotype (genetic information) remains the same.

Over the last 60 years, advances in genetic science have revealed the inadequacies of classical Darwinism—so today, most biology, geology, and psychology textbooks teach a modern form of evolution—neo-Darwinism, that evolution is the result of positive random mutations sorted by natural selection. This 'modern synthesis' of Darwinian evolution attempts to incorporate modern advances in genetics but has utterly failed to find unequivocal mechanisms (biological processes) for evolution. (Natural selection and positive random mutations are disucssed in the following sections.)

Nevertheless, our educational system and the world continue to march to the anthem, "Darwin said it, I believe it, and that settles it."[9] This is one of the great paradoxes of our time—why does mankind continue to blindly rally to the mantra of evolution?

Sketch by Anton Zakharov

Consider the following quotes:[10]

"This evolutionist doctrine is itself one of the strangest phenomena of humanity ... a system destitute of any shadow of proof, and supported merely by vague analogies and figures of speech. ... Now no one pretends that they rest on facts actually observed, for no one has ever observed the production of even one [transitional form]. ... Let the reader take up either of Darwin's great books, or Spencer's 'Biology,' and merely ask himself as he reads each paragraph, 'What is assumed here and what is proved?' and he will find the whole fabric melt away like a vision. ... We thus see that evolution as a hypothesis has no basis in experience or in scientific fact, and that its imagined series of transmutations has breaks which cannot be filled."—Sir William Dawson, *The Story of Earth and Man*, New York: Harper and Brothers, 1887, pp. 317, 322, 330, 339.

"We have all heard of *The Origin of Species*, although few of us have had time to read it. ... A casual perusal of the classic made me understand the rage of Paul Feyerabend. ... I agree with him that Darwinism contains 'wicked lies'; it is not a 'natural law' formulated on the basis of factual evidence, but a dogma, reflecting the dominating social philosophy of the last century."—Kenneth J. Hsu, "Sedimentary Petrology and Biologic Evolution," *Journal of Sedimentary Petrology*, vol. 56, September 1986, p. 730.

"The irony is devastating. The main purpose of Darwinism was to drive every last trace of an incredible God from biology. But the theory replaces God with an even more incredible deity—omnipotent chance." —T. Rosazak, *Unfinished Animal*, 1975, pp. 101–102.

The reality is that evolutionary doctrine is the continuous rebellion of man against his Creator as proclaimed in Romans 1:18–32. Beneath the cover of sophisticated pseudoscientific language, evolutionary doctrine is the same pagan defiance that man has waged against God since the beginning of time.

Natural Selection and Extinction

Most of us have heard the term "survival of the fittest." It refers to natural selection, a mechanism for change that occurs when organisms with favorable genes suitable for a specific environment are able to survive, reproduce, and pass these genes on to the next generation through a process known as adaptation.

Contrary to popular belief, natural selection or survival of the fittest is *NOT evolution.* It is merely a genetic process that selects gene traits from an "existing gene pool" best suited for specific environmental conditions. A gene pool is genetic information *present* in the population for a "kind" of animal or plant. (The original gene pool comprised a large or wide variety of genetic information *present* in the original population.) Most people are unaware that natural selection (or survival of the fittest, adaptation, or speciation) is a ***thinning out process*** which leads to *loss of genetic information.*

What do we mean by "kinds" of animals and plants? In Genesis 1, the phrase "after his kind" or "after their kind" occurs ten times and refers to animals and plants as they reproduce on earth. The Bible clearly teaches that God created "fixed kinds of animals and plants," each to reproduce after its own kind. Genesis 1:11–25, for example, refers to many kinds of plants and trees bearing fruit, sea creatures, birds, livestock, insects, and wild animals. We know there are great variations (speciation) within a kind of animal or plant, but kinds have never evolved or merged into other kinds.[11] What we observe today are distinct groups of animals and plants (i.e., mankind, bear kind, horse kind, dog kind, cat kind, and so forth) with no in-between or transitional forms found anywhere in the world.

To understand natural selection, let's first briefly review the genetic reproductive process. Living things are programmed genetically to pass on their information by making copies of themselves. The genetic code (DNA) of the father is passed on by the sperm gamete, and the genetic code (DNA) of the mother is passed on by the egg gamete.

In the human body, each cell has 23 pairs of chromosomes (or a total of 46 chromosomes)—and, statistically, the number of different genetic variations of eggs a mother can produce or sperm a father can produce is more than 8 million (2^{23} for each gamete). During the initial phase of fertilization (a process known as meiosis), each parent copies half of their genetic information (or half of their chromosomes), which allows the offspring to have the same number of chromosomes as their parents (46 chromosomes). When fertilization occurs, 70 trillion ($2^{23} \times 2^{23}$ = 70 trillion) different zygotes, or fertilized eggs or variations are possible! See Appendix B.

Natural selection or adaptation (commonly known as "survival of the fittest," speciation, or variation, and sometimes referred to as microevolution) is the genetic process within a "kind" of animal population which selects gene traits from an *existing gene pool* best suited for a specific environment.

Kind (group): The classification of living things is class, order, family, genus, and species, whereby order or family may have represented the biblical "kinds" or groups.

Type (form): Term used to depict fictitious transitional animals (or plants) evolving from one form to another—none have ever been observed within living populations.

Gene shuffling (reshuffling or genetic recombination) of the *same information* in response to a specific environment will also result in much variation in any animal or plant "kind" population—thus speciation and variability.

Most people are unaware that natural selection (or survival of the fittest, adaptation, or speciation) is a thinning out process which *leads to loss of genetic information.*

Natural selection is similar to artificial breeding of dogs, cats, horses, and other domesticated animals by humans. To help understand the mechanism of natural selection, let's briefly take a look at artificial breeding of dogs. Dogs with medium-length hair will have puppies with long, medium, and short hair. If a breeder mates dogs with only the longest hair, and does this multiple times, eventually a "new" long-haired dog will arise whose descendants will all have long hair. This can apply to any physical characteristic. How does this happen?

The short- and medium-haired dogs are weeded out of the artificial breeding process. Selection does not change the total capacity of DNA available to store information, but the selection process will eventually allow long-haired genes to occupy positions formerly held by short-haired or medium-haired genes. So the long-haired dogs have *less variety of information* (that is, there is a *loss of genetic information*) than their medium-haired ancestors, which have information for both short and long hair.

Original Gene Pool

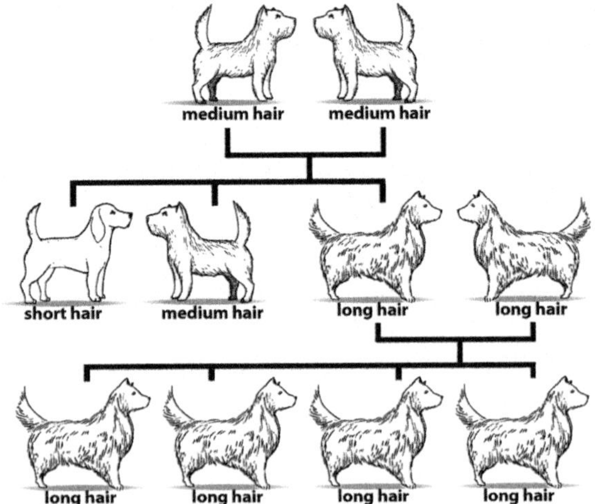

Sketch by Anton Zakharov

Each subgroup or species carries only a fraction of the original gene pool of the original kind of animal. Gene positions are now held exclusively by animals with specific traits or characteristics. That's why, starting with a

Chihuahua and Great Dane
Drawing by Patrick Gerrity

Pekingese or a Chihuahua, a breeder will never be able to produce a German Shepherd or a Great Dane—the necessary genetic information is no longer in the smaller breed!

"Nature" can also naturally select dogs and other animals by "survival of the fittest"—this is nature's way of allowing animals to adapt to their environment. In a cold region, for example, long-haired dogs will more likely survive. Short-haired varieties would be unable to survive a cold climate, but long-haired dogs will have an advantage and continue to breed. Over many generations the long-haired dogs "pass on" their genetic information that allows them to survive (similar to artificial breeding of a new long-haired dog).

The original "dog kind" (wolves), which was created with a large gene pool or built-in variation with no original defects, could vary simply by recombination (reshuffling) of the original genetic information because of an extreme or unique environment. Such variations would give rise to different species of wolves (domestic dog, coyote, dingo, jackal, fox (swift, gray, red), African and Asian wild dogs, and so forth.)

Wolf variations (*Canis spp.*): Gray, Arctic, Tundra, Arabian, Mexican, Russian, Italian, Egyptian, Eurasian, Eastern, Great Plains, Northwestern, Indian, Iberian, Ethiopian, and Red. *Canis* also includes coyotes, dingoes, and jackals.

> **Canidae** is the biological family of carnivorous and omnivorous mammals that includes wolves, foxes, jackals, coyotes, and the domestic dog. This family is divided into wolf and dog animals of the tribe Canini and the foxes of the tribe Vulpini.

The natural selection process can be applied not only to dogs but to cats (domestic cat, lion, tiger, leopard, jaguar, puma, cheetah, bobcat, ocelot, etc.), bears (polar bear, American black bear, brown or grizzly bear, giant pandas, Asiatic black bear, sloth bear, spectacled bear, sun bear, etc.), cattle (domestic cattle, bison, yak, etc.), and all other animal (and plant) kinds or groups. All animal species (subgroups) today have developed great diversity: size and form, hair length and color distinctions, and mating calls or gestures which allow them to distinguish their species from other species, and they remain reproductively isolated (although many separate species can "interbreed" in captivity).

> Long hair versus short hair, change in dominate color, and other variations are NOT evolutionary changes but merely reshuffling of genes within the *existing* "gene pool" that was originally present in a certain "kind" of animal population. ***No one has ever observed uphill drift or the addition of genetic information to the gene pool. Genetic drift has always been "downhill"—consistent with the Second Law.***

Finches (sparrows) are another example of natural selection and are found throughout the world including the Galapagos and Hawaiian Islands. Finches first arrived at these islands with a wide variety of genetic information for beaks. If the main food source on an island was hard seeds, finches with genes for short-thick beaks were able to break the seeds apart and survive. If another island had fewer seeds but many grubs (larvae of beetles), finches with long thin beaks could easily prod deeper into the ground, grab their prey, and survive. On the island with hard seeds and fewer grubs, selection did not change the total capacity of DNA available to store information, but adaptation eventually allowed short thick beaks to occupy positions formerly held by long thin or medium beaks. On the island with fewer seeds and many grubs, adaptation eventually allowed long thin beaks to occupy positions formerly held by short thick or medium beaks.[12] With natural selection, there is always a loss of genetic information—thus, natural selection is not evolution.

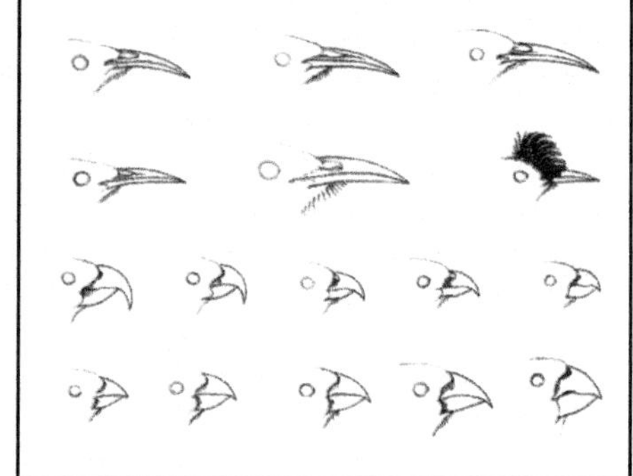

Beak shapes of the Drepanidinae, the finch family

In summary, natural selection is able to "cull and sort" genetic information within the *original* "kind" gene pool, and enables a new species to become unique while adapting to a different environment (niche or habitat)—it is the process of specialization to specific environmental conditions. Animals with "best suited" genes to an environment will continue to survive and animals with "least suited" genes will eventually die. Over many hundreds of years, natural selection results in many different species without new genetic information ever being added to the original gene pool—therefore, ***no evolution***.

Another commonly used example of natural selection is the DDT-resistant mosquito. Such a mosquito has adapted to an environment where DDT is present but, in the process, the population of mosquitoes has lost genes that were present in mosquitoes that were not resistant to DDT. Why? Because the information for resistance was in a few individuals in the population before the sprays were used in the environment. These resistant genes took the place of non-resistant genes along the genetic chain. Non-resistant mosquitoes, with a more diversified gene pool, died and resistant mosquitoes, with a less diversified gene pool, survived. The new population has lost genetic information from its original gene pool—that is, there is a "downhill drift" in genetic information.

Although adaptive specialization significantly improves the survival potential of the species, it makes the species prone to extinction if environmental conditions should ever change. For example, with a drastic change of climate—from a temperate antediluvian (pre-worldwide flood) environment to a world overwhelmed by a catastrophic flood followed by the Great Ice Age—dinosaurs and other animals such as woolly mammoths and saber-toothed tigers were unable to adapt and eventually became extinct. (See chapters 4 and 10, The Great Ice Age, and chapter 7, Rapid Burial versus Slow Deposition and Dinosaur Extinction, for more information.)

Extinction is the cessation of existence of a species or group of animals (or plants). Extinction is usually brought about by adaptation to a specific environment, following a drastic change in that environment.

Dinosaurs either drowned in the flood or those that were taken on the ark were overwhelmed with the post-flood climate, with its colder temperatures and lack of food. Remember that within a new species (subgroup), the original gene pool has been downsized (genetic code has less information), providing a survival advantage to a specific environment (e.g., antediluvian sub-tropical environment; see chapter 10, Dinosuars on Noah's Ark), but such adaptation would be a distinct disadvantage if climatic conditions should ever dramatically change (i.e., severe winter conditions or ice age following the flood).

"There is no known law of nature, no known process and no known sequence of events which can cause information to originate by itself in matter."[13] —Werner Gitt, *In the Beginning Was Information* (1997).

Most examples of evolution are nothing more than adaptation to their particular environment. If evolution were ever to succeed from molecules to man, the process would need a way to create new, complex genetic programs, or information—it needs new information to arise by natural "uphill" progression. There would have to be a mechanism to produce information from disorder by chance, and then continually add new information so new types could evolve—contrary to the Second Law. "Uphill" progression of the gene pool has NEVER been observed in nature.

"I suppose that nobody will deny that it is a great misfortune if an entire branch of science becomes addicted to a false theory. But this is what has happened in biology: for a long time now people discuss evolutionary problems in a peculiar 'Darwinian' vocabulary—'adaptation,' 'selection pressure,' 'natural selection,' etc.—thereby believing that they contribute to the explanation of natural events. They do not, and the sooner this is discovered, the sooner we shall be able to make real progress in our understanding of evolution. I believe that one day the Darwinian myth will be ranked the ***greatest deceit*** in the history of science."[14] [Italics and bold added]

Mutations - Deformities and Disease

Genetically and thermodynamically, it is impossible for an amoeba or protozoan to become a human being. Observed changes in living things simply head in the wrong direction to support evolution (see chapter 2). So what is the evolutionist to do? In evolutionary doctrine, the role of generating new information is consigned to positive random mutations—that is, mutations that are supposedly beneficial to life. (See section, Are There Beneficial Mutations?) In public schools and universities, students are being taught neo-Darwinism—that is, positive random mutations sorted by natural selection are responsible for the evolution of species, but this idea is *simply not true*.

What are mutations? Mutations are random errors or defects in the cell's DNA chemical structure—and such errors may affect whole chromosomes or just one gene. As mentioned in the previous section, when fertilization occurs, 70 trillion different variations of fertilized eggs are possible. With these kinds of odds, the more distantly related the parents, the less likely they will have the same mistakes in their genes. Genetic mistakes are inherited—the next generation makes a copy of the defective DNA, so the defect is passed on. Somewhere down the line another mistake happens, and the mutational defects accumulate. This is known as the problem of *increasing mutational load* or "genetic burden," which is consistent with the Second Law of Thermodynamics.

The more closely related the parents (as in inbreeding of cousins, or even worse, siblings), the more likely they will have "similar mistakes" in their genes because their genes may have been inherited from the same parents. Often, the good gene or trait overrides the bad gene so disease or deformity does not occur. However, if a child should inherit the same bad gene from both parents, it would result in two bad gene copies and a potentially serious defect leading to disease or deformity. Genetic drift has always been "downhill."

Chromosomes are comprised of DNA which consists of thousands of genes that carry genetic instructions or blueprints used in the development and functioning of all known living organisms. All organisms are programmed genetically to pass on copies of these instructions, or genetic code.

Mutations are random errors or defects in the cell's DNA chemical structure—and such errors may affect whole chromosomes or just one gene. If a gene is damaged, incomplete, or missing, the altered information for building and maintenance may lead to disease.

Genetic mistakes are inherited—the next generation makes a copy of the defective DNA, so the defect is passed on. Somewhere down the line another mistake happens, and the mutational defects accumulate. See Appendix B for more information.

Biblical Account - The Edenic Curse

In the beginning God created man (male and female) in His image (Genesis 1:26-28, 5:1-2). There were no genetic mistakes. Man was given dominion over the entire world and God upheld and *continuously restored* everything (Colossians 1:17 and Hebrews 1:3), and all that God made was "very good" (Genesis 1:31). This was the beginning of Creation and the First and Second Laws (*unrestrained* universal decay) did not apply.

When sin entered the world (because of rebellion by the first man, Adam—Genesis 3:6; see Prologue), God *cursed the world* (Genesis 3:14–19), and the perfect creation began to *degenerate—that is, suffer death and decay* (Romans 5:12, 6:23, and 8:22). Creatures under man's rule, though morally innocent, shared in God's judgment. The Curse ended Creation and initiated the unfettered reign of the First and Second Laws. There was no death in the animal kingdom until Adam sinned—and with the original sin (Genesis 3), so entered death as verified in Romans 6:23 (KJV), *"For the wages of sin is death; but the gift of God is eternal life through Jesus Christ our Lord."*

The law prohibiting marriage between close relatives (Leviticus 18–20) was given at the time of Moses (1526 BC), or about 2,478 years after Adam and Eve. As long as marriage was one man to one woman for life, there was no defiance of God's law! When Abraham (born 2166 BC) married his half-sister (Genesis 20:12), God blessed this union to produce the Hebrew people through Isaac and Jacob.

The degeneration and dying process started slowly because of zero mutational load with Adam and Eve. Cain was the firstborn to Adam and Eve and would have received perfect genes from his parents because the effects of sin and the Curse were nonexistent in the beginning. According to Genesis 5:5, Adam lived 930 years. Although the life span of Eve is unknown, they would have had many children that intermarried without producing deformed offspring.

Over the course of 2,000 years, mutational defects increased from one generation to another. The law prohibiting marriage between close relatives (Leviticus 18–20) was given at the time of Moses (1526 BC), or about 2,478 years after Adam and Eve. Today, our society forbids siblings to marry because their children would have a high risk of deformity.[15] We are in a world that "groans and suffers" (Romans 8:22, NAS) because of the Edenic Curse.

Physical and Spiritual Death: In Romans 6:23 (KJV), it states, *"the wages of sin is death"*—a physical and spiritual death. Physical death is necessary for the atonement of sin. In Hebrews 9:22, NIV, *"In fact, the law requires that nearly everything be cleansed with blood, and without the shedding of blood there is no forgiveness."* As described in Hebrews 2:14-18, the death of Jesus Christ on the cross and his resurrection brought reconciliation (restored harmony or brought to terms) between God and mankind.

In John 11:25-26 (NIV), Jesus said, *"I am the resurrection and the life. He who believes in me will live, even though he dies; and whoever lives and believes in me will never die."* To avoid spiritual death, we must personally accept the free gift of salvation by recognizing the Lord Jesus Christ as our personal Savior. This is a recurrent theme throughout the New Testament. See John 3:16, 5:24, 14:6; Romans 10:9, 13; and Ephesians 2:8-9.

The spirit is that part of humans that is able to love and experience God directly—it allows humans (and angels) to have a relationship with the Creator. The soul and spirit are connected but separate (Hebrews 4:12)—the soul is who we are, but the spirit connects us with God. Spiritual death results in eternal separation from God. See section, Uniqueness of Human Beings.

6,000 Years of Genetic Burden (4000 BC - AD 2000)

Genetic burden or mutational load began sometime after the dawn of man around 6,000 years ago. Degeneration has produced genetic mistakes and increasing mutational load in all living things. Today, we know that environmental factors such as excessive x-rays, ultraviolet light, radioactive substances, or certain chemicals may cause genetic defects. In most cases, genetic decay and increasing mutational load since the time of Adam and Eve are the *primary source of diseases and deformities, and eventual death.*

> In the beginning God created man (male and female) in His image and gave man dominion over the entire world (Genesis 1:26-28, 5:1-2) and, therefore, all of creation shared in God's judgment. The Edenic Curse ended Creation and initiated the unfettered reign of the First and Second Laws.

There are thousands of genetic defects and inherited diseases known in humans—and they may vary widely. Some defects are slight—for instance, a patch of skin loses its pigmentation without explanation, or some people let their hair grow over their ears to hide one that is lower or bigger than the other—or we have noticed, especially with celebrities on television, that a person's nose may be slightly out of alignment.

Genetic mutations are also the cause of some of the worst diseases known to man. "Genome research has already exposed [mutations] in these instructions [DNA] that lead to heart disease, cancer, and neurological degeneration."[16] Genetic mistakes account for most cancers including breast cancer, hereditary diabetes, hemochromatosis, sickle-cell anemia, cystic fibrosis, thalassemia, phenylketonuria, and more than 4,000 other genetic diseases including autism and Down syndrome in children. These diseases are crippling and often fatal.

Bacteria, Viruses, and Insects

Other diseases are the result of mutations that appear within bacteria and viruses. Such creatures are everywhere in abundance and are essential for life on earth. In the beginning (creation) all bacteria and viruses were benign and beneficial in their original form. Some were independent and others had beneficial symbiotic relationships with animals and humans. At the time of the Fall (Genesis 3:6), the degeneration process and ensuing mutational load began to produce genetic mistakes and increasing mutational load in all living things.

Bacteria are microscopic one-celled organisms that are essential to ecology and our life support system. They recycle dead animal and plants; help digest food in human and animals; help clean up the environment by stabilizing pH in soil; recycle nutrients such as carbon, nitrogen and sulfur; and help cleanse our water supply. But if bacteria are infected with mutated genes, they will cause disease.

> Bacteria carrying mutated genes cause diseases ranging from minor skin infections to tuberculosis, pneumonia, cholera, diphtheria, bacterial meningitis, bubonic plague, leprosy, and more.

Viruses are extremely small, non-living entities or proteins that carry genes from one plant or animal to another and replicate upon entering a living cell or bacteria. (A bacteriophage, or phage, is a virus that infects and replicates within bacteria. Phages are used in medicine today to help control bacterial growth.) A common misconception is that viruses serve exclusively as an infectious agent. Researchers are just now beginning to understand that viruses serve genetic functions essential for life. Many viruses exist in host cells without ever causing problems, but if a virus is carrying mutated genes, or there is an accidental movement of genes, the virus becomes an infectious agent (or pathogenic virus) that transmits disease.[17]

> Viruses carrying mutated genes can infect all types of life forms, from animals and plants to bacteria. Viral diseases range from colds, cold sores, and warts to mononucleosis, herpes, shingles, hepatitis B, yellow fever, viral meningitis, chicken pox, mumps, rabies, polio, smallpox, viral pneumonia, AIDS, ebola, and some cancers. Auto-immune diseases associated with viruses include diabetes, lupus, multiple sclerosis, and rheumatoid arthritis.

Pathogenesis, or the development of disease, is evidence of 'something gone wrong'—that is, genetic corruption or mutations. Over time bacteria, viruses, fungi (ringworm and yeast infections), microscopic protozoa (amoeba), parasitic worms (roundworms such as hookworms, whipworms and pinworms; and flatworms such as tapeworms and flukes), plasmodia (malaria causing parasite), and many more infectious agents have mutated. These parasitic agents lost their ability to live independently or symbiotically, and became harmfully dependent on the host—animals and humans.

Insects have beneficial ecological functions such as pollination, aid the decomposition of plants and animals, soil aeration, food for birds, and production of useful products such as honey, beeswax, and silk, but insects are not immune to genetic breakdown and mutational load. Inhospitable insects may include recluse spiders, leeches, millipedes, mites, bed bugs, fleas, and crab louse. Self-defense mechanisms (bites and stings) in the post-Fall world are the result of natural selection, or adaptation due to necessity (predator-prey relationships). Some insects became parasitic as a result of mutations.

> Mutations overwhelmingly cause loss of genetic information. There is no evidence of progressive evolution.[18] Also, all ancient viruses found in 'ancient amber' (fossilized tree resin) are fully formed and functional. See next section, Are There Beneficial Mutations?

Although the eight individuals (Noah, his wife, three sons and their wives) who entered the Ark at the time of the Flood (2385 BC) carried some mutational genetic load, they did not carry all the disease-causing parasites and microbes that we see today. Most, if not all, pre-flood pathogenic organisms died in the flood. But over the next 4,000 years (following the flood), there was plenty of time for bacteria, virus, fungi, protozoa (amoeba), worms, and insects to degenerate into today's pathogenic organisms. What about insects? Insects would have easily survived the flood on massive amounts of floating debris including the ark.

Human Intelligence

As previously mentioned, many people are under the misconception that our generation is the healthiest and most intelligent that has ever lived. This is *not the case*. Modern technology of this generation does not mean that we are the most intelligent or the healthiest—in fact, quite the contrary. We have accumulated technological knowledge while, during the same period, our brains and bodies have endured 6,000 years of mutations and degeneration.

In reality, the observed trend in nature is toward decay, not improvement. Accumulation of damaging effects of random mutations, and "downward drift" of genetic information in living organisms over the many centuries all destroy the validity of organic evolution (as one would expect in light of the Second Law).

For more information, see chapter 2, De-Evolution (and end note #18) and chapter 7, Upside Down Evolutionary Vision of Ancient Man; and the book, *The Genius of Ancient Man - Evolution's Nightmare* by Don Landis (Ed.), Master Books (2012).

Also refer to the book, *World Winding Down*, by Carl Wieland (2012, Creation Book Publishers, Powder Springs, GA). Dr. Wieland provides an easy to understand description of the Second Law of Thermodynamics (i.e., the relentless tendency of all things to become more disorderly with time) and its effect on the human genome.

Are we as intelligent as our early ancestors? "New research shows that mutations are making humans less intelligent and less able to relate emotionally...and in the past 3,000 years...every person has sustained two or more mutations [2.5 to 6 mutations in intellectual deficiency genes} harmful to intellect or emotional functioning."[19] Other genetic studies indicate similar findings.

As described in Genesis, early humans such as Adam, Methuselah, and Noah lived longer than 900 years while still retaining their youthful physical and cognitive functions. This is because mutational load at the time of creation was ZERO—God created humans and animals without flaws (Genesis 1:31). The degeneration and dying process started slowly allowing people to retain their youth and live longer in the beginning. See graph, chapter 2, Entropy and Heat Death.

Over the next 6,000 years life spans gradually decreased as disease and mutations accumulated, until reaching today's average age of 78 years in developed countries. Degeneration and death is not a natural process—it is an abomination. It is not what was originally intended when God created human and animal life.

Are There Beneficial Mutations?

Evolutionists maintain there must have been "beneficial" mutations, or positive random mutations, on occasion to allow uphill drift of genetic information. Although there are small numbers of mutations that make it easier for an organism to survive in an extreme environment, these by definition are "equivocally beneficial." ***None*** are "unequivocally" beneficial or "uphill" in the sense of adding new genetic information to the gene pool.

> **Equivocally beneficial mutations** can add "new functions" to the gene pool in a survival sense in the form of decay—**not new "uphill" increases in genetic information.** Positive random mutations, or equivocally beneficial mutations, are often mistaken by evolutionists as uphill drift.
>
> **Unequivocally beneficial mutations** that may add new genetic information to the gene pool is "nonexistent in nature." If evolution were true, we should find scores of information adding mutations, but **none have ever been found.**

Let's look at a few examples of what may be considered "equivocally" beneficial mutations:[20]

> Wingless beetles that live along the seacoast survive better than winged beetles because they are less likely to be blown away by the wind; or fish in dark caves are able to survive if a mutation causes loss of eye development because without eyes they are not prone to eye disease or injury; or bacteria are able to resist antibiotics because of a deformity in cell wall proteins that interferes with their pumping mechanism; thus, they are unable to pump antibiotics into their system.

> There are always a few individuals in the population with a mutation that allows them to survive under specific environmental conditions. Over time, these individuals dominate the population but there is a **loss of genetic information** (similar to the dog example in the previous section).

Corruption of DNA information necessary for wings or for eyesight or for chemical pumping into the bacteria's interior is *genetic decay* ("downhill" drift), not addition of genetic information to the gene pool. Such defects can add "new functions" to the gene pool in a survival sense in the form of decay or loss—but these are not examples of new "uphill" increases in genetic information—information that may lead to new genetic coding for new structures. There aren't any! There is a loss of genetic information. Gene mutations are overwhelmingly degenerative and constitute further proof of the Second Law. The bottom line is that genetic mutations do not offer any help for evolutionary doctrine.

In an article titled "Mutations: The Raw Material for Evolution?" Dr. Barney Maddox, M.D., explains that most mutations cause the worst diseases known to man (see previous section)—yet biology textbooks focus on very rare equivocally beneficial mutations.[21]

> Virtually all the 'beneficial mutations' known are only equivocally beneficial, not unequivocally beneficial. In bacteria, several mutations in cell wall proteins may deform the proteins enough so that antibiotics cannot bind to the mutant bacteria. This creates bacterial resistance to that antibiotic. Does this support evolutionary genetic theory? No, since the mutant bacteria do not survive as well in the wild as the native (nonmutant) bacteria.
>
> Biology textbooks in theory present positive and negative mutations to students as though these were commonplace and roughly equal in number. However, these books fail to inform students that *unequivocally positive mutations are unknown to genetics, since they have never been observed* (*or are so rare as to be irrelevant*). [Italics added] (p. 11)

> The ratio of "beneficial mutations" (equivocally beneficial) to harmful mutations has been estimated at about 0.000041![22] So assuming an "unequivocal" mutation—a mutation that is nonexistent in nature—at the very least "the next 10,000 mutations...would each be fatal or crippling, and ...**would bring the evolution process to a halt.**" (p. 12) [Bold added]

> While instructing students that harmful mutations were more numerous than 'beneficial' mutations, this textbook failed to disclose that even equivocally beneficial mutations (which still have a downside) are extremely rare (about one in 10,000), and that *unequivocally beneficial mutations are nonexistent in nature.* [Italics added] (p. 13)

Dr. Maddox concludes that "observational (i.e., scientific) evidence, as seen in medical research every day, leads one to be skeptical of the claims of evolutionary biology. How does science explain that mythical first bacterial cell three billion years ago? Did it transform itself—by random mutations in the DNA—into all the 'wondrous profusion' of life forms (one million species), and all their wondrous functional organs, over an imaginary time period? The evidence says no." (p. 13) For more information, see Appendix C, Equivocal and Unequivocal Beneficial Mutations.

In an article titled "Superbugs Not Super After All," Dr. Carl Wieland, M.B., B.S. (medical doctor and surgeon, Australia) provides a similar explanation concerning the mechanisms for bacterial resistance to antibiotics.[23] A "superbug" may be resistant to a variety of antibiotics in several different ways.

- First, resistant bacteria may already be in the population similar to the presence of certain mosquitoes resistant to DDT. The information for resistance was already in a few individuals in the population before the antibiotic was invented (in the case of synthetic antibiotics) or first used clinically. An antibiotic may kill all the bacteria except for a few resistant varieties in the population. Fortunately, in most cases, the body's immune systems will eliminate the remaining few "before [they] can multiply and cause harm, so resistance will not become a problem. However, if this doesn't happen...their offspring will obviously also be resistant."
- Second, bacteria can directly transfer their resistance to others. "One germ inserts a tiny tube into another, and a little loop of DNA called a 'plasmid' transfers from one to another. This sort of gene transfer, which can obviously pass on information for resistance to a drug, can even happen between different species of bacteria." Note that the "information for the resistance must already exist in nature before it can be passed on... this is information transfer, not information creation."
- Third, some bacteria can become resistant through mutations. For example, there are "chemical pumps" in bacteria that effectively move "nutrients from the outside through the cell wall into the germ's interior." Certain antibiotics kill after being pumped into the bacterium's interior in this way. However, if these bacteria inherit a defective gene by way of a mutation that interferes with the pumping mechanism, the loss or reduced efficiency of the pump due to the defect will cause or enhance resistance to such an antibiotic. This is an *equivocally beneficial mutation* (similar to wingless beetles or fish without eyes mentioned earlier). Although these bacteria may not be able to survive in normal circumstances, this defect "actually gives it [the bacterium] a survival advantage" in the presence of the antibiotic. "All such mutations appear to be losses of information, degenerative changes."

Supergerms are "not really super at all—they are actually rather 'wimpy' compared to their close cousins." Unfortunately, supergerms tend to "thrive in hospitals because all the antibiotics and antiseptics being used there keep wiping out the ordinary bacteria which would normally out compete, wipe out and otherwise keep in check these 'superwimps' [supergerms]." This is a big problem for most hospital environments and is one reason why antibacterial soap is not recommended by some medical practitioners. The bottom line is supergerms are "less hardy, and less fit to survive outside of the special conditions in hospitals...and where a mutational defect causes resistance, the survival advantage is almost always caused by a loss of information."

Dr. Lee Spetner of Johns Hopkins University has written a fascinating book called *Not by Chance! Shattering the Modern Theory of Evolution*. Dr. Spetner concludes that "***Not even one mutation*** has been observed that adds a little information to the genome. That surely shows that there are not the millions upon millions of potential mutations the theory demands. There may well not be any. The failure to observe even one mutation that adds information is more than just a failure to find support for the theory. It is *evidence against* the theory. We have here a serious challenge to neo-Darwinian theory."[25] [Italics and bold added] In reviewing Spetner's book, Professor E. Simon, Department of Biology, Purdue University, says: "It is certainly the most rational attack on evolution that I have ever read."[26]

> In no case is there any evidence of an information-adding "uphill" change. Insect adaptations to pesticides and bacterial adaptations to antibiotics provide no evidence that living things evolved from simple to complex by adding new information progressively over millions of years.
>
> "If evolution from goo-to-you were true, we would expect to find countless information-adding mutations. But we have not even found one. Even worse...mutations damage instead of improve the genome of an animal, meaning that mutation causes the breakdown of a species, not its evolution."[24]

> Extremely rare equivocally beneficial mutations can add "new functions" to the gene pool in a survival sense in the form of genetic decay, not new "uphill" increases in genetic information. (See explanation at start of section.) So, is there any evidence of new information being generated by random mutations? An example often cited by evolutionists is nylon-degrading bacteria.
>
> In 1975, Japanese scientists discovered that bacteria, *Pseudomonas* and *Flavobacterium*, degrade nylon compounds and produce new enzymes.[27] (These bacteria were originally described by scientists in 1872 and 1889.) Such findings were viewed by some as an example of new information produced by random mutations and natural selection—but there are substantial empirical evidences to doubt such claims.[28]
>
> In spite of these bacteria being "so ubiquitous, so prolific and so readily adaptable," Dr. Don Batten concludes there is "no evidence of directional change" over a period equivalent to "tens of millions of years of human generations..." The bottom line is, "stasis rules, not progressive evolution." In other experiments designed to detect evolution in fruit fly DNA (after 600 generations) and in *E. coli* DNA (after 40,000 generations), they became either normal, mutant, or dead—but none evolved.[29]
>
> Suppose some trivial cases of increased information occurred amid the billions of mutations throughout earth's history. Does this provide a mechanism or platform for evolution? The answer is *No*—the accumulation of mutations (information-losing processes) is *so overwhelmingly negative* (crippling or fatal) that the so-called evolutionary process would come to an *immediate halt*. See information at start of section, article by Dr. B. Maddox, M.D., titled "Mutations: The Raw Material for Evolution?"

Race and Human Morphology

Racial Groups

In our modern world, anthropologists classify people into racial groups, such as the Caucasoid (European), Mongoloid (Chinese, Eskimo), Native Americans, Ethiopian (later termed Negroid or black Africans), Australoid (Australian Aborigines), and Malayan race.[30] These racial groups have notably different morphological features such as skin and hair color. As previously mentioned, variations in human groups or races are not evolutionary changes but merely reshuffling of genes within the "existing gene pool" that was present in the original group or human "kind" (that is, Noah and his family) in response to the environment.

Modern geneticists agree that all human races are extremely close biologically, which is consistent with all racial characteristics being present in one small ancestral human "kind" population (common gene pool at the beginning of early civilization). For example, many people do not realize that there is just one skin coloring pigment in mankind. The shade of brown depends on how much a person produces the substance called melanin.[31]

If a person has a lot of melanin (darker skin) and lives in a northern country with little sunshine, it will be difficult to absorb enough vitamin D, which would cause rickets.[32] If, on the other hand, a person has too little melanin (lighter skin) and lives in a lower-latitude country with lots of sunshine, skin cancer may develop. Consequently, there is a natural selection or "thinning out" of the gene pool of a population "kind" in response to a specific environment.

> Darker skin is able to dissipate UV radiation as heat and lighter skin is able to absorb more vitamin D through exposure to sunlight.

Following the worldwide flood and onset of the Great Ice Age (see chapters 4 and 10), the descendants of Noah eventually moved to remote areas of the world (see chapter 10, The Postdiluvian Period), and the pressure of severe winter conditions affected the balance of genes. During this period of history, physical characteristics of subgroups of people began to adapt to their specific environment. For example, people of higher latitudes began to acquire pale skin and people of lower latitudes began to acquire dark skin.

But this is not always the case. The Inuit (Eskimos) have brown skin yet live where there is not enough sun, and native South Americans have lighter skin yet live near the equator. Such examples validate that natural selection does not create new information (that is, expand gene pools)—if the gene pool of the subgroup does not comprise the variation gene, natural selection cannot create the variation (e.g., the dark skin gene that is lacking in native South Americans, or the light skin gene that is lacking in Eskimos). Obviously, genetics (the original gene pool) of each group is more important than the environment.

There is just one race of people and that is the "human race," which came into existence following Noah's Flood. The first man, Adam, from whom all other humans are descended, was created with the best possible combination of genes and the largest gene pool—for skin color, intelligence, and so forth. Noah and his family were probably mid-brown, possessing genes for both dark and light skin, and the world's population today is still predominantly mid-brown.

Chapter 10 provides a more detailed discussion of racial groups and early civilization. Also see Dr. Carl Wieland's book, *One Human Family: The Bible, Science, Race & Culture* (2011, Creation Book Publishers, Powder Springs, GA) for a captivating and informative review of this subject.

Genetics of Humans and Apes

Over the last several decades geneticists have shown that DNA sequences of humans and apes, especially chimpanzees, are comparable. Similarity of DNA sequences has ranged from 94% to 99%, depending on the particular study, although 95% is considered a better estimate overall. When the chimpanzee genome was mapped in 2005, scientists announced a 96% similarity,[33] and then another study in 2006 revealed a 94% genetic similarity.[34]

Chimpanzee

Evolutionists claim that DNA similarity and existence of similar structures are caused by a common ancestor and, therefore, prove evolution. For example, evolutionists maintain that various appendages are similar (i.e., forelimbs of mammals, wings of bats or birds, and arms of primates). So are such comparisons meaningful? Does this mean that humans evolved from a common ancestor and are directly related to chimpanzees? Absolutely not—such a leap is absurd! There is no proof of anything. A much stronger argument can be made that such characteristics would be expected with Divine Creation.

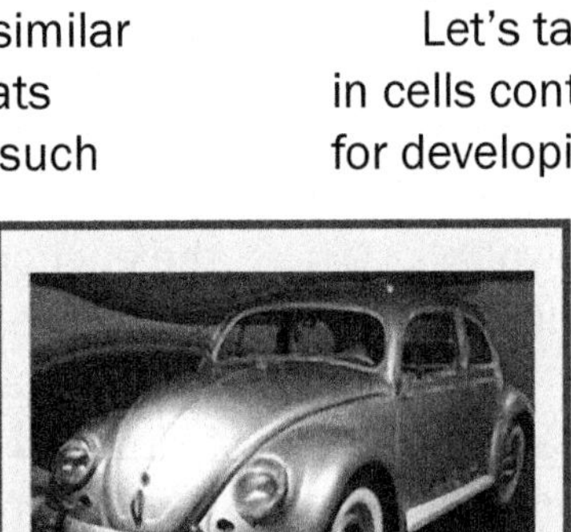
Volkswagen Beetle

When an engineer is given a design problem, he tries to find an "optimal" design to achieve the greatest efficiency and beauty. Designed machines do not occur by random chance. For example, many different automobiles comprise similar parts (for instance, wheels, brakes, axles, bearings, shocks, steering, front and rear lights, seals, belts and hoses, gears, levers, and pistons, transmission, drive shaft, radiator, fuel pump and tank, electrical wiring, and battery.)

Porsche 356, the first Porsche sports car

An example is the Porsche sports car and Volkswagen Beetle car, both German automobiles. They both have air-cooled, rear-mounted engines, two side doors, trunk in the front, and many other obvious similarities. The Volkswagen was initially called the *Porsche 60* by Ferdinand Porsche—and Porsche's chief designer, Erwin Komenda, was responsible for the design and style of the car. "Why do these two very different cars have so many similarities? Because they had the same designer."[35] Similarity of humans and chimps—morphologic (appearance) and internal (organs)—shows a common designer, not a common ancestor.

Let's take a look at genetic similarity. DNA in cells contains all the information necessary for developing an organism. If two organisms look similar, we would expect similarity to be reflected in their DNA. Apes are partially upright with clasping hands at the end of their forelimbs, and there is likely just one optimal way to design internal organs functioning interdependently—brain, heart, lungs, stomach, teeth, tongue, eyes, ears, skin, muscles, bones, a nervous and vascular system, and all the other organs, all working together. Genetic similarity is the result of structural and functional similarity. The genetic code is so enormously complex that it is impossible such information could have arisen by random chance, all working interdependently without a Creator.

"But in fact God has arranged the parts of the body, every one of them, just as he wanted them to be." (1 Corinthians 12:18, NIV)

Let's take a closer look at the 94% to 96% genetic similarity between apes and humans. Using the same method of determining similarity (that is, comparing DNA sequences), people are 88% the same as rodents (mouse, rat, squirrel, or beaver), 60% the same as chickens, and 80% the same as sea squirts.[36] In this context, a 95% similarity with chimps isn't so remarkable. Using the evolutionists' chosen method of genetic comparison (that is, comparing sequences), who does this make us closer to after chimps? Rats and sea squirts.

Additionally, does similarity of DNA sequences mean that two DNA strands have the same meaning or function? The answer is No. Millions of DNA insertions and deletions along a 3-billion nucleotide sequence may cause much of the genetic difference between human and animal groups, but are not included in these percentages of similarity. One study considered insertions and deletions in its research and found a 86.7% similarity between humans and chimps.[37] It is one thing to map the human genome but deciphering and translating the genetic code (i.e., chromosomes, genes, proteins and amino acids, and other translation machinery) is a completely different matter.

Let's look at just a few examples of genetic dissimilarities between humans and chimps that are not reflected in the comparison of DNA sequences. Chimps have 24 pairs of chromosomes and humans have 23 pairs.[38] It has been shown that chimps produce only 29% of the same proteins as humans.[39] If 29% are the same, then 71% are different!

Although 'DNA sequences' are similar (just as they are in rodents, sea squirts, and chickens), deciphering the genetic code is another story. In addition to protein differences, the chimp genome is larger (about 12%) than the human genome, possesses genes not present in the human genome, and lacks genes present in the human genome. Also, there are many other unknowns in the chimpanzee genome, such as the neglected "informationally-rich" so called 'junk-DNA' (non-protein-coding regions).[40]

Obviously, the complex translation machinery is not yet fully understood—scientists have just scratched the surface in deciphering this information. The comparison of DNA sequences does not mean what evolutionary scientists claim—in fact, much of this research is "flawed and biased." In his article "Human Chimp Similarities," Dr. J. Tompkins states that "some of the most critical DNA sequence is often omitted...[and] only similar DNA sequences are selected for analysis...[thus] similarity becomes biased on the high side."[41]

In a recent high-profile article in the journal *Nature* (January 2010), scientists have contradicted long-held claims of DNA similarity between humans and chimps.[42] This new research found that "chimpanzee and human Y chromosomes are remarkably divergent in structure gene content." But even in the light of contrary data, researchers make up stories (e.g., rapid evolution and evolution by gene loss and decay) in an attempt to validate evolutionary doctrine. In response to this research, J. Tompkins and B. Thomas comment, "when all aspects of non-similarity—sequence of categories, genes, gene families, and gene position—are taken into account, it is safe to say that the overall similarity was lower than 70 percent."[43]

The bottom line is humans and apes were created separate and distinct. To put it simply, in the real world, genetic drift has always been downhill consistent with the Second Law—not uphill as many evolutionists would have you believe. No one has ever observed "uphill" drift or new, genuine increases in genetic information.

Uniqueness of Human Beings

Forms of Life and Love

Many of the obvious similarities and differences between humans and apes are morphologic—that is, similarities of appearance. What makes humans unique is not necessarily morphologic or genetic, but intangible differences such as creativity, culture, abstract reasoning, moral judgments, and a spiritual connection to our Creator.[44]

The Bible makes the claim that humans alone are "created in the image of God." In Genesis 1:26-27 (NIV), *"Then God said, 'Let us make man in our image, in our likeness, and let them rule over the fish of the sea and the birds of the air, over the livestock, over all the earth, and over all the creatures that move along the ground.' So God created man in his own image, in the image of God he created him; male and female he created them."*

What makes humans unique from all the other creatures? No other species of animal, including apes, is able to create and understand images of art such as paintings and sculpture, and appreciate the beauty of nature. In abstract thinking, for example, humans constantly invoke unobservable images to explain events or why things happen whereas other creatures operate in the observable world of tangible things—that is, the world that can be seen.

The ability to make "moral judgments" is another characteristic that is only found in humans. Even apes are unable to make moral judgments about the behavior of other animals. No other species of animal, including apes, displays a sense of conscience—that is, pride, remorse, shame, and guilt.

Another major difference is the spiritual connection between humans and the Creator. Both Adam and Eve had a personal relationship with God. Such a personal relationship is not described for any other animal species. It is the presence of a "spirit" that separates humans from animals. As stated in Genesis 2:7 (NIV), *"...the Lord God formed the man from the dust of the ground and breathed into his nostrils the breath of life, and the man became a living being."*

There are three forms of life that God has created in this universe:[45]

1. Body only. Examples are reptiles, amphibians, fish, invertebrates and other lower forms of life.
2. Body and Soul. Examples are birds and mammals.
3. Body, Soul, and Spirit. Examples are humans and angels.

The soul is the eternal part of a living being, separate from the body, and it is essential to consciousness and personality—the soul may be synonymous with the mind, will, and emotion. (Consciousness is awareness; personality is a psychological portrait of a person; mind is intellect that includes thought, perception, memory, emotion, and imagination; will is the ability to make choices free from limitations (free will); and emotion is associated with mood, temperament, personality, disposition or habit, and motivation.) Only birds and mammals display these characteristics, which is why humans can form close relationships with birds and mammals.

The spirit is that part of humans that is able to love and experience God directly—it allows humans (and angels) to have a relationship with the Creator. The soul and spirit are connected but separate (Hebrews 4:12)—the soul is who we are, but the spirit connects us with God. In 1 Thessalonians 5:23 (NAS), *"Now may the God of peace Himself sanctify you entirely; and may your spirit and soul and body be preserved complete, without blame at the coming of our Lord Jesus Christ."* And in Romans 8:16 (NAS), *"The Spirit Himself bears witness with our spirit that we are children of God."*

In addition to these forms of life, there are four forms of love: friendship (platonic), romantic, natural affection (familial), and agape. The first three can be understood in terms of shared mutual benefit that is found in all animals. Agape is a self-sacrificing love found only in humans. It is described in John 3:16 (NIV), *"For God so loved the world that he gave his one and only Son, that whoever believes in him shall not perish but have eternal life."* The word translated "love" in this verse is *agape*. When 1 John 4:8 says "*God is love*," the Greek New Testament uses the word agape to describe God's love.

Human Nature and the Moral Law

When we discuss the laws of nature we usually refer to the laws of chemistry, physics, and biology. Such laws include thermodynamics, gravitation, motion, fluid dynamics, buoyancy, elasticity, partial pressure, biogenesis, and heredity. But there is another law that is just as valid—a Moral Law that was examined by C. S. Lewis in his book, *Mere Christianity*—a universal law of decent behavior or morality—a law of "right and wrong" behavior. Unlike other natural laws, the Moral Law is a law that human beings are free to disobey.[46] (See Nature of Mankind in the Prologue.)

Since the rebellion by Adam, humanity has had an inherited or inborn tendency to sin and rebel—to oppose the Creator's sovereignty over their lives. In the days of Noah (4004 BC - 2385 BC), mankind was rebellious and exceedingly wicked—and with this came eventual consequences—the Flood that destroyed all of humanity. Rebellion and wickedness have been continually present throughout the history of mankind.

In opposition to this tendency, the Moral Law motivates humans to make personal sacrifices that may lead to suffering, injury, and even death without the possibility of personal benefit. Such behavior is in direct conflict with apes (nonhuman primates) and other animals. According to Lewis, man has two impulses: the first is to provide help and the second is to avoid danger, but the overriding drive is to help—not to run away. (p. 9-10) Although we are free to disobey (free will) this inner voice of conscience, the motivation to do the right thing is real. The Moral Law will ask me to save a drowning man with personal sacrifice, even if he is an enemy or stranger. (p. 10)

The reality is, not one person keeps the Moral Law all the time and many of us never keep the law at all—and in the act of not keeping the law, most humans try to shift the blame or alleviate their wrong behavior with all types of excuses. Quarrelling between two individuals, for example, is "appealing to some kind of standard of behavior which he expects the other man to know about." (p. 3) Lewis sums it up: "human beings, all over the world, have this curious idea that they ought to behave in a certain way...[but] they do not in fact behave in that way." (p. 8) The question is, what lies behind this law? (p. 13)

As Lewis explains, "If there was a controlling power *outside* the universe, it could not show itself to us as one of the facts inside the universe—no more than the architect of a house could actually be a wall or staircase or fireplace in that house. The only way in which we could expect it to show itself would be inside ourselves as an influence or a command trying to get us to behave in a certain way. And that is just what we do find inside ourselves. Surely this ought to arouse our suspicions?" (p. 24) [Italics added]

God wants man to behave in a certain way, urging man to do what is right and just, and if he does not, he is made to feel uncomfortable. (p. 23) One must conclude that this must be a holy and righteous God who despises evil. This is certainly proclaimed throughout the Bible. If God is outside the universe—outside the natural world—the tools of science cannot find God. Ultimately, the decision to acknowledge and accept God is based on personal choice and faith (Matthew 17:20; Romans 1:17, 3:22, 3:28; and 2 Corinthians 5:7, NIV, "*We live by faith, not by sight.*")

Summary

The major problems for evolutionists are the lack of any evidence for spontaneous generation of life from nonlife and lack of any credible genetic mechanism for an expanding, more complex gene pool. Biological research has been unable to show or agree how evolution occurs on a genetic level or even to provide examples of evolution occurring today. What we observe today are distinct groups of plants and animals (mankind, horse kind, dog kind, cat kind, etc.) with no transitional forms anywhere in the world.

Gene mutations are *overwhelmingly degenerative* and although evolutionists claim uphill drift was the result of natural selection and random mutations, *all scientific evidence is to the contrary*. No one has ever observed "uphill" drift or new, genuine increases in genetic information.

> In most cases, genetic decay and increasing mutational load since the time of Adam and Eve are the *primary source of diseases and deformities, and eventual death*. Pathogenesis (disease) is consistent with the 2nd Law of Thermodynamics (chapter 2).

Diversity of species within "kinds" of animals and variations in human racial groups are not evolutionary changes but, rather, a "thinning out" process that leads to a loss of genetic information. Speciation (natural selection) is a genetic process within populations that selects favorable gene traits from an "existing gene pool" best suited for a specific environment. Genetic drift has always been "downhill"—consistent with the Second Law.

What about morphologic similarity among animals? Evolutionists claim that similar structures are caused by a common ancestor. The stronger argument is that all animals had the same designer—similarity of structures shows a common designer, not a common ancestor. Additionally, the uniqueness of humans (apart from all other animals) and the moral law point directly to a holy and righteous God who creates every "kind" of life.

Consider the following quotes:[47]

Origins of Life

"If one believes in evolution, then one has to also account for the origin of life—the very first step. Without this, the whole subject of evolution hangs on nothing." —Stephen Grocott, *In Six Days*, 2000, p. 148. (Ph.D., Organometallic Chemistry)

"The question 'How did life originate?' which interests us all, is inseparably linked to the question **'Where did the information come from?'** Since the findings of James D. Watson and Francis H. C. Crick, it was increasingly realized... that the information residing in the cells is of crucial importance for the existence of life. Anybody who wants to make meaningful statements about the origin of life, would be forced to explain how the information originated. All evolutionary views are fundamentally unable to answer this crucial question." —Dr. Werner Gitt, *In the Beginning Was Information*, 1997, p. 99. (Gitt was a director and professor at the German Federal Institute of Physics and Technology.)

"All of us who study the origin of life find that the more we look into it, the more we feel that it is too complex to have evolved anywhere. We believe as an article of faith that life evolved from dead matter on this planet. It is just that its complexity is so great, it is hard for us to imagine that it did." —Harold C. Urey, quoted in *Christian Science Monitor*, January 4, 1962, p. 4.

"In fact, evolution became, in a sense, a scientific religion; almost all scientists have accepted it and many are prepared to *bend* their observations to fit with it. To my mind, the theory does not stand up at all. If living matter is not, then, caused by the interplay of atoms, natural forces, and radiation, how has it come into being? I think, however, that we must go further than this and admit that the only acceptable explanation is Creation." —H.S. Lipson, "A Physicist Looks at Evolution," *Physics Bulletin*, 1980, vol. 31, p. 138.

"The origin of life is still a mystery. As long as it has not been demonstrated by experimental realization, I cannot conceive of any physical or chemical condition [allowing evolution] ... I cannot be satisfied by the idea that fortuitous mutation ... can explain the complex and rational organization of the brain, but also of lungs, heart, kidneys, and even joints and muscles. How is it possible to escape the idea of some intelligent and organizing force?" —Merle d'Aubigne, "How Is It Possible to Escape the Idea of Some Intelligent and Organizing Force?" In *Cosmos, Bios, Theos*, Margenau and Varghese (Eds.), p. 158.

"Scientists have no proof that life was not the result of an act of creation, but *they are driven by the nature of their profession to seek explanations for the origin of life that lie within the boundaries of natural law.* They ask themselves, 'How did life arise out of inanimate matter? And what is the probability of that happening?' And to their chagrin they have no clear-cut answer, because chemists have never succeeded in reproducing nature's experiments on the creation of life out of nonliving matter. Scientists do not know how that happened, and furthermore, they do not know the chance of its happening." [Italics added] —Robert Jastrow, *The Enchanted Loom: Mind in the Universe*, 1981, p. 19.

"Often a cold shudder has run through me, and I have asked myself whether I may have not devoted myself to a fantasy." —Charles Darwin, *The Life and Letters of Charles Darwin*, 1887, vol. 2, p. 229.

Genetic Code - DNA and RNA

"One cell division lasts from 20 to 80 minutes, and during this time the entire molecular library, equivalent to one thousand books, is copied correctly." —Werner Gitt, professor of physics and technology, *In the Beginning Was Information*, 1997, p. 90.

"But where the first RNA came from is a mystery; it's hard to see how the chemicals on early Earth could have combined to form the complicated nucleotides that make up RNA." —John R. Davenport, "Possible Progenitor of DNA Re-Created," *Science Now*, November 16, 2000, p. 1.

"The problem is not as simple as might appear at first glance. Attempts at engineering—with considerably more foresight and technical support than the prebiotic world could have enjoyed—an RNA molecule capable of catalyzing RNA replication—have failed so far." — C. de Duve, "The Beginning of Life on Earth," *American Scientist*, 1995. (De Duve is a Nobel Prize-winning biochemist.)

"If the DNA of one human cell were unraveled and held in a straight line, it would literally be almost one meter long and yet be so thin that it would be invisible to all but the most powerful microscopes. Consider that this string of DNA must be packaged into a space smaller than the head of a pin and that this tiny string of human DNA contains enough information to fill almost 1,000 books, each containing 1,000 pages of text.... For compactness and information-carrying ability, no human invention has even come close to matching the design of this remarkable molecule." —John P. Marcus, *In Six Days*, 2000, pp. 174-175. (Ph.D., biochemistry)

"There is no known law of nature, no known process and no known sequence of events which can cause information to originate by itself in matter." —Werner Gitt, *In the Beginning Was Information*, 1997, p. 107 (Professor of physics and technology)

"If they eventually make a computer as small as a cell with a huge information storage capacity like DNA, and I scoff and claim: 'You didn't do that! It just came about by accident,' they will rightly consider me a fool." —Thomas F. Heinze, *How Life Began*, 2002, p. 107.

"Common sense says that the amazing complexity of life cannot arise out of random process." —Lee Spetner, *Not By Chance!* 1997, p. 75. (Ph.D. in physics from MIT, years of research in genetics at Johns Hopkins University)

"... Among all the mutations that have been studied, there aren't any known, clear, examples of a mutation that has added information." —Lee Spetner, *Not By Chance!* 1997, p. 131.

"I have carefully studied molecular, biological, and chemical ideas of the origin of life and read all the books and papers I could find. Never have I found an explanation that was satisfactory to me. The basic problem is with the original template (be it DNA or RNA) that would have been necessary to initiate the first living system that could undergo biological evolution. Even reduced to the barest essentials, this template must have been very complex indeed. For this template and this template alone, it appears it is reasonable at present to suggest the possibility of a creator." —Henry Margenau, *Cosmos, Bios, Theos,* 1992. (Margenau is professor of physics and natural philosophy at Yale and author of over 200 research articles and 14 books.)

"Darwinian evolution only has chance mutations at its disposal. Because no 'advance thinking' can possibly be allowed, there is no way that the nucleotides can arrange themselves in a 'predefined code,' since this assumes prior knowledge. Thus, the very existence of the DNA-coded language stalls evolution at the first hurdle." —Andrew McIntosh, *In Six Days*, 2000, p. 160. (Ph.D., mathematics)

"The likelihood of life having occurred through a chemical accident is, for all intents and purposes, zero. This does not mean that faith in a miraculous accident will not continue. But it does mean that those who believe it do so because they are philosophically committed to the notion that all that exists is matter and its motion. In other words, they do so for reasons of philosophy and not science." —Robert Gange, *Origins and Destiny*, 1986, p. 77. (Ph.D., a research scientist in the field of cryophysics and information systems)

"It was fashionable in the middle part of the twentieth century to attribute biological information and complexity to chance plus time. However, as our understanding of the enormous biochemical complexity associated with the origin of life and the development of more complex forms has matured, appeals to chance have gradually lost credibility." —Walter Bradley, "Design or Designoid," *Mere Creation: Science, Faith & Intelligent Design*, 1998, p. 41. (Ph.D., mechanical engineering, and author of over 100 publications in material science)

"But no one has demonstrated how RNA could have formed before living cells were around to make it. According to Scripps Research Institute biochemist Gerald Joyce, RNA is not a plausible candidate for the first building block of life 'because it is unlikely to have been produced in significant quantities on the primitive earth.' Even if RNA could have been produced, it would not have survived long under the conditions thought to have existed on the early Earth." —Jonathan Wells, *Icons of Evolution: Science or Myth*, 2000, p. 23. (Ph.D., molecular and cell biology, Berkeley)

notes: Chapter 3

1. de Duve, Christian (September–October 1995). The beginnings of life on earth. *American Scientist*, 428. Also cited in Morris, J.D. (2003). *Is the Big Bang Biblical?* Green Forest, AR: Master Books, 78–79.

2. Thaxton, Bradley, and Olsen. (1984). *The Mystery of Life's Origin: Reassessing Current Theories*, NY: Philosophical Library, 80; as cited in Riddle, M. (2010). Chapter 6: Can Natural Processes Explain the Origin of Life. Retrieved February, 2012, from http://www.answersingenesis.org/articles/nab2/natural-processes-origin-of-life.

3. Morris, H. (December 1997). *That Their Own Words May Be Used Against Them*. Green Forest. AR: Master Books, 55.

4. Riddle, M. (2010), op. cit., 8. Also see Ham, K. (Ed.) (2008). *The New Answers Book 2*. Green Forest, AR: Master Books, 72.

5. Fox News.com. Leading atheist philosopher concludes God's real. Associated Press article, December 9, 2009. Retrieved May 2010, from http://www.foxnews.com/story/0,2933,141061,00.html. Also see EveryStudent.com. Is God real? Why the DNA structure points to God. Retrieved April 2010, from http://www.everystudent.com/wires/Godreal.html; and http://www.healthy-elements.com/atheists.html. (Dr. Antony Flew, Professor of Philosophy, and former atheist, author, and debater.)

6. Truth and Science Ministries, In Search of the Truth. Retrieved April 2008, from http://www.truthandscience.net/quotes.htm. Also see Morris, H.M. (1997). *That Their Words May Be Used Against Them*. Green Forest, AR: Master Books, 51, 58.

7. Paley, William (First published 1802; reissued 2008). *Natural Theology*. (Mathew D. Eddy and David Knight, Ed., 2006). Published as an Oxford World's Classics Paperback, 2006. New York: Oxford University Press Inc. Also retrieved from http://ourworld.compuserve.com/homepages/rossuk/Paley.htm.

8. Schultz, Duane P. (October 1981). *A History of Modern Psychology*. Section Titled: *The Evolution Revolution: Charles Darwin (1809–1882)*, Harcourt Brace Jovanovich Inc., 114. Also retrieved from http://www.soulcare.org/Creation/Evolution.html.

9. Quote has been used in numerous articles about evolution. Source unknown.

10. Evolutionism: Is evolution a religion? Northwest Creation Network, Mountlake Terrace, WA. Retrieved December 2007, from http://nwcreation.net/evolutionism.html (contact@nwcreation.net).

11. Whitcomb, J.C. and Morris, H.M. (1961). *The Genesis Flood*. Phillipsburg. NJ: The Presbyterian and Reformed Publishing Co, 66.

12. Sarfati, J. (2011). *Refuting Evolution 2*. Atlanta, GA: Creation Book Publishers, 71-74.

13. Truth and Science Ministries, In Search of the Truth. Retrieved April 2008, from http://www.truthandscience.net/quotes.htm

14. Løvtrup, Søren (1987). *Darwinism: The Refutation of a Myth*. New York, NY: Croom Helm, 422; as cited in Morris, H.M. (1997). *That Their Words May Be Used Against Them*. Green Forest, AR: Master Books, 117.

15. Catchpoole, D., Sarfati, J., and Wieland, C. (2008). *The Creation Answers Book*. (D. Batten, Ed.). Atlanta, GA: Creation Book Publishers, 132-139.

16. The Human Genome Project, Announcement from the University of Texas Southwestern Medical School, May 6, 1993; as cited in Maddox, B. (September 2007). Mutations: The raw material for evolution? *Acts & Facts*, 36 (9), Dallas, TX: Institute for Creation Research, 11. Copyright © 2007 Institute for Creation Research, used by permission. Maddox, B. (2007). Mutations: The raw material for evolution? *Acts & Facts*, 36 (9): 10-13.

17. Bergman, J. (April 1999). Did God make pathogenic viruses? *Journal of Creation*, 13 (1): 115-125; and Herrmann, B. and Hummel, S. (1994). *Ancient DNA*. Springer, New York, NY.

18. Ibid.

19. Stanford Scientist: Humans are getting less intelligent, *Creation*, 35 (2), 7-8. Also see creation.com/Sanford; creation.com/time-genetic; and creation.com/dna-repair; see medicalxpress.com, 12 November 2012. Also see Crabtree, G.R. (2013). Our fragile intellect. *Trends in Genetics*, 29 (1), 1-3; as cited in Thomas, B. (May 2013). Is Mankind Getting Dumber. *Acts & Facts*, 42 (5), 17.

20. Wieland, C. (2001). *Stones and Bones*. Green Forest, AR: Master Books, 24-25. Also see Wieland, C. (December 2003). Beetle bloopers. *Creation*, 19 (3): 30; and Wieland, C. (August 9, 2000). New eyes for blind cave fish? http://creation.com/new-eyes-for-blind-cave-fish. Also see end notes 21 and 23.

21. Maddox, B. (September 2007). Mutations: The raw material for evolution? *Acts & Facts*, 36 (9), Dallas, TX: Institute for Creation Research, 13. Copyright © 2007 Institute for Creation Research, used by permission. Maddox, B. (2007). Mutations: The raw material for evolution? *Acts & Facts*, 36 (9): 10-13.

22. Sanford, J. (2005). *Genetic entropy and the mystery of the genome*. Lima, NY: Elim Publishing, 26, as cited in Maddox (2007).

23. Wieland, C. (December 1997). Superbugs not super after all. *Creation*, 20 (1): 10–13. Retrieved September 2008, from http://creation.com/superbugs-not-super-after-all. Also see Wieland, C. (1994). Antibiotic resistance in bacteria. *Journal of Creation*, 8 (1): 5–6. Permission received on August 21, 2009. (Dr. Wieland is Managing Director of Creation Ministries International (Australia) and founding editor of *Creation* magazine. CMI is the distributor of *Creation* magazine and the *Journal of Creation*.)

24. Sarfati, J. (2011). *Refuting Evolution 2.* Powder Springs, GA: Creation Book Publishers, 160; and Williams, A. (2008). Mutations: Evolution's engine becomes evolution's end. *Journal of Creation*, 22 (2): 60-66; Sanford, J. (2008). *Genetic Entropy and the Mystery of the Genome.* Waterloo, NY: FMS Publications.

25. Spetner, Lee (1996). *Not by Chance! Shattering the Modern Theory of Evolution.* Brooklyn, NY: The Judaica Press, 160. Also see Sid Galloway. (1998, updated 2006). Questions & evidence for evolutionists as well as "progressive" creationists. Retrieved May 2008, from http://www.soulcare.org/Creation/Evolutionary_Questions.html.

26. Spetner, Lee (1996), op. cit., back book cover. Also see Galloway, Sid. (1998, Updated 2006). Questions & evidence for evolutionists..., op. cit.; and Sid Galloway (June 2003). Evolution: A racist religion..., op. cit.

27. Kinoshita, S., Kageyama, S., Iba, K., Yamada, Y., and Okada, H. (1975). Utilization of a cyclic dimer and linear oligomers of ε-aminocapronoic acid by *Achromobacter guttatus* K172. *Agric. Biol. Chem.,* 39 (6):1219–1223. Note: *A. guttatus* K172 syn. *Flavobacterium* sp. K172.

28. Batten, D. (December 2003). The adaptation of bacteria to feeding on nylon waste. *Journal of Creation,* 17 (3): 3-5. Retrieved December 2011, from http://creation.com/the-adaptation-of-bacteria-to-feeding-on-nylon-waste.

29. Thomas, Brian (2012). Four scientific reasons that refute evolution. *Acts & Facts*, 19. (Note: Some bacteria accessed citrate for food, but this was equivocally beneficial; that is, the new function resulted from genetic decay, not the addition of genetic information to the gene pool.)

30. Race (classification of human beings) (January 5, 2009). In Wikipedia, the free encyclopedia. Retrieved September 2008, from http://en.wikipedia.org/wiki/Race_(classification_of_human_beings).

31. Melanin exists in the plant and animal kingdoms, where it serves as a pigment. In humans, melanin is found in skin and hair, in pigmented tissue of the iris, in the adrenal gland, in the inner ear, and in brain tissue, and it is the primary determinant of skin color. Melanin is able to dissipate more than 99.9% of the absorbed UV radiation as heat. Wikipedia, the free encyclopedia.

32. Rickets is a softening of the bones in children, potentially leading to fractures and deformity. It is among the most frequent childhood diseases in many developing countries. The predominant cause is a vitamin D deficiency, primarily from inadequate sun exposure; additionally, lack of adequate calcium in the diet may lead to rickets. Wikipedia, the free encyclopedia.

33. Boyle, A. (September 1, 2005). Chimp genetic code opens human frontiers. *MSNBC Science & Technology*. Retrieved from http://www.msnbc.msn.com/id/9136200/.

34. Minkel, J.R. (December 19, 2006). Human-chimp gene gap widens from tally of duplicate genes. Retrieved June 2009, at http://www.scientificamerican.com/article.cfm?id=human-chimp-gene-gap-wide.

35. Batten, Don (2001). Does the DNA similarity between chimps and humans prove a common ancestry? Retrieved July 2008, from www.christiananswers.net/q-aig/aig-c018.html. See *Christian Answers Network* website, www.christiananswers.net.

36. Gunter, Chris, and Dhand, Ritu (September 1, 2005). The chimpanzee genome. *Nature,* 437 (7055).

37. Anderson, Daniel (2006). Decoding the dogma of DNA similarity. Retrieved July 2008, from http://creation.com/decoding-the-dogma-of-dna-similarity. Also see Anzai, T. et al. (2003). Comparative sequencing of human and chimpanzee MHC class I regions unveils insertions/deletions as the major path to genomic divergence. *Proceedings of the National Academy of Sciences, USA,* 100 (13): 7708–7713; Nelson, Warren J. (2004). Human/chimp DNA similarity continues to decrease: counting indels. *Journal of Creation*, 18 (2): 37–40. Retrieved from http://creation.com/human-chimp-dna-similarity-continues-to-decrease-counting-indels.

38. The human-ape connection. Retrieved July 2008, from http://www.cs.unc.edu/~plaisted/ce/apes.html.

39. McBrearty, S. and Jablonski, N. (September 2005). Initial sequence of the chimpanzee genome and comparison with the human genome. *Nature*, 437 (7055), abstract.

40 Thomas, B. (March 2010). Evolution's best argument has become its worst nightmare: How functional transposons refute "Junk DNA" and human evolution. *Acts & Facts,* 39 (3), Dallas, TX: Institute for Creation Research, 16-17; Anderson, op. cit. and Batten, D. (2005). No joy for junkies. *Journal of Creation*, 19 (1): 3. Retrieved June 2009, from http://creation.com/no-joy-for-junkies.

41. Thompkins, J. (June 2009). Human-chimp similarities. *Acts & Facts*, 38 (6), 12-13.

42. Hughes, J.F. et al. (2010). Chimpanzee and human Y chromosomes are remarkably divergent in structure gene content. *Nature*, 463 (7280): 536-539.

43. Thompkins, J. and Thomas, B. (April 2010). New chromosome research undermines human-chimp similarity claims. *Acts & Facts*, 39 (4), 4-5.

44. Deem, Rich (September 2007). Man, created in the image of God: How mankind is unique among all other creatures on earth. Retrieved July 2008, from http://www.godandscience.org/evolution/imageofgod.html.

45. Ibid.

46. Lewis, C.S. (1952). *Mere Christianity*. New York, NY: Harper Collins Publishers, chapters 1-4. Also see Collins, Francis S. (2006). *The Languages of God.* New York: Free Press, chapter 1.

47. Truth and Science Ministries, In Search of the Truth. Retrieved April 2008, from http://www.truthandscience.net/quotes.htm.

Chapter 4

Catastrophic Phenomena in Geology

"Then God said, 'Let the waters below the heavens be gathered together into one place, and let the dry land appear'; and it was so." —Genesis 1:9, NAS

The doctrine of uniformitarianism maintains that geological and other physical processes operating in the world today have remained constant throughout earth's history. It is the belief that God has never intervened and judged the world through a worldwide flood—primarily because we are told by atheistic academia and evolutionists that this is the case.

Catastrophism maintains that normal geological and physical processes of the earth have been interrupted by a cataclysmic worldwide flood. According to the Bible, there have been two great worldwide upheavals since the beginning of time: Original Creation (Genesis 1) and Noah's Flood (Genesis 6-8). Because there has been no worldwide catastrophic event in recent times, most people believe there is no reason for concern—that everything will remain constant and life will continue on as before.

Uniformitarianism (uniformity theory) maintains that geological, meteorological, and other physical processes have remained constant during earth's history. It is the belief that existing natural processes are sufficient to account for all past changes in our world and the universe. Essentially, the doctrine assumes that conditions on earth have never been interrupted by worldwide cataclysmic events. This principle states that scientific laws and processes operating today operated in the past.

Catastrophism maintains that normal geological and physical processes of the earth have been interrupted by a cataclysmic worldwide flood as described in Genesis 6-8. Does a worldwide flood covering the mountains seem far-fetched from a geologic and hydraulic standpoint? There is a logical explanation and an impressive array of scientific evidence for a global flood and recent creation of the earth.

The apostle Peter describes these skeptics, *"Know this first of all, that in the last days mockers will come with their mocking, following after their own lusts,* [they ridicule the Bible] *and saying, 'Where is the promise of His coming?* [they deny the Second Coming of Jesus the Messiah] *For ever since the fathers fell asleep, all continues just as it was from the beginning of creation.' For when they maintain this* [that is, they maintain uniformitarianism and evolution], *it escapes their notice* [they 'deliberately forget'] *that by the word of God the heavens existed long ago and the earth was formed out of water and by water, through which the world* [cosmos] *at that time was destroyed* [Noah's flood], *being flooded with water* [they deny the Genesis account of creation and the worldwide flood]. (2 Peter 3:5–6, NAS)

The Apostle Peter begins by describing the last days—the days which we are now living—a time when people will ridicule the Bible, deny the Second Coming of Jesus the Messiah, and believe in evolution and uniformity theory. Peter continues by describing two catastrophic events that cannot be explained based on the natural day-in and day-out processes of today's world: The first was the creation of the world and the second was Noah's Flood.

Peter continues, "*But the present heavens and earth by His word are being reserved for fire, kept for the day of judgment and destruction of ungodly men.*" (2 Peter 3:7, NAS) The phrase *"the present heavens and earth by His word are being reserved"* implies normal, uniformitarian processes throughout most of world history following the Flood except for local or regional earthquakes, hurricanes, tornadoes, volcanic eruptions, and tsunamis. The Bible allows for various types of astronomical (Genesis 8:22, Jeremiah 33:20-25), geological, and meteorological processes (Amos 1:1, Zechariah 14:5, Ecclesiastes 1:6, Matthew 16:2-3), which are not related to worldwide catastrophes.

While many advocates of uniformity doctrine allow for local or regional catastrophes, they do not allow for supernatural worldwide catastrophes as described in Genesis 6-8 or the final destruction of the world by fire [nuclear fire] as described in 2 Peter 3:7 and in Revelation 16 and other prophetic books of the Bible. In fact, the scientific community cannot allow for any supernatural interventions by God because uniformity doctrine justifies evolution, which in turn justifies atheism.

Science does not accept the term "supernatural" or the possibility of supernatural intervention in its vocabulary—it is considered outside the boundary of science and our three-dimensional world. Most scientists hold to evolutionary doctrine for fear of being ridiculed, as one secular scientist said, "any hint of teleology [i.e., intelligent design] must be avoided."[1]

Uniformitarianism assumes conditions on earth have never been interrupted by worldwide cataclysmic events—but the reality is, geological landforms found throughout the world are the result of tectonic forces vastly more intense than those observed today. Such landforms include:

- enormous mountain ranges with no denudation (erosion) (chapters 4, 5, and epilogue),
- huge monoliths and gigantic igneous batholiths (chapters 4, 5, and epilogue),
- vast areas of igneous and metamorphic rock exposed at the surface (continental shields) (chapter 5 and epilogue),
- ancient super volcanoes (chapters 4, 5, and epilogue),
- volcanic rock, known as basalt, comprises most of today's oceanic crust (chapters 4, 9, and Appendix D).
- mountain overthrust (low angle reverse thrusting) (chapter 5 and epilogue),
- rock layers and formations most often uplifted, inverted, tilted, folded, or missing (chapters 5, 7, epilogue, and Appendix G),
- dry falls and canyons, ancient dry lake basins and enclosed salt lakes, and overfit valleys and raised river terraces (chapter 5 and epilogue),
- tightly bent rock strata (chapters 5, 6, and epilogue),
- sharp contacts between rock layers and formations supposedly many millions of years old with no evidence of erosion, soil layers, or animal activity (chapters 4, 6, and epilogue).

For many distinguished geologists who have researched geologic landforms and catastrophic processes, evidence of a global flood is **indisputable**.

Phases of the Catastrophic Worldwide Flood

In the early 20th century, Alfred Wegener, a German meteorologist, noted that the continents (including the continental shelves) fit together as a single supercontinent. This antediluvian (pre-flood) landmass is commonly called Pangaea, from the Greek root word for "all lands." The northern part of Pangaea is called Laurasia and the southern part is called Gondwanaland.

Although no one could have observed the separation of Pangaea into the present-day continents, the evidence which supports the splitting of this ancient supercontinent is substantial. This includes not only the physiographic fit of the continents but also the alignment of major fault zones when the continents are placed together. The questions are, how long did it take, and when did the splitting occur?

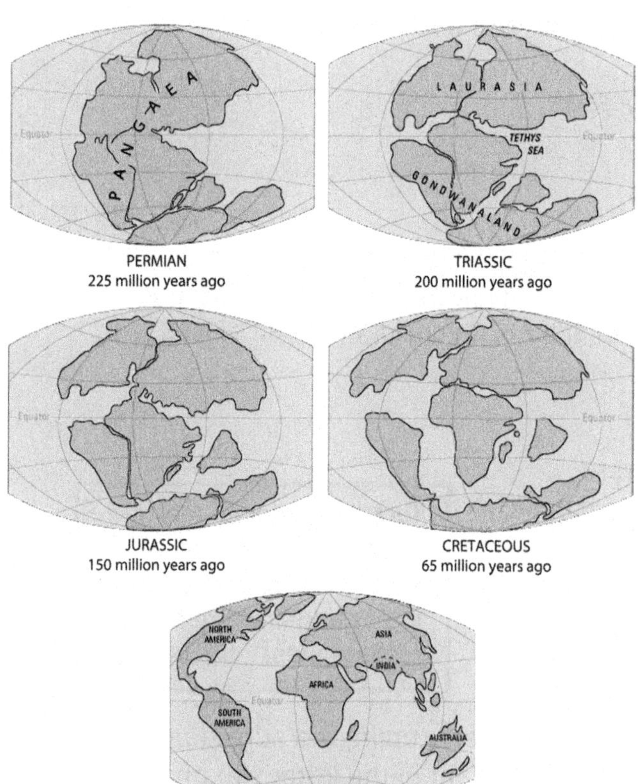

Map of Pangaea

> Evolutionists typically believe that the earth is about 3 to 5 billion years old—and that a super-landmass called Pangaea existed about 1 to 2 billion years ago—and this landmass began to slowly drift apart about 200 million years ago.[2]

> **Pangaea** is the primeval supercontinent composed of all the major landmasses.

From a biblical and young earth perspective, the original ancient continent was created by God as a single landmass about 6,000 years ago: *"Let the waters below the heavens be gathered into one place, and let the dry land appear."* (Genesis 1:9, NAS) By the end of the 1,600-year antediluvian period just before the flood (see chapter 10), it is believed that the population was more than 250 million people and the society was sophisticated, perhaps comparable to the early Egyptian culture.

Also, climatic and topographic conditions were much different from our current world. Although the supercontinent had mountains, rivers and seas, *its topography was much less prominent than we know today*[3]—the oceans weren't so deep and the mountains weren't so high (i.e., high hills and plateaus). There were other significant differences. The climate was similar to today's temperate regions with moderate seasonal variations (Genesis 1:14, 8:22). A milder environment can be verified by fossils of tropical plants found in polar regions.[4] (See chapters 7 and 10.)

The separation or splitting apart of this ancient landmass took place about 4,400 years ago, during a catastrophic worldwide flood—a global event described in Genesis 6-8. This catastrophic shifting of landmasses and flooding can be separated into three phases (each phase overlapping into the next) which took place within a span of just one year—NOT millions of years as maintained by uniformitarian geologists.

Three Phases of the Worldwide Flood

1. **Splitting of the Supercontinent (Pangaea)**
   Beginnings of Catastrophic Earthquakes* and Tsunamis
   Beginnings of Catastrophic Volcanoes*

2. **Uplift of Ocean Basins, Worldwide Flooding, and Sediment Deposition**
   Catastrophic Flooding and Sediment Deposition
   Sandstone Turbidites - Rapid Formation of Layered Sediments
   Limestone - Rapid Formation of Lime Mud Layers
   Coal and Oil - Uprooted Forests Rapidly Buried and Sealed

3. **Subsiding Ocean Basins, Continental Uplift, and Torrential Drainage**
   Uplift of Mountains and Plateaus
   Torrential Drainage and Rapid Erosional Processes
   Relict Erosional Landforms

* Catastrophic earthquakes and volcanoes continued throughout phases 1-3.
According to Scripture, Phases I and 2 lasted 150 days and Phase 3 lasted 164 days, and the earth was completely dry after 370 days. See chapter 8, Chronology of Noah's Flood.

Events leading to our current land features were catastrophic rifting and subduction, uplift of ocean basins, flooding of the continents, sedimentary deposition, burial of uprooted forests (creating the coal and oil we find today), mountain and continental uplift, and torrential erosional drainage of the floodwaters. Horizontal movement—seafloor spreading and continental drift—was the main tectonic force during the first phase of the flood event, and vertical movement (uplift and subsidence of ocean basins, and mountain uplift or orogeny) was predominant in the latter two phases of the flood.

Tectonics refer to geologic structures within the earth's lithosphere and the forces and movements that created these structures.

Phase Diagram of the Worldwide Flood

(Cross-section of earth's crust, looking south to north)

Phase One

Splitting of the Supercontinent (Pangaea)
Horizontal Movement Predominant
Beginning of Earthquakes, Volcanoes, and Tsunamis

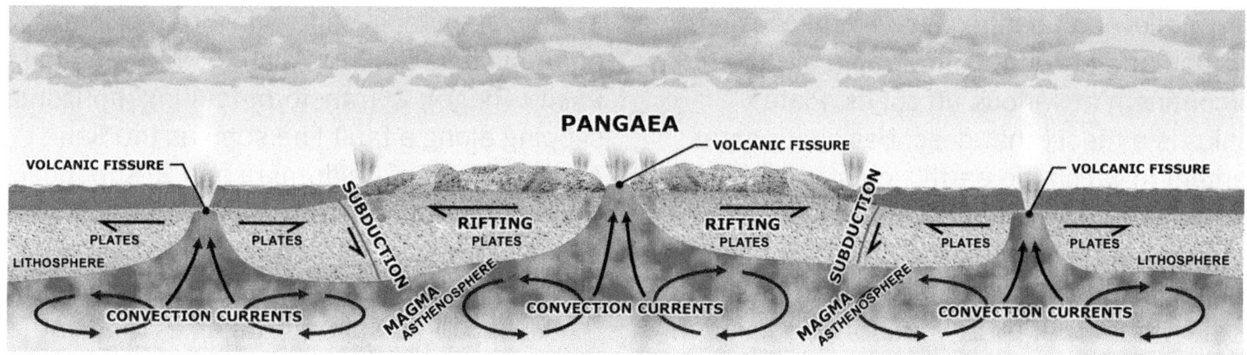

Phase Two

Uplift of Ocean Basins, Worldwide Flooding, and Sediment Deposition
Enormous Tsunamis – Water Spills Over Land

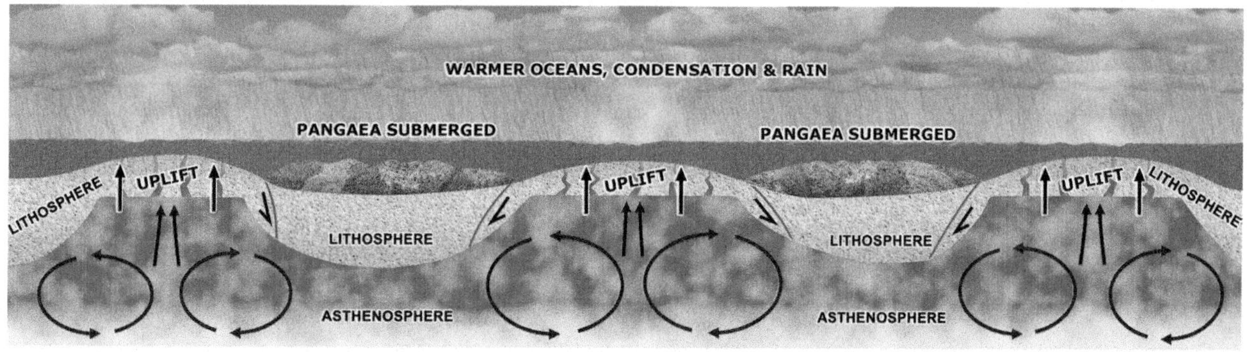

Phase Three

Subsiding Ocean Basins, Continental Uplift, and Torrential Drainage
Vertical Movement Predominant

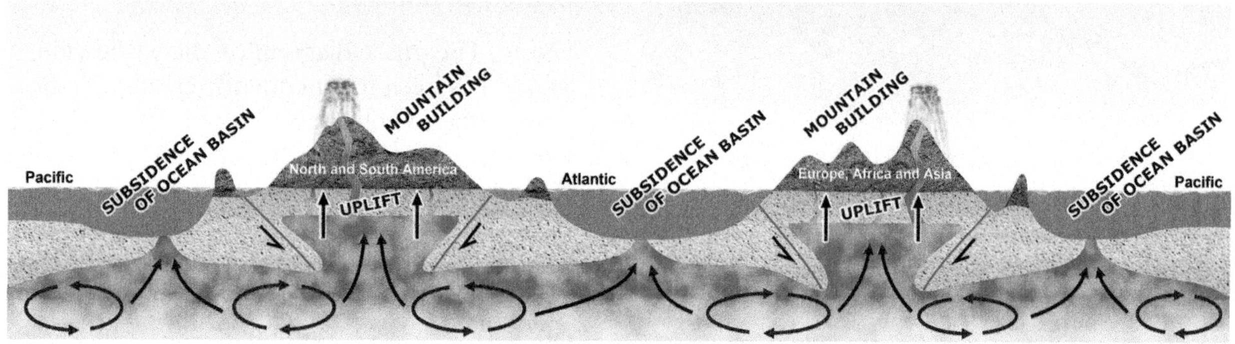

Drawings created by Roger Gallop and drawn by Chris Gallop
See Pangaea Flood Video at www.CreationScienceToday.com.

Phase One

Splitting of the Supercontinent

During the initial phase of this worldwide event, rapid "plate tectonics" is believed to be the mechanism causing separation and horizontal movement of the supercontinent at various rift zones. Plate tectonics is a theory that describes continental movement in which the earth's crust, or lithosphere, is segregated into crustal plates floating on a semi-fluid (plastic) asthenosphere.

Seismic and tectonic activity would have occurred primarily at plate boundaries as plates moved relative to one another. As the plates separated, hot asthenosphere rock material (magma) erupted along the rift lines in the form of violent volcanoes and earthquakes, and magma rose to fill the gap left by the spreading plates. It is most interesting to note that dense volcanic rock, known as basalt, comprises most of today's oceanic crust.

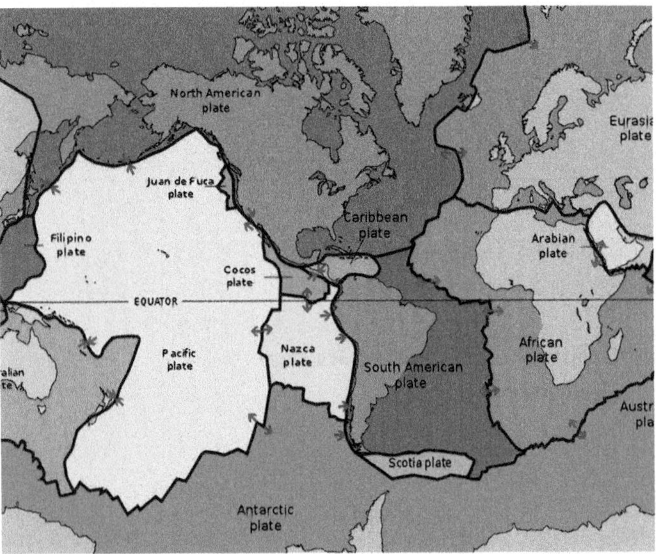

Boundary Map of the Tectonic Plates

Deformation at the boundaries of the plates is caused by three types of horizontal motion: 1) rifting (splitting, separating, or moving apart such as the Mid-Atlantic Ridge and East Pacific Ridge); 2) transform faulting (horizontal slipping along a fault line such as the San Andreas Fault of California); and 3) subduction or compressional deformation, when one plate descends beneath another. Examples are the Cocos Plate beneath South America and the Pacific Plate beneath Japan. Rifting and subduction are illustrated in the phase diagram.

> **Asthenosphere** is the portion of the upper mantle of the earth where magma forms. See Appendix D.

Dr. John Baumgardner, a geophysicist at the Los Alamos National Laboratory in New Mexico, used a two-dimensional finite model to simulate "thermal runaway, catastrophic plate tectonics" associated with rifting and subduction.[5] This model explains the tectonic mechanisms for rapid continental separation (horizontal rifting), catastrophic processes such as massive earthquakes and super-volcanic eruptions, and a worldwide flood resulting in the erosional landscape and geologic features we observe today. Various features of Baumgardner's tectonic modeling have been independently validated by other research scientists.[6]

Note: The phase diagram of the worldwide flood has been independently created by the author, Roger Gallop, and is not affiliated with Dr. Baumgardner's research and computer modeling.

> **Rifting** (horizontal movement) was the primary operating force early in the flood. Geomagnetic field reversals parallel to mid-ocean floor rifts are evidence of rapid sea floor spreading. See chapters 8 and 9.

> **Subduction** is the tectonic process in which a crustal plate descends beneath another plate and dives into the asthenosphere. **Runaway subduction** is "rapid" subduction associated with the worldwide catastrophic flood. See phase diagram of worldwide flood.

In the phase diagram, convection currents are shown in the asthenosphere. These currents are one of the major modes of large-scale heat and mass transfer in a fluid. The less dense liquid magma in the lower mantle would rise while denser rocks in the upper mantle would sink, creating slow, concentric currents within the mantle. This movement of warmer and cooler rocks creates pockets of circulation called convective cells.

Convection current is a circular current in the asthenosphere (consisting of semi-fluid rock and magma) formed when heated materials rise and cooler materials sink. See phase diagram.

Today, these convection currents move mantle rocks only a few centimeters a year but the circulation of these convective cells was likely the driving force behind the rapid movement of crustal plates over the asthenosphere during the worldwide flood. No one has ever observed these convection currents in the earth's mantle, and the Bible is silent about plate tectonics.

Beginnings of Catastrophic Earthquakes and Tsunamis

This was the beginning of worldwide catastrophic earthquakes and tsunamis. Whenever an underwater earthquake occurs, a huge volume of water is displaced and a shock wave races through the water for thousands of miles, often at speeds exceeding 200 miles per hour. In the deep oceans this energy wave does little damage but once it reaches shallow waters, it may wrap around islands and shorelines and becomes a wall of water from ten to hundreds of feet high (depending on the severity of the quake). Initially, the water will pick up anything in its path of destruction and then deposit its load of sediment and sea creatures.

Recent examples are the Sri Lanka-Sumatra tsunami which struck on December 26, 2005, and the Japanese tsunami which struck on March 11, 2011. Although devastating, killing hundreds of thousands of people along the coastlines, these tsunamis were small compared with those which occurred during Noah's Flood.

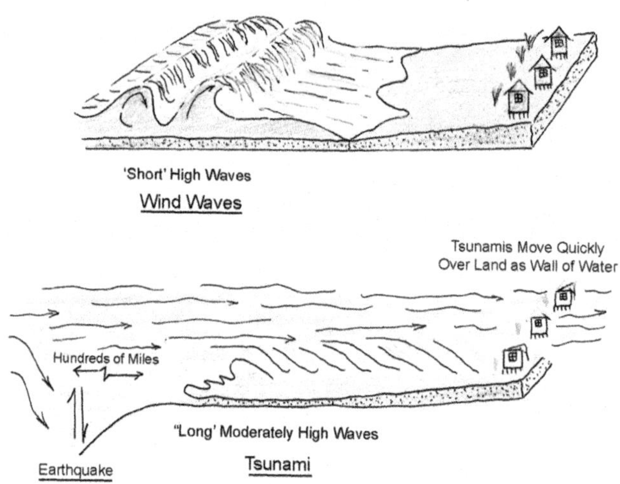

Top drawing: Normal wind waves.
Bottom drawing: Tsunami. *Sketch by Roger Gallop*

Tsunami is a series of massive waves created when a body of water is rapidly displaced, usually in the open ocean. A tsunami has a smaller amplitude (wave height) and a very long wavelength (often hundreds of kilometers long) as compared with a "wind wave" (usually with a larger amplitude but much smaller wavelength).

Tsunamis may be no taller than normal wind waves but are more dangerous. If a portion of a seafloor collapses during an earthquake, sea level falls with it. Water fills the low spot and overcompensates, creating the tsunami. In open oceanic waters, the crest may be 1–5 meters high but 62 to 186 miles long (100–300 km) and moving 200 to 300 miles per hour (322–500 km/hr). In Noah's Flood, gigantic tsunamis were likely hundreds of feet high.

Beginnings of Catastrophic Volcanoes

Accompanying earthquakes and tsunamis are volcanoes which spew and spill hot molten rock, gases, and ash under intense pressure. There are perhaps 4,000 to 5,000 volcanoes in the world today, about 1,900 of which are considered active, and possibly another 1,500 to 2,000 extinct volcanoes. Some well-known examples are the Hawaiian Islands and their principal volcanoes: Kilauea, Mauna Loa, and Mauna Kea; and in Washington State, Mount Rainier and Mount St. Helens. See chapter 5, Columbia Plateau, and chapter 6.

The eruption of Mount St. Helens in Washington on May 18, 1980, was one of the most significant geologic events in the United States in recent times. The islands of Hawaii are a series of shield volcanoes that have built up from the ocean floor. Some of the islands are composites of extinct volcanoes, while Maui's Haleakala and four of the big island of Hawaii's volcanoes—Kilauea, Mauna Loa, Loihi, and Hualalaia—have erupted in the past 2,000 years. As with earthquakes, about 80 percent of all volcanoes exist within the "Ring of Fire" of the Pacific Ocean, the boundaries of crustal plates.

Although recent volcanoes have been destructive, they are not comparable to volcanoes of the past. For example, one of the largest super volcanoes lies beneath Yellowstone National Park. A super volcano refers to a volcano that produces the largest

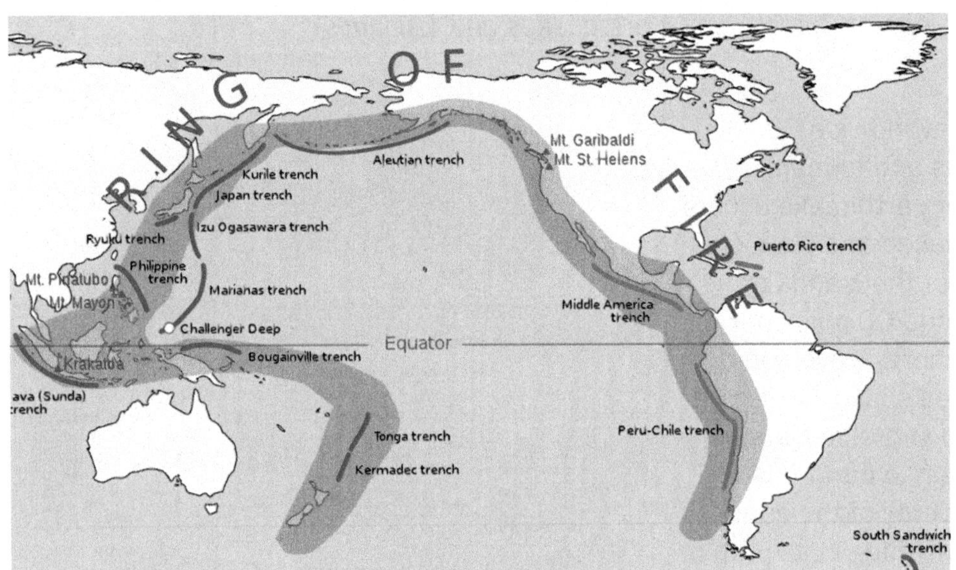

Pacific Ring of Fire

The 1980 Eruptions of Mount St. Helens, Washington

Hawaii — A Chain of Shield Volcanoes

and most voluminous eruption on earth. Yellowstone formed a 35-mile by 40-mile caldera (top of volcano; circular volcanic depression) and was likely tens of thousands of times more powerful than any known volcano today.

Volcanologists believe that current activity at Yellowstone has been increasing and the huge caldera has actually bulged and deflated during recent times. When magma builds up, it rises toward the surface where it pushes against the bottom of the caldera. Although most volcanologists believe that this volcano is in no danger of erupting anytime soon, recent bulging and deflating, and many hot springs, bubbling mud pots, and erupting geysers are worrisome to many geologists, park attendants, and park visitors.

The northeastern part of Yellowstone Caldera, with the Yellowstone River flowing through Hayden Valley and the caldera rim in the distance

Although secular geologists maintain that the volcano erupted 600,000 years ago, the reality is, this volcano and many other ancient super volcanoes (e.g., Valles caldera in New Mexico; Long Valley caldera in east-central California; Tampo caldera in New Zealand; Aira caldera in southern Japan; Lake Toba in North Sumatra, Indonesia; the recently discovered world's largest volcano, Tamu Massif, located beneath the Pacific Ocean about 1,000 miles east of Japan, and ancient super volcanoes under Utah-Nevada, and in Siberia, Russia)[7], including igneous batholiths such as the Columbia Plateau (described in chapter 5) were actually formed during the year of the Flood just 4,400 years ago.

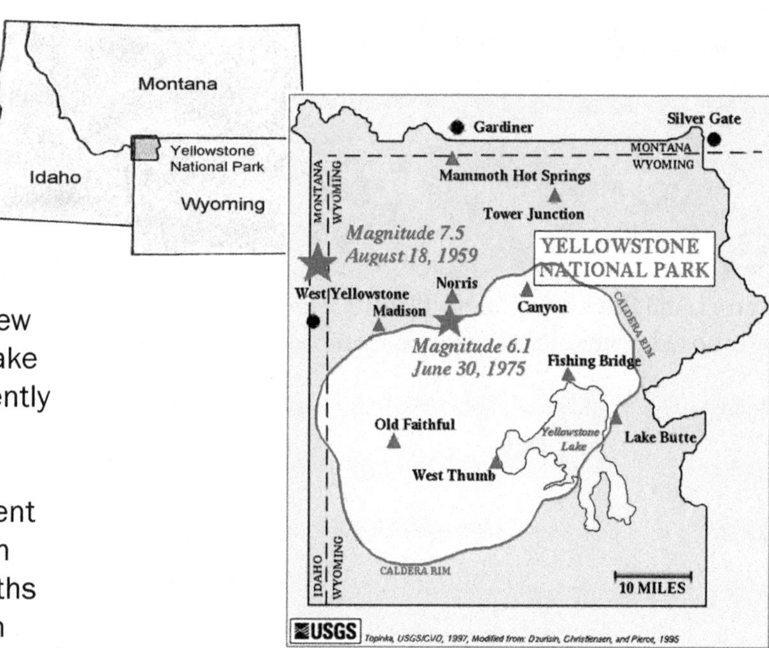

Catastrophic earthquakes, volcanic eruptions, and immense tectonics (horizontal and vertical earth movement) increased throughout phases 2 and 3 of the Flood.

El Capitan, Arizona, SR 163 1500 foot high, eroded volcanic plug or neck (when magma hardens within a vent of a volcano—consists of volcanic breccias). Surrounding sediment was swept away by floodwaters. *Photos by Roger Gallop*

Dr. Tas Walker provides a similar description of the Glass House Mountains (volcanic plugs) and surrounding lands in Brisbane, Australia ("Signs point people in the wrong way," *Creation,* 34 (2), 2013, p. 55.)

Pyroclastic Rocks, Colorado Plateau. Pyroclastics are rocks composed primarily of volcanic materials.

Phase Two

Uplift of Ocean Basins, Worldwide Flooding, and Deposition

During the initial phase of the global flood, plate tectonics caused the separation of the supercontinent at various rift zones. Rifting (horizontal movement) was the primary operating force early in the flood event but, in phase two of the flood, runaway subduction played a major role in ocean basin uplift.

As the event progressed, hot rock material (magma) erupted throughout the oceanic bottom in the form of violent cataclysmic volcanoes and earthquakes. As horizontal movement accelerated, the sea floor rapidly lifted because of increased temperature and pressure.[8] As described in Genesis 7:11 (NIV), *"on that day all the springs of the great deep burst forth"*—a catastrophic event that caused gigantic tsunamis and massive flooding of the continents.

Basalt is a dense, volcanic rock (type of igneous rock) which is usually dark gray or black and fine-grained. It forms most of the oceanic crust and is extruded by volcanoes or through fissures.

Along the ocean floor, super volcanoes generated massive eruptions which vaporized prolific amounts of seawater producing geysers of superheated steam along these zones of separation. This thermal steam condensed throughout the atmosphere as intense rainfall which accounts for much of the *"rain* [that] *fell on the earth forty days and forty nights"* as stated in Genesis 7:12 (NIV). See chapter 10, The Antediluvian Period and The Great Flood, for more information.

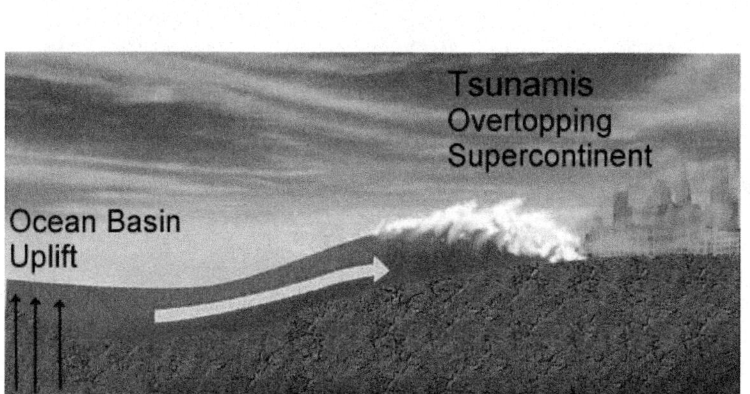

Ocean Basin Uplift and Worldwide Flooding

Skeptics insist that there was not enough water in our present oceans to cover all the mountains of the earth that exist today. Although the supercontinent had mountains, its topography was much less prominent—the oceans weren't so deep and the mountains weren't so high (likely no more than 2,500 feet high). It is these skeptics who are mistaken. See phase diagram of worldwide flood at the beginning of chapter and flood video at www.CreationScienceToday.com.

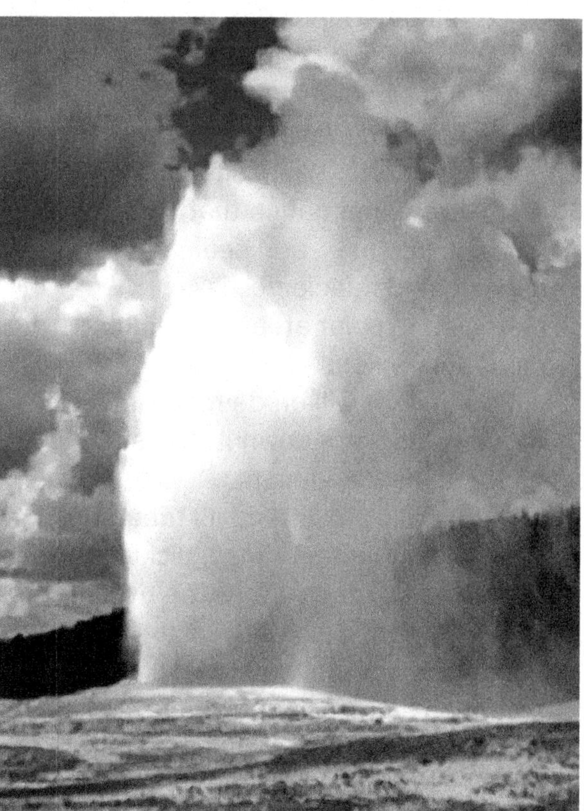
Old Faithful Geyser during an eruption

Catastrophic Flooding and Sediment Deposition

According to Scripture, the flood began when Noah was 600 years old (second month) and ended when Noah was 601 years old (first month). The flood lasted almost a year. During this event, much of the world's massive sediments were laid down as mud deposits.

Sediment deposits early in the flood period were "dumped in a heap" by tsunamis, but later turbidites were laid down in rapid succession in an underwater sea environment. A succession of earthquake events over a short time resulted in layered pancake-like strata. The lowest of these sediment layers was deposited during the early and middle phase of the flood and layers near the top were laid down late in the flood.

Sediment deposition today occurs by the action of water, wind, and ice, but *nearly all sedimentation (deposition) occurs underwater.* Wind deposits rarely result in sedimentation that hardens into rock, and sediment deposition by glacier ice is restricted to polar areas and high mountain regions. When water flows rapidly it will erode rock, and when water slows down it deposits its suspended sediment.

Visualize a river and its tributaries flowing through a mountain range, picking up dirt and pebbles along the way. As the river enters a lake or ocean, it slows down and the sediments are deposited, the larger rocks and pebbles first, then sand, silt, and finally the finer materials such as clay (sometimes referred to as deltaic deposits). An example is the Mississippi River entering the Gulf of Mexico, forming a great delta.

On a year-in, year-out basis normal flow rates and energy levels are steady, and not much geologic work occurs—that is, little erosion and deposition take place. However, on rare high-energy occasions, sedimentation (and erosion) can occur very rapidly because of undersea earthquakes. These underwater sediments are often deposited catastrophically by mud slides or avalanches in a horizontal, successive pancake-like position, similar to sediment layers of the Grand Canyon, Appalachian Mountains, and formations found throughout most of the world.

The Mississippi River Delta

Catastrophic sedimentary deposition, including great fossil graveyards with vast numbers of creatures violently mixed together, are found worldwide. Rapid deposition occurs within the mainland when high-energy floods suddenly slow—for example, in the 2005 Sri Lanka-Sumatra tsunami. (For more information, see chapter 7, Rapid Burial versus Slow Deposition.)

Sandstone Turbidites – Rapid Formation of Layered Sediments

Turbidites are pancake-like layered formations that result from catastrophic underwater avalanches or mudslides. Such avalanches distribute vast amounts of sediment into an ocean environment. Within the last century, turbidite deposits have been formed by massive slides at locations where rivers form large deltas. Often, continental slopes fail in response to earthquake shaking or excessive sedimentation load.

The concept of sediment deposits being "dumped in a heap" by tsunamis can be easily understood, but the formation of layered turbidites is a more difficult concept and has not been understood until recently. It was not until the 1950s that oceanographers and geologists were able to observe and sample underwater turbidite formations off the coast of Newfoundland.

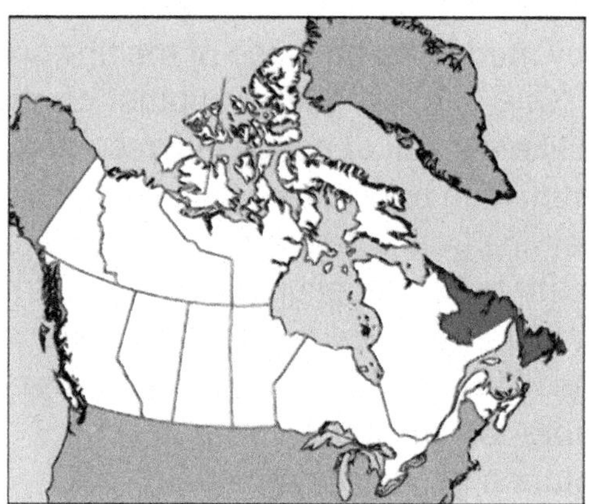

Map of Canada with Newfoundland in red

In 1927 a major earthquake occurred off the coast of Newfoundland, resulting in rapid pancake deposition. As Dr. John D. Morris explains, this earthquake "...caused a great volume of loose sediments on the continental shelf to slide as an underwater avalanche. The mud raced down the slope at 60 miles per hour and eventually came to rest on the ocean bottom. This underwater mudflow behaved much as wet cement behaves flowing off a cement truck. When it finally lost its energy and came to a halt, it had covered an area of 40,000 square miles with a deposit two to three feet thick. In the 1950s and 1960s geologists were able to study this deposit with the help of submarines. Through this study they were able to define a 'turbidite' as the result of catastrophic water currents."[9]

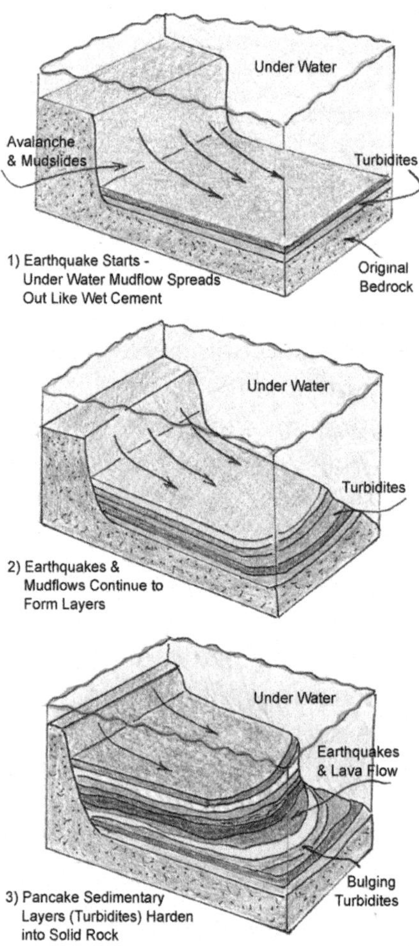

Turbidite Formation
Sketch by Roger Gallop

> Sediments are deposited underwater by oceans, rivers, lakes, or flood events, often as long, flat layers known as turbidites. Such layers can be seen in mountains and canyons throughout the world—evidence of a worldwide flood.

Tapeats Sandstone
Turbidites – Pancake Layers
Little Colorado River Gorge, Arizona
Photos by Roger Gallop

Examples of sedimentary pancake layers can be seen throughout the Grand Canyon (see chapter 5) and Little Colorado River Gorge. Tapeats Sandstone, for example, is hard rock (distinctively brown) and resistant to erosion found near the bottom of the canyon. In the past, uniformitarian geologists maintained that these layers (collectively, the 100 to 300 foot high Tapeats Sandstone formation) resulted from slow and gradual deposition on the bottom of a calm ocean.

Geologists have begun to realize that Tapeats Sandstone and other layered formations were due to a series of underwater flows of sand and mud. This sandstone, either exposed at the surface or often observed by oil and gas drillers, exists as pancake layers covering much of central North America and western Canada. As explained by Dr. Morris, "It would have taken a great underwater event to create such a formation."[10] In the latter phases of the worldwide flood, after massive amounts of sediment were deposited, turbidites were laid down in rapid succession in an underwater environment following massive earthquakes and uplift.

It is interesting to note that pancake sedimentary layers were also formed in association with the 1980 eruption of Mount St. Helens. While the eruption and ash fallout were devastating, the greatest destruction was caused by mud flows because of melting snow and ice. "One mud slide after another covered the area like a stack of pancakes, resulting in a sediment pile up to 600 feet thick in places."[11] Layers varying from a few millimeters to more than one meter thick accumulated in a matter of seconds to just a few minutes. Trapped inside these layers are many polystrate trees (see chapter 6) which are now petrifying.[12] Uniformitarian geologists thought it took millions of years to deposit such thicknesses of sediments, but "under ideal conditions, it can happen within days."[13]

Turbidites eventually turned into sandstone in the presence of overburden and cement (dissolved calcium carbonate, silica, and iron). Sediment layers of the Grand Canyon, deposited by catastrophic water action, look essentially the same as rock formations found worldwide.

Limestone – Rapid Formation of Lime Mud Layers

In today's world, calcareous lime muds (calcium carbonate, $CaCO_3$) found on sea floors typically form by the disintegration of carbonate-containing sea creatures and by slow precipitation of calcium carbonate. According to most geology textbooks, lime muds accumulate at a rate of about one foot per 1,000 years in shallow, tranquil seas. Because evolutionists believe that "the present is the key to the past"—that is, bio-physical processes have remained constant throughout earth's history—they believe it is impossible for young earth advocates to explain massive limestone and chalk formations found throughout the world. But visual and empirical evidences show otherwise.

Extremely well-preserved and diverse fossils, commonly known as Lagerstätten, are found within limestones of the Santana Formation of northeast Brazil and within similar formations worldwide.[14] These are examples of fossilization of animals that were rapidly covered in lime mud of fine texture. Visual evidence of rapid lime mud burial of the Redwall Limestone of the Grand Canyon, for example, includes large underwater sand waves (cross-bedding indicative of underwater sand dunes formed in ocean currents), preservation of fast-swimming nautiloids and other delicate fossil remains, and the absence of coral and sponge reef formations expected to be found in ancient tranquil seas[15]—all evidence of rapid deposition. See Appendix E.

So what really happened? The answers are found in the visual examination of limestone and geochemical mechanisms for rapid deposition of calcium carbonate. Primary mechanisms for rapid limestone formation are: 1) aggregation or flocculation of fine-grain crystals of calcite or aragonite in a high energy environment, 2) warming of seawater by underwater volcanoes and superheated geysers on a massive scale, in turn, resulting in direct precipitation of dissolved calcium carbonate, and 3) precipitation of dissolved calcium carbonate due to the depletion of carbon dioxide in the atmosphere—all mechanisms the result of the worldwide flood. A discussion of these mechanisms is found in chapter 6, Origin of Immense Limestone Formations and Chalk Beds, and Appendices E and F.

Coal and Oil – Uprooted Forests Rapidly Buried and Sealed

Uniformity theory assumes that coal and oil were formed by heat and pressure of organic matter in inland swamp environments with periodic burial by water deposited sediment over many millions of years. No one has ever observed the supposed gradual change of peat into coal and oil under such conditions—theorists *merely assume* the process. Observational and empirical evidence, however, points to a rapid formation of coal and oil as vast forests were uprooted and buried beneath massive amounts of heated sediment during a catastrophic worldwide flood. This phenomenon is explained in chapter 6, Origin of Coal and Oil.

Phase Three

Subsiding Ocean Basins, Continental Uplift, and Drainage

Sometime during the latter phase, "runaway subduction," that is, the rapid sinking or plunging of crustal plates beneath the continents, would have created tremendous volcanic and tectonic activity. Sedimentary layers (laid down during phase 2) would have been uplifted, folded, crumpled, and buckled while still soft and pliable (bendable), and intruded from below by massive igneous rock, forming batholiths and mountain ranges we see today.

Batholith is a large emplacement of igneous intrusive rock that forms from cooled magma deep in the earth's crust. It usually comprises granite or similar rock.

Rifting (horizontal movement) was the primary operating force early in the flood but, in the latter stages, **runaway subduction** likely played a major role in mountain building processes (vertical movement).

Such activity (mountain uplift) would have *balanced* the collapse of cooling oceanic basins, in effect forming new basins to collect receding floodwaters. Continual cooling of ocean bottoms would have increased the density of the new oceanic crust (mainly basalt from lava flows and sediment) causing further deepening of ocean basins.

Psalm 104:8 (NAS) translates as, "*The mountains rose; the valleys sank down*" implying that vertical earth movements (orogeny) were the main tectonic forces operating at the close of the flood. This period was the time of predominate mountain uplift to current elevations.

Every mountain range today looks different; some are angular, some are rounded, some are extremely high (over 29,000 feet), and others are low-lying (3,000 feet or less).

Much evidence of folding of stratified layers during the deposition period can be observed in all mountain ranges. The Alps, the Himalayas, the Appalachians, and the Rocky Mountains are considered folded mountains which were crumpled and buckled by compressive forces. Domed mountains, such as the Black Hills of South Dakota, were pushed from below and fault-block mountains, such as the Grand Tetons in Wyoming and the Adirondacks in New York, were pushed up on one side.[16] See chapter 5, Mountain Chains.

Mountains of today were formed toward the end or latter phase of the flood with subsidence of ocean basins, collision of tectonic plates, and continental uplift. In support of this, the layers that form the summit of Mount Everest (Himalayas) and other mountain ranges are *composed of fossil-bearing, marine limestone.*

Sierra Nevada Batholith, Half Dome, Yosemite, CA
See map in chapter 5, Mountain Chains

Psalm 104:6-8 (NAS) describes the receding waters that had stood above the mountains, "*Thou didst cover it with the deep as with a garment; the waters were standing above the mountains. At Thy rebuke they fled; at the sound of Thy thunder they hurried away. The mountains rose; the valleys sank down to the place which Thou didst establish for them.*" See chapter 10.

NASA International Space Station photograph of the Himalayas, looking south from over the Tibetan Plateau. The perspective is illustrated by the summits of Makalu [left (8,462 meters; 27,765 feet)] and Everest [right (8,850 meters; 29,035 feet)] at the heights typically flown by commercial aircraft.

View of the majestic Mount Everest from the Rongbuk valley

Mount Everest, the highest peak on earth at 29,035 feet, is part of the Himalayan range in Asia. The mountain is known to have a wide band of marine limestone near its summit—such rock had its origin beneath water. Marine fossils such as shells and coral are common on the earth's high mountain summits. Igneous and metamorphic rock masses are also interspersed within mountains. **Mount Ararat** is a 17,000-foot mountain described in the Bible (Genesis 8:4) as the resting place of Noah's ark.

Sand Dollar, Museum of Natural History, San Diego, California. *Photo by Roger Gallop.*

The museum sign states: "This sand dollar once lived on a Miocene seafloor, but was found as a fossil on the side of a mountain. How did it get there? Plate movements pushed up the ancient seafloor with buried fossils, and formed mountains. Lifted up, the mountains eroded, revealing fossils." (Coyote Mountains, Imperial County, California)

These mountains, with marine limestone, shells, and other marine fossils, had their origin in pre-flood tropical waters. Mountain chains were pushed up AFTER the floodwaters covered the earth and deposited their sediment.

Torrential Drainage and Rapid Erosional Processes

Some of the most dramatic geologic features on earth are due to erosion rather than mountain building. Such features include mesas, buttes, monoliths, peneplains, arches, spires, hoodoos, balanced rocks, and entrenched meanders. See photos on the next page and in chapter 5, Erosional Features of the Colorado Plateau and Rapid Continental Drainage – Relict Landforms.

During normal seasonal events, erosional processes such as rain and wind, freezing and thawing, chemical weathering, and even the activities of plants and animals, can do steady erosional work over time. These effects are hardly noticed on a daily or yearly basis but sometimes erosional events can be very fast, as witnessed by regional tsunamis, hurricanes, earthquakes, and volcanoes. So what are the mechanisms for rapid erosion?

Hydrologic mechanisms of floodwaters that rapidly erode even the most hardened sediment rock are cavitation, plucking, and kolking. When waters flow faster than 20 miles per hour, cavitation produces shock waves which strike the bottom rock. Plucking is when loose chunks of underlying sedimentary bedrock are ripped up and dragged along the bottom, pulverizing other rock. Kolking, or underwater whirlpools, can lift large pieces of bottom material. These mechanisms work together during high-energy flood events. Although normal weathering can do steady work over time, the action of major high-energy events is far greater over a short period of time.

> Erosion can occur either 1) by wind, rain, frost, and water while exposed as dry land, or 2) during torrential drainage and scouring processes in a high energy water environment. During Phase III of the global flood, water velocity varied, often exceeding 60 mph.
>
> For many distinguished geologists who have researched geologic landforms and catastrophic processes, overwhelming evidence supports the latter. Erosional relict landforms we see today are a testament to rapid deposition, tectonic uplift, and torrential drainage over a very short period of time. See chapter 5.

Secular geologists argue that various rock formations, such as those found in the Grand Canyon, underwent periodic tranquil inundation (period of sediment deposition) and uplift or ocean recession (exposure; period of erosion or non-deposition) over a span of millions of years.

Creation scientists maintain that the earth's features (based on empirical and observational evidence) are the result of a worldwide flood and torrential drainage over a very short period of time. Many geologists today are beginning to realize that erosional features throughout the world, including the Grand Canyon and the Colorado Plateau, are primarily the result of catastrophic drainage.

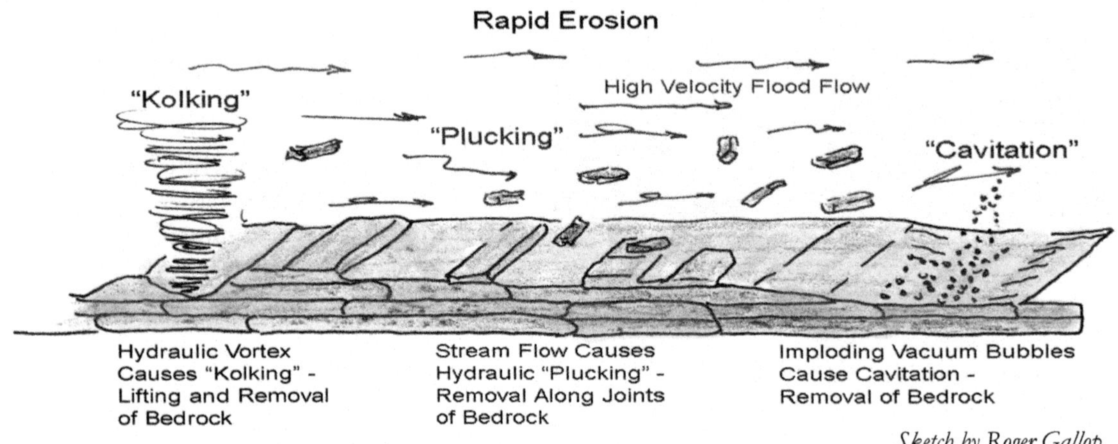
Rapid Erosion
High Velocity Flood Flow
"Kolking"
"Plucking"
"Cavitation"
Hydraulic Vortex Causes "Kolking" - Lifting and Removal of Bedrock
Stream Flow Causes Hydraulic "Plucking" - Removal Along Joints of Bedrock
Imploding Vacuum Bubbles Cause Cavitation - Removal of Bedrock

Sketch by Roger Gallop

Relict Erosional Landforms

When a continental area is flooded and then uplifted, torrential drainage and erosion will take place. Once the water is gone, isolated landforms will remain—evidence of rapid water erosion on a major scale. Many relict landforms of the Western United States and throughout much of the world include such features as plateaus, mesas, buttes, spires, and hoodoos sculptured by catastrophic floodwaters draining from the newly uplifted continents.

Mesas, buttes, and spires are erosion resistant relict features that were left standing while the ground around them eroded away. Other relict features found worldwide include dry falls, dry lake basins, deeply incised meanders, underfit rivers and overfit valleys, and sandstone monoliths surrounded by great peneplains. Such landforms stand as monuments to a worldwide flood and rapid drainage. See chapter 5 and photos.

Butte in Sedona, Arizona
Photos by Roger Gallop

Bryce Canyon NP, Utah
Spire with Balanced Rock

Bryce Canyon NP, Utah

Arch at Bryce Canyon National Park, Utah

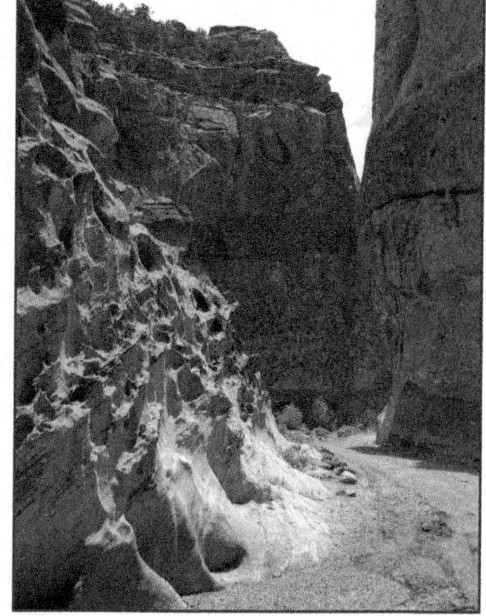

Capitol Reef NP, Utah
"Deep twisting, water carved, sheer walled canyons."

The Great Ice Age

The Ice Age was a time of global cooling when glaciers descended into lower latitudes of North America, Europe, Asia, and South America. "In North America, ice covered much of Canada and as far south as Kansas. Northern Europe was likewise blanketed with ice."[17] The most obvious effect was the removal and storage ("locking up") of great quantities of water in the form of glacial ice which, in turn, would have significantly lowered sea level.

As maintained by secular uniformitarian geologists and standard geology textbooks, glacial ice sheets formed about 10,000 to 15,000 years ago during the Pleistocene epoch. Accordingly, they believe this epoch was marked by four widespread glacial periods (Illinoian, Kansan, Nebraskan, and Wisconsin). There is evidence the latest ice age, the Wisconsin, formed vast continental glaciers—hundreds if not thousands of feet thick over thousands of square miles—but there is little or no evidence of the first three glacial periods.

Evidence of the supposed three earlier glaciations (the Illinoian, Kansan, and Nebraskan) are deposits of gumbotil, supposedly a mature weathered clay soil comprising small stones (tills), but these supposed glaciations do not comprise typical features common to the latest glacial stage, the Wisconsin. Interestingly, the Wisconsin ice sheets show no evidence of gumbotil formations but do include such features as tills, striations, moraines, kames, drumlins, eskers, kettle lakes, and erratic boulders. "These are found only in connection with the supposed last glacial maximum and its retreat, the so-called Wisconsin stage."[18]

One of the reasons it is difficult for secular geologists to account for earlier glaciations is because *these glaciations NEVER existed.* "The evidence for more than one glaciation, whether in the Pleistocene, the Permian, the Precambrian, or any other geologic system, is utterly inadequate."[19] There was only one Great Ice Age—not four!

So what caused the Ice Age and how cold did it actually get? The Ice Age has always been a problem for uniformitarian geologists. While there is abundant evidence for continental glaciation, the cause has always remained a mystery. Dozens of hypotheses have been proposed—for example, global cooling, decrease in the sun's intensity, prolific volcanic activity—but all have serious problems. None has been accepted because of uniformitarian constraints; that is, slow and gradual processes (geological and meteorological) to explain the cause of the Ice Age simply do not work. No hypothesis has been able to explain such intense changes—none except a catastrophic worldwide flood.

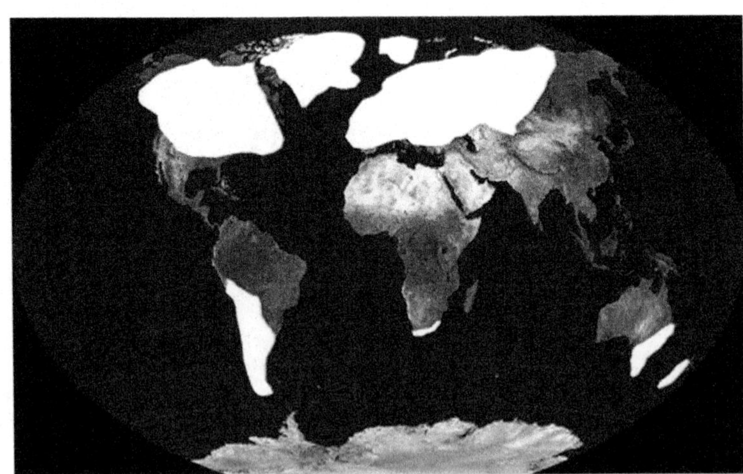

World map has been modified to show extent of glacial ice during the Great Ice Age.

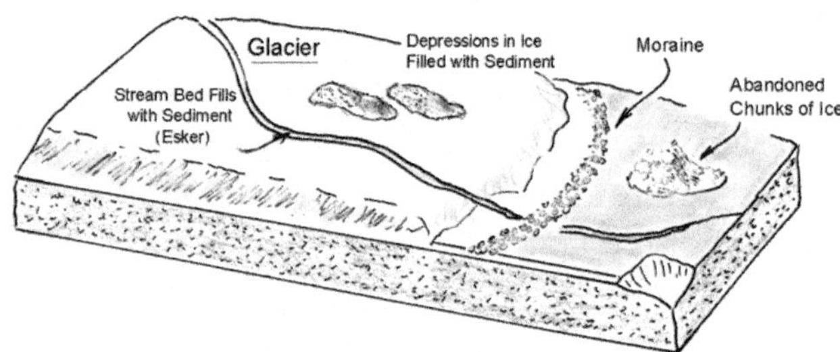

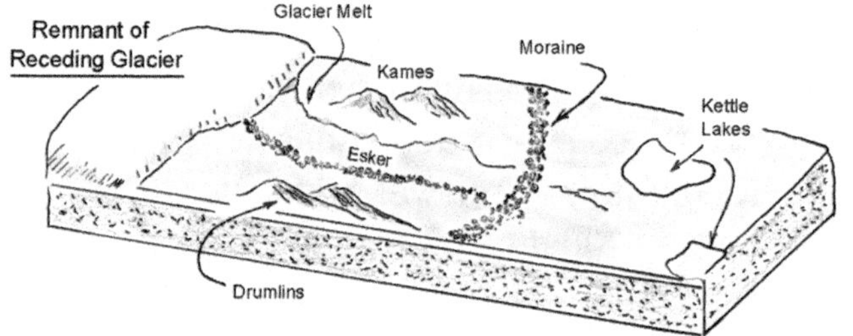

Glacier Landforms
Sketch by Roger Gallop

Gulkana Glacier, South-central Alaska, Oct. 5, 2003

Wolverine Glacier, South-central Alaska
Sept. 10, 2003

South Cascade Glacier, Washington, Oct. 5, 2000

Cause of the Great Ice Age

As explained by Drs. J. C. Whitcomb and H. M. Morris, "Glacial geologists have never answered the cogent criticisms of Sir Henry Howorth, president of the Archaeological Institute of Great Britain at the close of the nineteenth century, who amassed a tremendous amount of evidence that most of the supposed ice sheet deposits may have been formed by a great flood sweeping down from the north. Howorth was not defending Genesis, in which he was not a believer, but only was concerned to show the scientific inadequacy of the glacial theory."[20] The most widely accepted theory is the "solar-topographic" hypothesis of the Yale glacial geologist, Dr. R. F. Flint. This theory explains "glaciations in terms of the worldwide mountain uplifts... combined with assumed fluctuations in incoming solar radiation."[21]

In agreement with these hypotheses, the biblical deluge offers a rational explanation for the cause of the Ice Age. The combined effect of a worldwide flood, uplift of the continental landmass, and torrential drainage would have induced great snow and ice accumulations. Glaciation as caused by the flood would more likely be one glacial event, not four.

What are the climatic requirements? As stated by Dr. John D. Morris, the "obvious requirements for ice buildup are more snowfall and less snow melt. But how does this happen? Merely tweaking [sic] modern conditions will not alter the earth's environment to that extent. It takes more than cold temperatures, for if things get too cold, the air can't contain much moisture and it doesn't snow much. And so the puzzle remains."[22] According to Dr. J. D. Morris and Michael J. Oard, the key ingredients to more snowfall are:[23]

- Warmer oceans due to volcanism; thus
- Increased evaporation; thus
- Moderately warm, wet winters for more snowfall (just below freezing);
- Cold mud-slick land surfaces reflecting radiation (no solar absorption); thus
- Moderately cold summers for snow accumulation (just at or above freezing).

When the flood ended, the oceans were much warmer because of increased heat and pressure within the earth's mantle (see phase diagram at the beginning of chapter). Volcanism and superheated geysers would have added a tremendous amount of heat to the oceans. "This warmth would yield a continual pump of warm moisture into the atmosphere—thus warm, wet winters. Thus, more evaporation."[24]

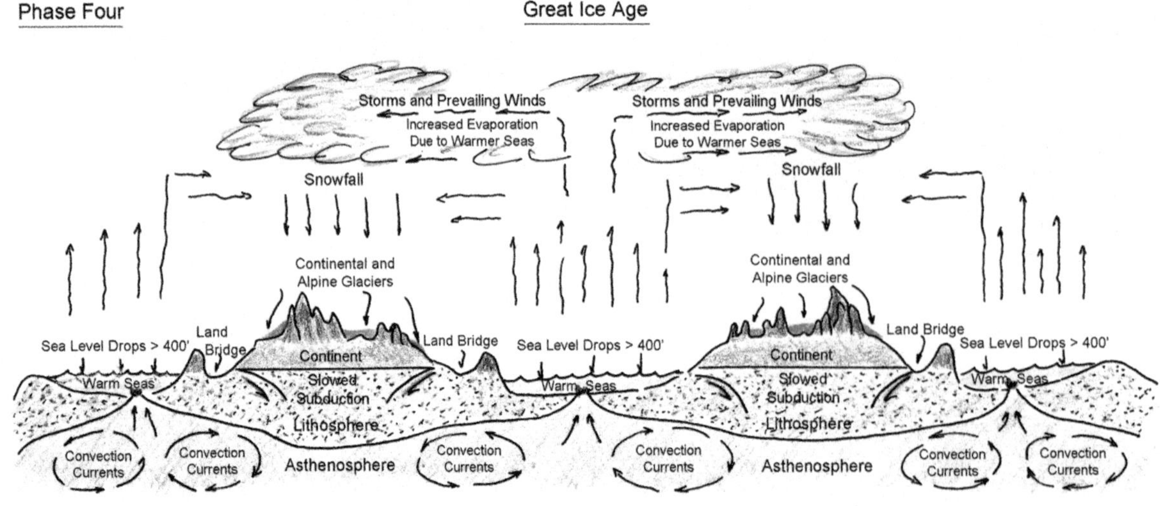

Sketch by Roger Gallop

Intense temperature gradients and strong winds (Genesis 8:1) would have transported evaporated moisture from the ocean to the continents, where it would condense as snow—and this was likely the first occurrence of snowfall since the time of creation. Moderately warm winters (below freezing) allowed for continuous snowfall.

The continental land surface following the flood was a "mud slick, and would have reflected solar radiation without absorbing much heat."[25] Also, volcanism during the flood would have saturated the atmosphere with "aerosols and dust, bouncing incoming solar radiation back into space." This in turn would have resulted in moderately cold summers, allowing for nonstop accumulation of snow and ice.

"Once an ice sheet got started, it would probably grow rapidly and extensively."[26] Most geologists and meteorologists agree that these conditions would cause an ice age, but uniformitarian doctrine cannot allow for such an abrupt change.

Retreat of Ocean Waters

During the Ice Age, vast amounts of water were removed from the oceans and stored in polar regions in the form of great ice caps. The amount of water confined in these great glaciers is unknown, but many geologists believe the amount was enormous. Evidence of this fact is the greatly lowered sea levels during this post-flood period. We have evidence that "ocean levels were at least 400 feet lower than at present, possibly much more, as shown by such features as the continental shelves, sea-mounts, submerged canyons and terraces, etc."[27]

Such evidence requires an explanation outside the doctrine of uniformitarianism; that is, a worldwide catastrophic flood event. Creation scientists believe there was only one Ice Age—an ice age caused by the flood. The lowered sea level at that time made it possible for animals to migrate over land bridges for many centuries. Men and animals could freely cross the Bering Strait which separates Asia and North America. Many animals such as kangaroos or the arboreal koalas in Australia, and unique species of Madagascar,[28] would have migrated slowly from Mt. Ararat to remote areas of the world.

With the return of ocean waters to their present levels, animals such as marsupials would have become isolated and protected from competition and predation.[29] Unique animals found in other remote locations simply mean these are the only places where these animals still survive.

Specialized animals found throughout the world required unique environments for survival. Specialization resulted in a loss of genetic information, from thinning out of the gene pool, or by degenerative mutations. Examples are China's giant panda with preference for bamboo shoots or Australia's koala with preference for eucalyptus leaves. (See chapter 3, Natural Selection and Extinction.)

End of the Great Ice Age

Geophysical and paleontological evidence indicates the "warming [of the world climate] which occurred at the close of the Wisconsin glacial time was extremely abrupt."[30] The Great Ice Age would last until oceanic volcanism subsided and oceans cooled, and trees and shrubs were reestablished on the newly formed continents. This likely would take place within 200 to 500 years following Noah's Flood. The end of the Ice Age ushered in the beginning of our modern hydrologic cycle (phase 5)—and seasonal and temperature fluctuations as we experience today.

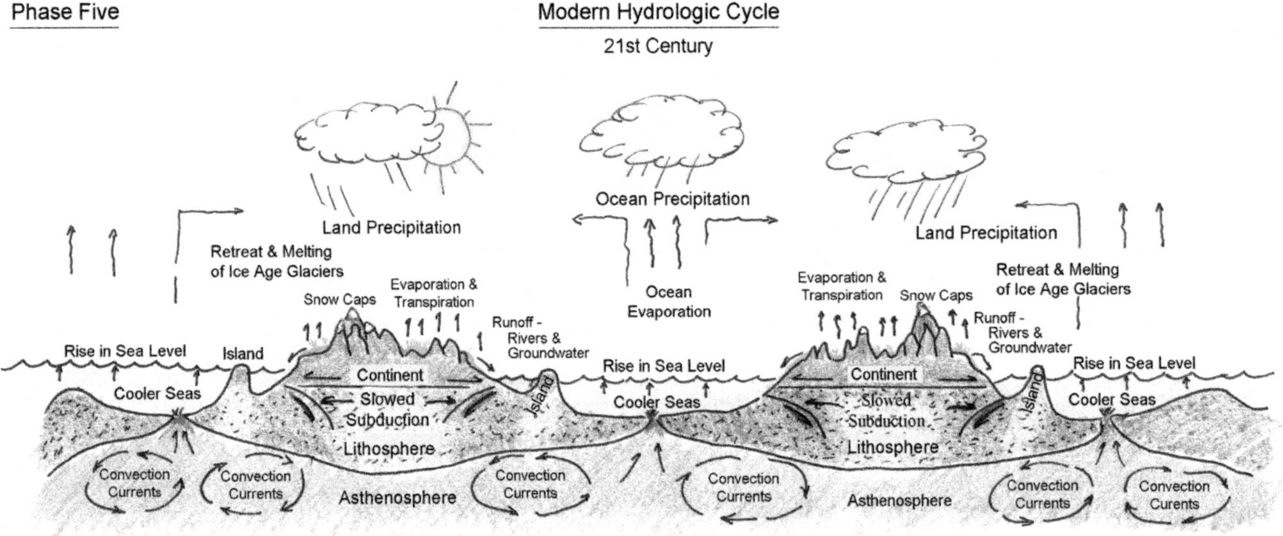

Sketch by Roger Gallop

Summary

Splitting of the original supercontinent, Pangaea, took place about 4,400 years ago during a worldwide flood as described in Genesis 6-8. This event is separated into three phases which took place within a span of just one year—NOT millions of years as maintained by uniformitarian geologists. There is an abundance of physical evidence in support of such an event—evidence further described in chapters 5, 6, and 9. Also, slow and gradual processes (geological and meteorological) cannot explain the cause of an ice age—yet, a worldwide flood provides the only reasonable explanation.

Evolutionary doctrine does not take into account a worldwide flood because it implies Divine creation and a young earth. Secular geologists believe that existing physical processes are sufficient to account for all past changes in our world and the universe. There is no basis for this belief other than the assumption of evolution as the only "scientific" mechanism for the existence of man and all living things—the "modern scientific establishment has bound itself to a single system of interpretation...and alternatives must be rejected out of hand."[31]

notes: ***Chapter 4***

1. de Duve, Christian (September–October 1995). The beginnings of life on earth. *American Scientist*, 428. Also cited in Morris, J.D. (2003). *Is the Big Bang Biblical?* Green Forest, AR: Master Books, 78-79.

2. Thompson, G.R., and Turk, J. (1991). *Modern Physical Geology*. Philadelphia, PA: Saunders College Publishing, 359-360.

3. Dillow, Joseph C. (1982). *The Waters Above*. Chicago, IL: Moody Press, 141. Also see Whitcomb, J.C. (1988). *The World That Perished*. Grand Rapids, MI: Baker Book House, 41-46.

4. Ibid., 335–353, 377. Also see Whitcomb (1988), 77-80.

5. Baumgardner, J.R. (1994). Runaway subduction as the driving mechanism for the Genesis flood. *Proceedings of the Third International Conference on Creationism,* Pittsburg, PA, 63-75; and Catchpoole, D., Sarfati, J., and Wieland, C. (2008). *The Creation Answers Book*. (D. Batten, Ed.). Atlanta, GA: Creation Book Publishers, chapter 11.

6. Catchpoole, Sarfati, and Wieland (2008), op. cit., 168. Also see Weinstein, S.A. (1993). Catastrophic overturn of the earth's mantle driven by multiple phase changes and internal heat generation. *Geophysical Research Letters*, 2: 101-104; Tackley, P.J., Stevenson, D.J., Glatzmaier, G.A., and Schubert, G. (1993). Effects of an endothermic phase transition at 670 km depth on spherical mantle convection. *Nature*, 36: 699-704; and Moresi L., and Solomatov, V. (1998). Mantle convection with brittle lithosphere: Thoughts on the global tectonic styles of the earth and Venus. *Geophysical Journal International*, 133: 669-682.

Other evidence of runaway subduction is the presence of pseudotachylyte (PST), a dark-colored, glassy material, formed by frictional melting during high-speed rock movement. According to Dr. Tim Clarey, Ph.D., "the presence of PST is considered evidence of high-speed rock movement.....documented at many locations around the world, including several in subduction zone settings." Clarey, T. (2013). Runaway subduction and deep catastrophic earthquakes. *Acts & Facts*, 42 (1), Dallas, TX: Institute for Creation Research; and Clarey, T.L. et al. (2013). Superfaults and pseudotachylyte: Evidence of catastrophic earth movements. In *Proceedings of the Seventh International Conference on Creationism.* Horstemeyer, M., Ed. Pittsburg, PA: Creation Science Fellowship, Inc.

7. The Florida Times-Union, Jacksonville, Florida. Meeri Kim's article of The Washington Post titled "Massive volcano discovered submerged in the Pacific," appeared on August 8, 2013; Fox News.com. Becky Oskin's article, LiveScience, titled "Largest volcano on Earth found under the Pacific," September 6, 2013. Retrieved September 2013, from http://www.foxnews.com/science/2013/09/06/largest-volcano-on-earth-found/; and AIPGeNews. Retrieved December 2013, from AIPG eNews [aipg@multibriefs.com]. American Institute of Professional Geologists.

8. Austin, S.A., Baumgardner, J.R., Humphreys, D.R., Snelling, A.A., Vardiman, L., and Wise, K.P. (1994). Catastrophic plate tectonics: A global flood model of earth history. *Proceedings of the Third International Conference on Creationism*, Pittsburg, PA, 609-621; and Catchpoole, Sarfati, and Wieland (2008), op. cit., chapter 11.

9. Morris, J.D. (2000). *The Geology Book*. Green Forest, AR: Master Books, 33-34.

10. Ibid., 35.

11. Morris, J.D. (1994). *The Young Earth*. Green Forest, AR: Master Books, 107.

12. Fossilization occurs when mineral matter such as hot, silica-rich water flowing through volcanic deposits is injected into pore spaces of the animal or fallen tree. Water carrying dissolved silica penetrates the tissue, precipitates to form quartz and replaces the organic tissue. The resulting fossil is petrified wood.

13. Morris (1994), op. cit., 106.

14. Lagerstätte (May 2009). In Wikipedia, the free encyclopedia. Retrieved May 2009, from http://en.wikipedia.org/wiki/Lagerst%C3%A4tten.

15. Austin, S.A. (Ed.). (1994). *Grand Canyon: Monument to Catastrophe.* Institute for Creation Research, El Cajon, CA, 26-28.

16. Morris, (2000), op. cit., 24-25.

17. Morris, J.D. (2003). *Is the Big Bang Biblical?* Green Forest, AR: Master Books, 146.

18. Whitcomb, J.C., and Morris, H.M. (1961). *The Genesis Flood*. Phillipsburg, NJ: The Presbyterian and Reformed Publishing Company, 296.

19. Ibid., 302.

20. Ibid., 292-293.

21. Ibid., 293.

22. Morris (2003), op. cit., 146.

23. Ibid., 147; Oard, M.J. (1990). A post-flood ice-age model can account for quaternary features. *Origins*, 17 (1): 8-26; Oard, M.J. The ice age and the Genesis flood. Institute for Creation Research, El Cajon, CA. Retrieved from http://icr.org/articles/print/272/; Wieland, C. (December 1996). Tackling the big freeze: Interview with weather scientist Michael Oard. *Creation*, 19 (1): 42-43; Oard, M.J. (1990). An ice age caused by the Genesis flood. Institute for Creation Research, El Cajon, CA, 33-38; Oard, M.J. (1979). A rapid post-flood ice age. *Creation Research Society Quarterly*, 16 (1), 29-37; Oard, M.J. (1986). An ice age within the Biblical time frame. *Proceedings of the First International Conference on Creationism*, Pittsburg, PA, 2: 157-166; and Catchpoole, Sarfati, and Wieland (2008), op. cit., 205-208.

24. Ibid., 147.

25. Ibid.

26. Whitcomb and Morris (1961), op. cit., 294. Also see Brooks, C.E.P. (1949). *Climate Through the Ages*. (2nd Ed.), McGraw-Hill, 31-45.

27. Ibid. Also see Russell, R.J. (December 1957). Instability of sea level. *American Scientist*, 45: 414-430.

28. Madagascar, an island nation off the coast of southeastern Africa, is renowned for its unique and diverse species of wildlife, especially lemurs—primates found nowhere else on the planet. Madagascar is home to more than 250,000 species of which about 70 percent are unique to this island. http://www.wildmadagascar.org/home.html.

29. Placental mammals give birth to live young which are nourished before birth through an embryonic organ known as the *placenta*. Placental mammals include whales and elephants, rodents and bats, dogs and cats, as well as many farm and work animals such as sheep, cattle, and horses. Humans, of course, are placental mammals. http://www.ucmp.berkeley.edu/mammal/eutheria/placental.html.

Marsupials, commonly thought of as pouched mammals (like the koala, wallaby, and kangaroo), give live birth but the gestation period is much shorter than placental mammals. The young animal climbs from the mother's birth canal into the pouch where the young continues to develop. http://www.ucmp.berkeley.edu/mammal/marsupial/marsupial.html.

30. Whitcomb and Morris (1961), op. cit., 304; and Broecker, W.S., Ewing, M., and Heezen, B.C. (June 1960). Evidence for an abrupt change in climate close to 11,000 years ago. *American Journal of Science*, 258: 429. Also see Dillow, Joseph P. (1982). *The Waters Above*. Chicago, IL: Moody Press, 363, 377, 397.

31. Evolution's evangelists. (May 2008). *Acts & Facts,* 37 (5), Dallas, TX: Institute for Creation Research, 10. Copyright© 2008 Institute for Creation Research, used by permission. 2008. Evolution's evangelists. *Acts & Facts,* 37 (5): 10.

Chapter 5

Geomorphic Features of a Worldwide Flood

> *"The mountains rose; the valleys sank down to the place which Thou didst establish for them."*—Psalm 104:8, NAS

Mountain Chains

Near the end of the worldwide flood, tectonic plates collided because of runaway subduction, and huge mountain chains were created with massive uplifting, folding, and thrusting of rock formations. One of many examples of massive mountain building (orogeny) is the long, broad mountain chain of western North America known as the Cordillera, or Cordilleran Mountain Chain (fold and thrust belt). This chain includes the Rocky Mountains, the Grand Tetons of Wyoming, and the San Juan Mountains of southwestern Colorado and the Coast Ranges—almost all the mountains in the western United States. The chain extends from Alaska to Mexico to South America.

> Formations and layers comprise igneous, sedimentary, and metamorphic rocks. For more information, see Glossary and Appendix D.

Other mountain ranges include the Cascade Range of northern California, Oregon, and Washington; the Sierra Nevada Batholith (massive igneous intrusive rock formation) of California (see photo of batholith, chapter 4, Phase Three); and the Appalachian Mountains of the eastern United States. Other well-known mountain ranges throughout the world include the Great Himalayan Range between India and China, the Alps stretching from Austria to France, and the Andes Mountain Range along the western coast of South America.

Uniformity theory maintains that mountain building has occurred steadily over hundreds of millions of years, if not billions of years, but if this actually occurred, the removal of rock and sediment by erosion (denudation) should offset mountain uplift over this vast period of time. Clearly, this is not the case. Visual evidence today shows there has been little or no denudation as suggested by uniformity theory. See chapter 9, Continental Erosion and Ocean Sediments.

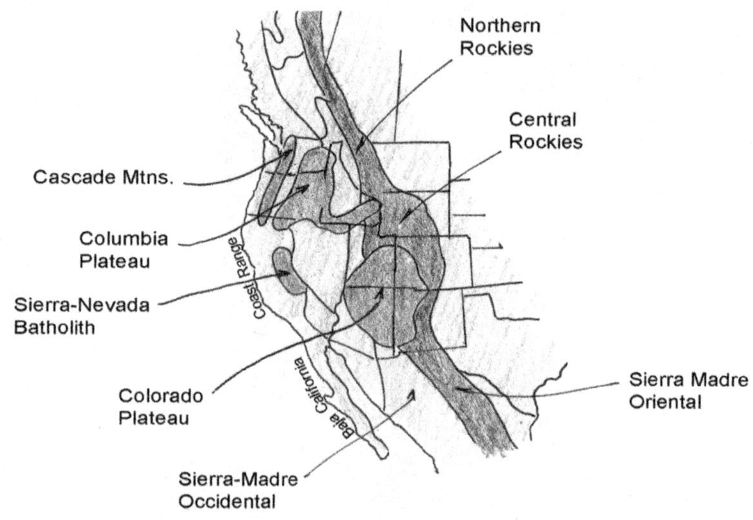

> Many secular geologists believe the Cordillera orogeny took place about 80 to 180 million years ago during the late Mesozoic Era while others believe that the Rocky Mountains were uplifted as part of the Laramide Orogeny about 50 to 70 million years ago, and the Colorado Plateau was uplifted sometime later about 5 to 10 million years ago.

In an effort to explain this lack of denudation (erosion), many secular geologists now believe these great mountain ranges were uplifted several miles during the late Pliocene and early Pleistocene epochs, a period of less than 5 million years. In terms of standard geologic time scale (see chapter 7), this is a relatively short time frame—in a "blink of the eye."[1] But, in fact, this 'blink of the eye' occurred during the worldwide flood (see chapter 4)—in less than one year! Visual evidence indicates that these great mountain chains underwent rapid and sudden uplift.

Metamorphic rock forms when rock (e.g., igneous, sedimentary, or other metamorphic rock) changes or recrystallizes in response to increased temperature and pressure.

Uniformity theory cannot explain mountain uplift with no denudation, and thrusting, tilting, and folding of rock formations *AFTER* the rocks have cooled and become brittle (non-pliable) over many millions of years. This poses a time problem for secular geologists. It is clear to most unbiased researchers (setting aside the presumption of uniformity theory) that geologic evidence of mountain uplift with no denudation is evidence of a young earth, and that thrusting, tilting, and folding (i.e., geosynclines) *occurred over a short, catastrophic period of time when magma rock was still soft (hot) and pliable*. (For more information, see chapter 6, Presence of Tightly Bent Strata Without Cracking, and chapter 9, Continental Erosion and Ocean Sediments.)

Northern Teton Range and Jackson Lake (Mt. Moran on left)

The Alps, view of Rigidalstock (left) and Titlis (center) from Mount Stanserhorn, Lucerne, Switzerland. *Photo by Roger Gallop*

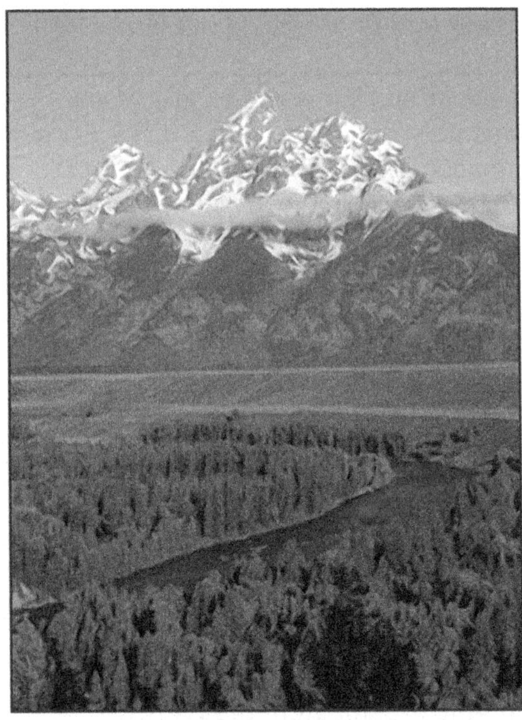

Cathedral Group of the Grand Tetons. Grand Teton National Park, Wyoming. Grand Teton and Snake River.

Visual evidence today indicates little or no denudation as suggested by uniformity theory. Uniformity theory cannot explain folding of rock formations after rocks have cooled and become brittle over hundreds of millions of years.

Additionally, mountain ranges comprise metamorphic rock (rock that has undergone increased temperature and pressure) deep within their core. Also, continental shields (e.g., Canadian, Amazonian, Baltic, African, Indian, China-Korean, Australian, and Antarctic shields) exposed at or near the earth's surface comprise vast areas of metamorphic and igneous rock. Although metamorphism has occurred in recent times on a very limited scale when hot magma intrudes into sedimentary rocks, *enormous areas of metamorphic rock found throughout the world are the result of far greater forces than those observed today.*

Uplift and Tilting. U.S. 15 west of Hurricane, Arizona. *Photo by Roger Gallop*

Folded limestone layers on Cascade Mountain in Provo Canyon, Utah

These mountain ranges were formed toward the end or latter phase of the flood with subsidence of ocean basins and collision of tectonic plates with massive folding and thrusting of rock formations as described in chapter 4. Mountain chains were rapidly pushed up *AFTER* floodwaters covered the earth and deposited their sediment. In further support of this, layers that form the uppermost parts of Mount Everest (Himalayas) and other mountain ranges are composed of fossil-bearing, water-deposited layers. Nevertheless, secular geologists continue to maintain a tight grip on old earth doctrine.

In the words of A. J. Eardley, "... the cause of the deformation of the earth's outer layers and the consequent building of mountains still effectively evades an [uniformitarian] explanation."[2]

Anticline (arch-shaped) with syncline (U-shaped) visible at far upper right in layered gneiss (metamorphic rock formed from pre-existing igneous or sedimentary rock) along NJ Route 23 near the rest area exit ramp (west of Butler, New Jersey). Note blurred image of man in foreground for scale.

Snow-dusted syncline in Provo Canyon, Utah

Columbia Plateau

The Columbia Plateau is a geologic region located between the Cascade Range and the Rocky Mountains, and is part of the Cordilleran Chain. It is a basaltic lava plateau (sometimes referred to as a batholith) of incredible thickness (6,000 feet or more; 1.8 km) stretching more than 63,000 square miles (160,000 km^2) across Washington, Oregon, and Idaho. This plateau is the result of giant volcanic eruptions producing a basalt lava plain known as the Columbia Basin. No geologic process known today could have produced this gigantic igneous formation.

Of the 5,000 volcanoes in the world, about 1900 are considered active, and possibly another 1,500 are extinct. See super volcanoes in chapter 4, Beginnings of Catastrophic Volcanoes.

Many geology textbooks state that this plateau formed by "rapid extrusion of magma from 14 to 17 million years ago."[3] Obviously, uniformity theory does not apply to this massive igneous structure and most igneous formations associated with mountain building (orogeny). "Some manifestation of catastrophic action [such as the tectonic activity of the flood described in Psalm 104:8—*the mountains rose; the valleys sank down to the place which Thou didst establish for them*] alone is sufficient."[4] Such events took place in the latter phases of Noah's Flood.

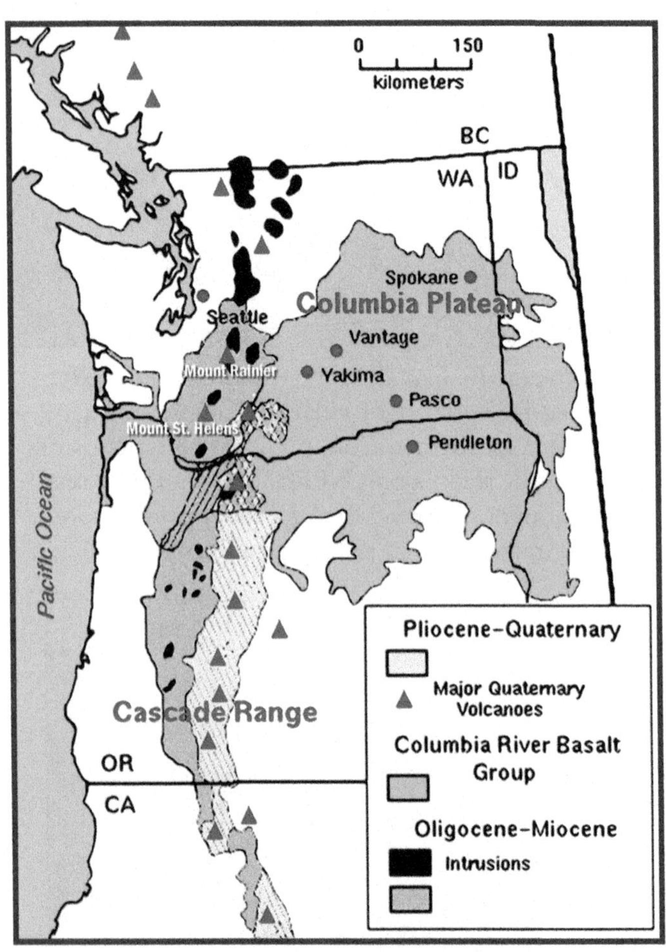

USGS Map of Columbia Plateau (green)

Moses Coulee is located in the northwest corner of the Columbia Plateau. The upper basalt is Roza Member, while the lower basalt is Frenchmen Springs Member. See section, The Missoula Flood – Channeled Scablands.

Moses Coulee cuts into the Waterville plateau in Douglas County, Washington

Mountain Overthrust

Another example of the failure of uniformity theory is the Lewis Thrust Fault located in northwestern Montana near Glacier National Park. Glacier National Park comprises mountain ranges referred to as the southern extension of the Canadian Rockies, which are an extension of the Canadian Cordillera. The Lewis Thrust is one of the world's great thrust faults—a 350-mile-long reverse fault between Alberta and Montana. The following diagrams depict a low angle reverse fault (thrust fault shown in many geology textbooks) and the Lewis Overthrust fault.

Most geology textbooks don't elaborate on the fact that the rocks in the upper half of the mountain range are Precambrian (much older rocks) while the lower rocks are Cretaceous shale, *at least 500 million years younger* (according to standard time scale). Because the principle of superposition would not allow the deposition of older rocks on top of younger rocks, secular geologists believe the older Precambrian rocks must have been pushed 50 miles or more *overtop* the younger Cretaceous shale, a more brittle sedimentary rock, despite the fact the fault is almost horizontal (dipping only three degrees).[5] This is *make-believe* thinking.

Principle of superposition states that in any undisturbed geologic sequence, older beds lie below younger beds. See chapter 7.

According to standard geologic time scale (see chapter 7), the Precambrian limestone is 570 million years old and the Cretaceous shale is 70 million years old.

U-shaped glacial trough of McDonald Valley in Glacier National Park, Montana

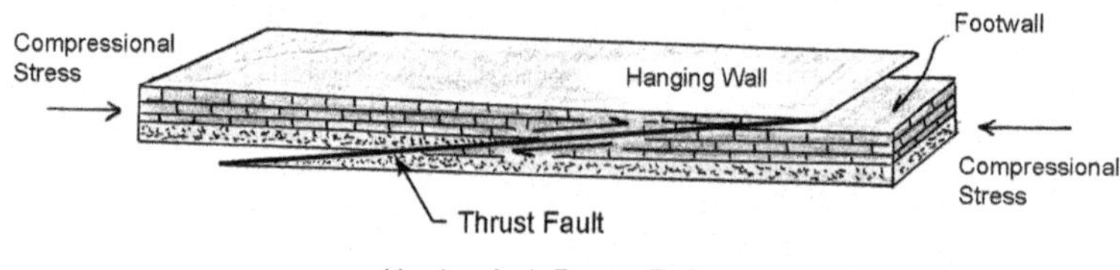

In a reverse fault, the hanging wall has been moved up relative to the footwall. A thrust fault is a special type of reverse fault that is *nearly horizontal.*
Sketch by Roger Gallop

"Lewis Overthrust"

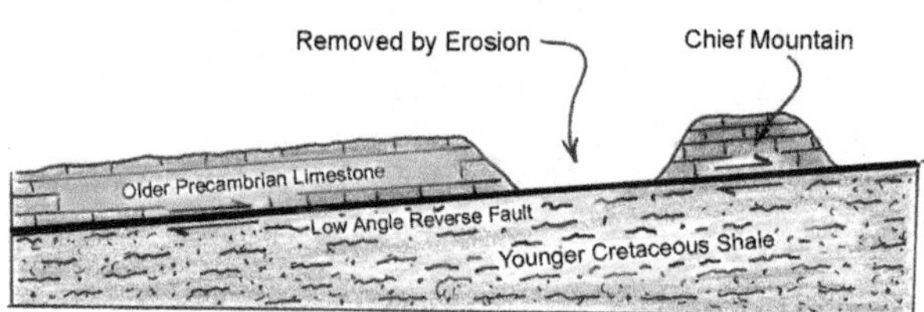

Chief Mountain and Formations. *Sketch by Roger Gallop*

Lewis Overthrust is not only a low angle reverse fault, but much *older* Precambrian limestone rests on top of much *younger* Cretaceous shale. How is this reasonably possible?

Chief Mountain at Glacier National Park, Montana

Based on known friction coefficients for sliding blocks, Dr. Henry Morris demonstrated the Precambrian rock formation could not be transported as a coherent block because of shear stress.[6] Others, such as Drs. M. K. Hubbert and W. W. Rubey, explained that "it would be impossible to move such a block" and confirmed that the existence of large overthrusts "appears to be a mechanical impossibility."[7] Low angle thrust faulting on such a massive scale would not be possible without the hydrodynamics of a worldwide flood as described in chapter 4. Either the older Precambrian limestone was deposited *after* the younger Cretaceous shale or the reverse thrust occurred *during* the flood.

Another overthrust is Heart Mountain located just north of Cody, Wyoming. Heart Mountain is an 8,123-foot (2,476 m) peak that protrudes from the floor of the Bighorn Basin. The mountain is composed of limestone and dolomite of Ordovician through Mississippian age (350 to 500 million years old according to standard time scale), but it rests on top of the Willwood Formation, alluvial rocks that are much younger of late Cretaceous to early Tertiary age (65 million years old according to standard time scale). Thus, rocks at the base of Heart Mountain are more than *300 million years younger* than the rocks at the summit.

How can this be? For over one hundred years geologists have tried to understand how these older rocks came to rest on top of much younger strata. It is clear to most objective researchers (setting aside the presumption of evolution and uniformity) that the older Mississippian and Ordovician rocks were deposited *after* the younger Cretaceous and Tertiary rocks or the massive reverse thrust was caused by the hydrodynamics of a worldwide flood.

Another contradiction in uniformity theory is the Matterhorn of the Pennine Alps located on the border between Switzerland and Italy. Similar to the previous examples, this is an "upside-down mountain" where sedimentary rocks and fossils are in reverse order. Near the top of the mountain, Cenozoic rocks (40 million years of age according to standard time scale) underlie rocks of Mesozoic age (200 million years according to standard time scale), all contrary to uniformity theory.[8]

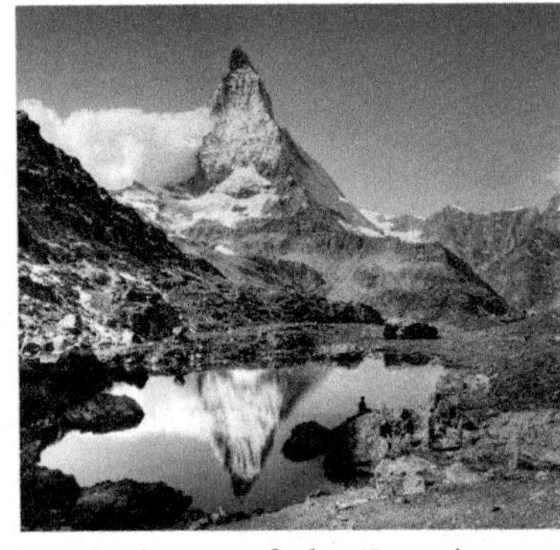
Matterhorn of the Pennine Alps on the border between Switzerland and Italy

Clearly, major geologic features with folding, tilting, and thrusting occurred over a short period of time when rock was soft and pliable. The thrust fault hypothesis including folding of igneous, metamorphic, and sedimentary rock over "millions of years" is *a futile attempt by secular geologists to explain evidence contrary to superposition and uniformity theory—that is, how older rock formations were deposited over younger formations.*

Other 'upside down' examples that are contrary to uniformity theory with no evidence of overthrust include the Franklin Mountains near El Paso, Texas, where Ordovician limestone (443 to 490 million years) lies on top of Cretaceous strata (65 to 144 million years); the Glarus Alps near Schwanden, Switzerland, where Permian limestone (248 to 290 million years) lies on top of Jurassic limestone (144 to 206 million years) which, in turn, overtops Eocene strata (34 to 55 million years); and the Empire Mountains in Southern Arizona where Permian limestone lies on top of Cretaceous strata.[9]

> Such geologic phenomena are immensely problematic for secular geologists. For more information, see chapter 7, Contradictions in Uniformity Theory, and chapter 4, Catastrophic Phenomena in Geology.

Such inconsistencies are found throughout the world. Based on the biblical worldwide flood event, such formations are consistent with catastrophic geologic processes (that is, earthquakes, volcanism, flood, and orogeny with folding, tilting, and thrusting of massive formations; and erosion and sedimentation) occurring within a short period of time.

The Colorado Plateau and Grand Canyon

The Colorado Plateau (also called the Colorado Plateau Province) is a physiographic region of the Four Corners of Colorado, New Mexico, Utah, and Arizona. The plateau covers an area of 130,000 square miles (337,000 km^2) and about 90 percent of the area is drained by the Colorado River and its main tributaries: the Green River, the San Juan River, and the Little Colorado River. The mostly flat-lying sedimentary layers are found in intermediate plateaus from 5,000 feet (1,524 m) to over 11,000 feet (3,353 m) above sea level.

In the southwest area of the plateau lies the Grand Canyon of the Colorado River, a colorful steep-sided gorge in the state of Arizona. It is 277 miles (446 km) long, ranges in width from 4 to 18 miles (6.4 to 29 kilometers), and reaches a depth of more than a mile (1.6 km). The South Rim of the canyon near the Grand Canyon Village, located just downstream of the Little Colorado River, has an elevation of 7,000 feet (2,134 m); the North Rim has an elevation of 8,000 feet (2,438 m); and the Colorado River has a bottom elevation of 2,400 feet (732 m).

The canyon is characterized by stacked, pancake-like sediment rock layers and formations about 4,000 feet thick. Formations (top to bottom) comprise Kaibab Limestone (~250 feet thick), Toroweap Sandy Limestone (~250 feet thick), Coconino Sandstone (~300 at the Grand Canyon Village to 1,000 feet thick about 100 miles to south), Hermit Shale (~300 feet thick), Supai Group (shale-limestone-sandstone; ~650 to 1,750 feet thick), Redwall Limestone (~500 to 800 feet thick), Temple Butte Limestone (~20 to 50 feet thick), Muav Sandy Limestone (limestone-sandstone-shale; ~350 to 1,000 feet thick), Bright Angel Shale (mudstone-shale-sandstone-sandy limestone; ~350 to 400 feet thick), and Tapeats Sandstone (~125 to 350 feet thick). The formation at the bottom of the Colorado River is Precambrian Cardenas Basalt rocks that existed before the flood as part of the supercontinent, Pangaea.

> **Formation** is a distinct body of sedimentary, igneous, or metamorphic rock with similar characteristics, and can be visually recognized in the field. A formation usually comprises a group of similar **layers** as shown in the Grand Canyon diagram.

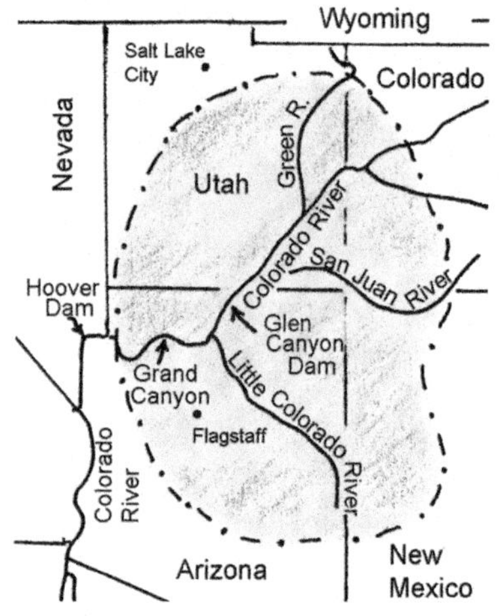

Colorado Plateau

Cross Section of Grand Canyon

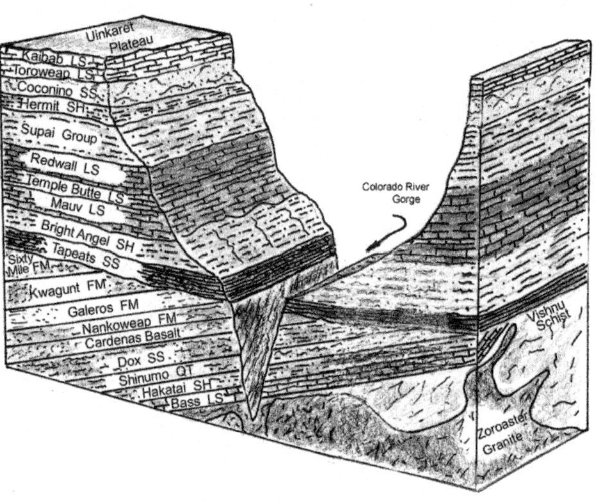

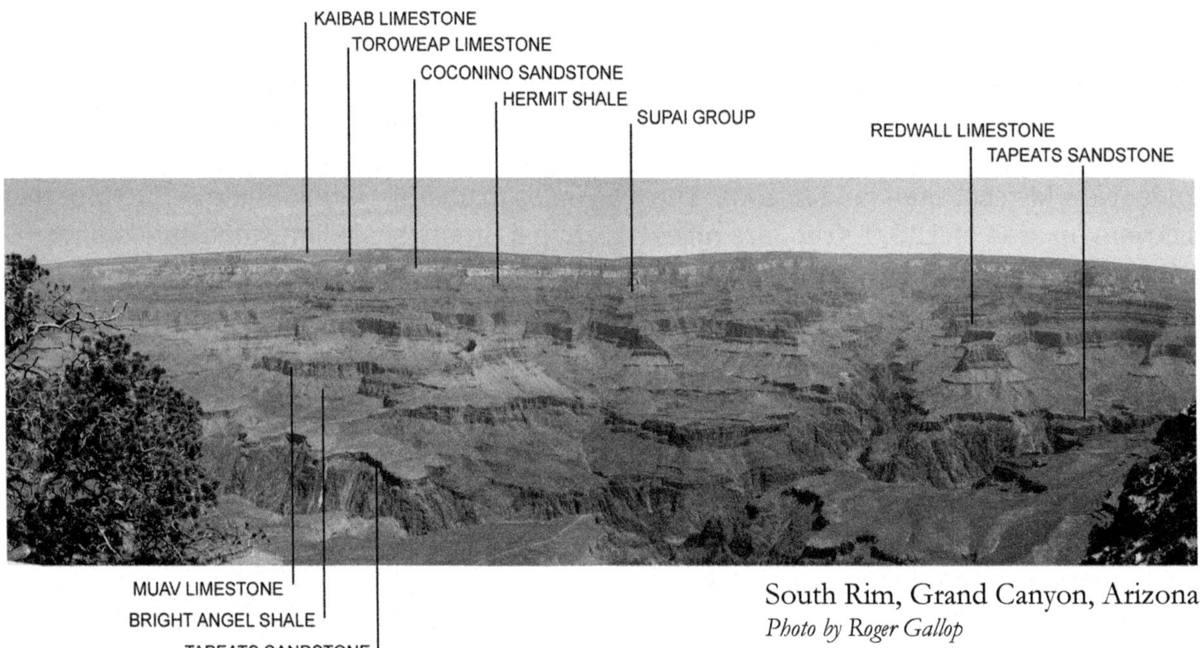

South Rim, Grand Canyon, Arizona
Photo by Roger Gallop

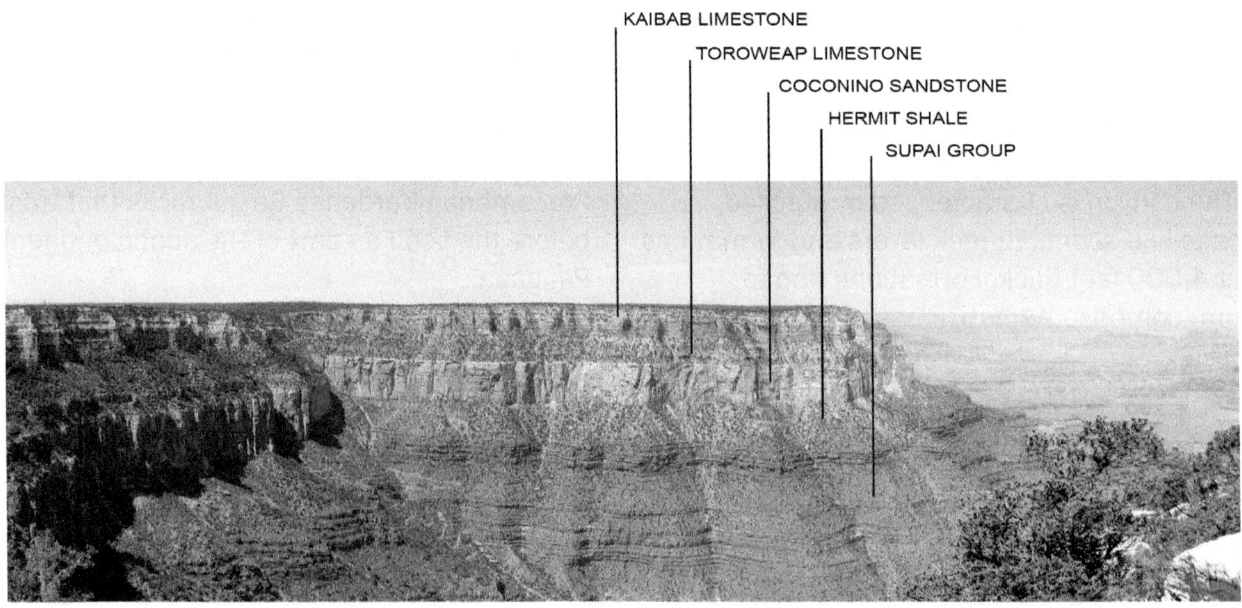

South Rim, Grand Canyon, Arizona
Photo by Roger Gallop

Kaibab Upwarp

According to old earth scenario (uniformity theory), the Grand Canyon's sandstone and limestone layers were formed by slow and gradual sediment deposition during alternating periods of inundation (usually described as a calm sea in many textbooks) and dry conditions over a span of 350 million years (during the Paleozoic Era; see cross-section of Grand Canyon). Uniformity theory maintains that Tapeats Sandstone (bottom sediment formation) of the Grand Canyon was deposited about 550 million years ago while the Kaibab Limestone (top formation) was deposited about 200 million years ago.

> Given these vast time periods, one would conclude the Tapeats Sandstone and overlying sedimentary layers would be hard and brittle after almost 350 million years of overburden and compression.

Then, according to secular geologists, these horizontal sediment layers of the Colorado Plateau representing vast periods of time were *slowly* "uplifted" almost a mile to their present elevations about 50 to 70 million years ago during the Laramide orogeny and formation of the Rocky Mountains (although some textbooks maintain the plateau was uplifted sometime later about 5 to 10 million years ago). A feature of the Grand Canyon associated with the Colorado Plateau is the *Kaibab Upwarp*, an elevated secondary plateau through which the Grand Canyon was supposedly carved by the Colorado River.

> Plastic, non-brittle folding of strata along the eastern edge of the Kaibab Upwarp indicates that layers were soft and pliable when bent, which is consistent with all layers being deposited quickly during a worldwide flood event.

Although most geology textbooks agree the Colorado Plateau was uplifted without much internal deformation, Dr. John Morris in his book, *The Young Earth,* points out that "most people don't know that on the edge of the plateau [*eastern edge of the Kaibab Upwarp*]... the rocks are, in places, standing in a near-vertical orientation."[10] Further, he explains that "the sandstone [Tapeats] appears to have been in a soft, unconsolidated condition when bending occurred. Nowhere are found elongated sand grains, or the cement which bound the grains together broken and recrystallized."[11] This is consistent with layers being deposited horizontally and then *bent quickly* while still soft.

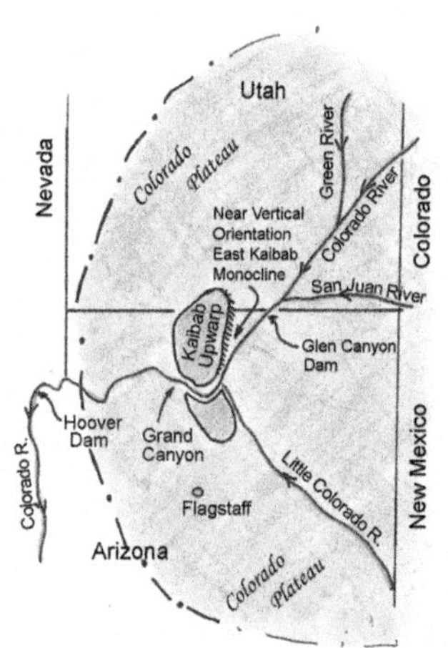

Colorado Plateau and Kaibab Upwarp

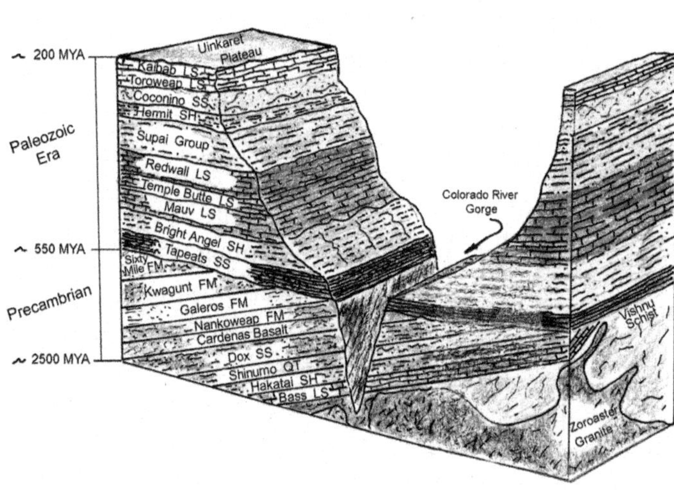

So how long does it take sediments (mineralized fragments such as quartz sand or silt) to turn into solid rock? Generally, it takes just a few years and certainly no more than 100 years for sandy deposits to turn to sandstone by deep burial with moisture and dissolved minerals (dissolved calcium carbonate, silica, and iron) that glue the grains together. The pressure of deep burial squeezes out the pore water and compacts individual grains, a process known as lithification. It does not take millions of years to form sediment rock.

Some old earth theorists maintain that rock formations can bend if buried deeply with enough heat and confining pressure on all sides. Says Morris, "...in a hard rock like the Tapeats sandstone, that sort of bending always results in elongated sand grains or broken crystals, neither of which have been found in these deformed Grand Canyon rocks."[12] Based on investigations by Morris and others, these sediment layers were still soft and pliable when bending occurred.

While many geologists continue to maintain a tight grip on uniformity theory while ignoring visual and empirical evidence, some geologists (including many of the secular "millions-of-years" variety) concede that the Grand Canyon was formed catastrophically and was not the result of gradual erosion by the Colorado River over millions of years.

Depositional Features

Nearly all geologists agree that sedimentary rocks found throughout the world were deposited horizontally underwater in an unconsolidated form. When the water receded, sediment layers hardened into rock as water was squeezed out and the individual grains were compressed. In almost every case, deposition was accomplished by water in oceans, rivers, streams, or lakes. *Wind deposits rarely result in sedimentation that hardens into rock* because of the lack of moisture and dissolved minerals (cement).

Sandstone Turbidites

According to young earth advocates, pancake-like sediment layers in the Grand Canyon and throughout the world were deposited underwater as turbidites during a catastrophic flood event. For example, Tapeats Sandstone near the bottom of the Grand Canyon exists as pancake layers covering most of the central portion of North America and western Canada.[13]

Turbidites are pancake-like sediment layers that have their origins in catastrophic underwater avalanches or mudslides—or a succession of underwater flows of sand. See chapter 4.

If many millions of years separated these various strata, as maintained by uniformity theory, each individual layer would show signs of erosion, bioturbation (physical disturbance by plant and animal life), soil layers, and other features representing vast periods of time between layers—but such evidence is missing.

Tapeats Sandstone, Grand Canyon, Arizona. Obviously, there are no time breaks between the deposition of these sediment layers. *Photos by Roger Gallop*

Many geologists now recognize that *most pancake rock layers are actually turbidites deposited underwater in a sea environment and uplifted during a catastrophic worldwide tectonic event!*

Tapeats Sandstone, Little Colorado Gorge, Arizona This shows no time breaks between the deposition of these sediment layers. *Photo by Roger Gallop*

Coconino Sandstone

The Coconino Sandstone Formation found near the top of the Grand Canyon is about 300 feet thick (thickens to 1,000 feet about 100 miles to the south) and extends over 100,000 square miles. According to flood geologists, the sharp contact between geologic formations, large cross-bedding and sharp angular bedding planes suggests the layers were deposited in deep, moderately fast-flowing water in days or months.[14]

The following photographs show the abrupt contact between Coconino Sandstone and Hermit Shale exposed along the Bright Angel Trail in Grand Canyon, Arizona. Also note the abrupt contact between Coconino Sandstone and Toroweap Limestone. According to uniformity theory, the time gap between these formations is 10 million years or more—again, this is wishful thinking.

Obviously, *sharp, undisturbed bedding planes (sharp contacts between layers and formations) indicate little or no time between the deposition of the layers and each formation*—otherwise, there would have been erosion, soil layers, and signs of plant and animal activity at the interface.

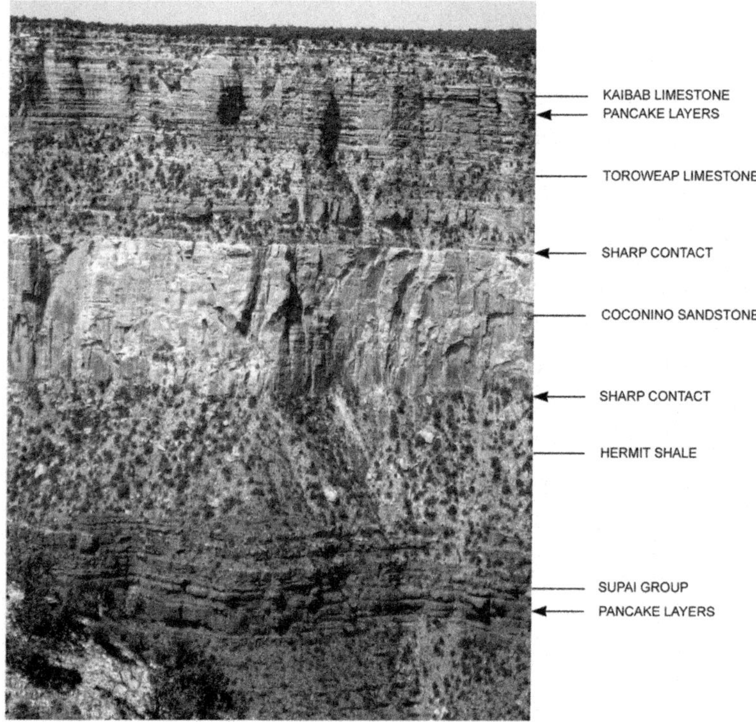

South Rim, Grand Canyon, Arizona
Photo by Roger Gallop

Sharp, undisturbed bedding planes between layers and formations are indicative of deposition over a short period of time—and such features are typically found worldwide.

In their article, "Startling Evidence for Noah's Flood," Drs. Andrew A. Snelling and Steven A. Austin present empirical and visual evidence supporting flood deposition of Coconino Sandstone.[15] These evidences are summarized below.

South Rim, Grand Canyon, Arizona. *Photo by Roger Gallop*

Coconino Sandstone and Hermit Shale, South Rim, Grand Canyon, Arizona. *Photo by Roger Gallop*

A common feature within the Coconino Sandstone is inclined bedding, or cross-bedding. For many years secular geologists have compared these cross-beds to eolian processes (Æolian)—that is, processes pertaining to the activity of the winds. These same geologists maintain the Coconino Sandstone formed over vast periods of time "in a windy desert by migrating dunes."[16]

Another sedimentary feature of Coconino Sandstone is "the large number of fossilized footprints." When compared with modern tracks in desert sands, it was readily assumed the Coconino footprints were produced by eolian processes—that is, these footprints were "covered up by wind-blown sand" and fossilized.[17] The most obvious question is: How can delicate sedimentary features (i.e., cross-bedding, animal tracks, fossil impressions; see chapter 6) be preserved in a dry desert environment?

Sand dunes in deserts rarely, if ever, result in cross-bedding that hardens or has become "frozen" into rock. Desert dune features are fragile and will soon erode and disappear because of exposure to wind erosion. Preservation of delicate features requires immediate burial, moisture, and cement (i.e., dissolved calcium carbonate, silica, and iron) and, in the case of animal fossils, no oxygen (anaerobic conditions); otherwise, decay will occur. Such necessities are not available in desert environments. Cross-bedding is not indicative of desert sand dunes, but rather, "underwater" sand dunes.

Experimental studies by Dr. Leonard Brand of Loma Linda University in California, in 1979, involved careful examination of 82 fossilized vertebrate footprints in the Coconino Sandstone along the Hermit Trail of the Grand Canyon.[18] He then examined another 236 footprints made by amphibians and reptiles in laboratory chambers. In his research, Dr. Brand concluded that the Coconino footprints did not support eolian processes but rather "submerged sand surfaces." Follow-up studies in 1991 confirmed the conclusions of his earlier research.[19] These findings were so significant that a report about the research appeared in *Science News* (1992, vol. 19) and *Geology Today* (May–June 1992, vol. 8, No. 3).

Wind-blown desert sands have also been refuted by Dr. Glen Visher, professor of geology at the University of Tulsa in Oklahoma, who is not a creationist geologist. In his research, Visher interpreted Coconino as a "submarine sand wave deposit" (dune-like waves) created by water, not wind.[20] Visher also points to a phenomenon called "parting lineation," which commonly forms on "sand surfaces during brief erosional bursts beneath fast flowing water." Although found in Coconino Sandstone, it is not known to exist in desert sand dunes. Such features are "evidence of vigorous water currents depositing sand which formed the Coconino Sandstone."

Cross-bedding, Coconino Sandstone, Grand Canyon, Arizona
Photo by Roger Gallop

A feature sometimes used by secular geologists in support of eolian processes is the "pitting" characteristic of sand grains. As studied under the microscope, sand grains from modern desert dunes often show pitted surfaces similar to that of Coconino Sandstone. However, research has confirmed that pitting of sand grains can occur "outside a desert environment"—for example, it has been shown to form during the cementation process.[21] Also, pitting is "not diagnostic of the last [geologic] process"—that is, it is more likely the result of water deposition of previously wind-blown sand.

Cross-bedding (inclined sand beds within the overall horizontal layer of sandstone) has been found to be "excellent evidence that ocean currents moved the sand rapidly as dune-like mounds called sand waves."[22] It has been demonstrated that not only is "sand wave" height related to the water depth but also water velocities can be determined.[23] The largest or highest sets of cross-bedding found in Coconino Sandstone are 30 feet (9 meters)[24] which imply "sand waves at least 60 feet (18 meters) high and a water depth of around 300 feet (between 90 and 95 meters). For water that deep to make and move sand waves as high as 60 feet (18 meters) the minimum current velocity would need to be over 3 feet per second (95 centimeters per second) or 2 miles per hour."[25]

Modern ocean currents at a depth of 300 feet or more cannot sustain such velocities (2 mph) except around restricted submarine landforms. Catastrophic flooding and torrential drainage from uplifted continents provide the only mechanism that can produce high-velocity ocean currents.

It is interesting to note that below Coconino Sandstone is Hermit Shale and above is Toroweap Sandy Limestone and, as agreed by most geologists, these sediment layers (Hermit and Toroweap) were deposited by water.[26,27] Uniformity theory maintains that these formations are separated in time by at least 10 million years. If the Coconino represents a desert (covering more than 100,000 square miles), then the submerged Hermit Shale had to be "uplifted, out of the water, to an elevation high enough and dry enough to be a desert" over millions of years.[28] Yet, the upper surface of Hermit Shale remains flat, with no evidence of erosion. How could this surface remain stagnant over this vast period of time? A simple and plausible explanation is that these layers were deposited quickly during the course of a worldwide flood.

It is *simply not possible* for such events to take place over tens of millions of years without evidence of erosion. *No surface on earth could remain dormant over this vast period of time.*

Limestone Formations

In today's marine environment, calcareous lime muds (calcium carbonate, $CaCO_3$) found on sea floors form by the physical breakdown of carbonate-containing sea creatures and by slow precipitation of calcium carbonate. Marine organisms, such as corals, clams, mussels, and snails, are able to extract calcium from seawater to build their skeletal structure. When these marine organisms die, their shells disintegrate and intermingle with bottom sediment and collect as calcareous ooze (lime muds).

According to most geology textbooks, accumulation rates of lime mud in shallow seas are about one foot per 1,000 years. Because evolutionists believe that "the present is the key to the past"—that is, physical processes have remained constant throughout earth's history, they believe it is impossible to explain the formation of massive limestone (including those in the Grand Canyon) by a worldwide flood.

> Massive limestone observed worldwide ranges from fine-textured calcite to turbidite-like mixtures of calcite, sand, and shell fragments.

There is substantial visual and empirical evidence for rapid and massive accumulation of lime muds in a high-energy water environment. It has been suggested that limestone was formed by "direct precipitation" of calcium carbonate during storm events. Microscopic examination reveals that modern lime mud particles aggregate into "pelletoids [which]... exhibit hydraulic characteristics of sand"[29] and settle out quickly.

As seawater warmed, pelletoids would have rapidly accumulated on the ocean bottom—and in response to earthquakes, massive slope failures formed turbidites (pancake-like strata) comprising lime mud, sand, and shell fragments. Bulging, pancake-like lime mud layers of Kaibab Limestone (see photos in chapter 6, Limestone Formations) indicate huge mudflow and turbidite formation.

> Warming of ocean waters from volcanic activity and superheated geysers associated with a worldwide flood would have directly precipitated massive quantities of lime mud deposits throughout the world. Today, calcium carbonate is often found precipitating near hot springs. In contrast, calcium carbonate is not found in sediments in the colder, deep marine environment.

Visual evidence of rapid lime mud burial of the Redwall Limestone, for example, includes large underwater sand waves, lack of coral and sponge reef formations (expected to be found in ancient tranquil seas), and preservation of fast-swimming nautiloids and other delicate fossil remains.[30] Also, extremely well-preserved and diverse fossils, commonly known as Lagerstätten, are found within limestone formations throughout much of the world.[31] These are examples of fossilization of animals that were rapidly covered in lime mud of fine texture. A discussion of this evidence is found in Appendix E.

What are the mechanisms for rapid deposition of limestone? Massive limestone formations found worldwide were derived from three primary processes: 1) aggregation or flocculation of fine-grain crystals of calcite or aragonite in a high energy environment, 2) warming of seawater and direct precipitation of dissolved calcium carbonate, and 3) precipitation of dissolved calcium carbonate due to the depletion of carbon dioxide in the atmosphere, all as a result of the flood. A discussion of these mechanisms is found in chapter 6, Origin of Immense Limestone Formations and Chalk Beds, and Appendix F.

Redwall Limestone (vertical formation wall near bottom of photos). Above Redwall is Supai Group (shale intermixed with limestone), then Hermit shale, and Coconino sandstone (vertical formation wall near top of right photo), and then Toroweap and Kaibab Limestones, South Rim of Grand Canyon. *Photos by Roger Gallop*

Erosional Theories of the Grand Canyon

According to uniformity theory, the Laramide orogeny lifted the Colorado Plateau, which includes the area of the Grand Canyon, and crumpled the Rocky Mountains. The canyon cuts directly through a huge uplifted region called the Kaibab Upwarp. Kaibab Limestone that caps this plateau is thought to be 200 million years old and the Colorado River has supposedly been cutting through this elevated region and creating the Grand Canyon for at least 50 million years. Once eroded, the sediments from the Colorado River and its tributaries were carried away and deposited somewhere downstream.

There are two uniformity theories regarding the formation of the Grand Canyon: the Antecedent River Theory and the Precocious Gully Theory. There is also one catastrophic theory: the Breached Dam Theory. These theories are thoroughly explained by Dr. Steven A. Austin in his book, *Grand Canyon: Monument to Catastrophe.*[32] A brief summary of these theories is presented in the following paragraphs.

The "Antecedent River Theory" proposes the Colorado River was already present in northern Arizona *before, or antecedent to*, the uplift of the Colorado Plateau. This theory has many unresolved problems. Based on empirical and observational geological data, suffice it to say the idea that the Grand Canyon slowly eroded over tens of millions of years is simply wrong and most canyon geologists have abandoned the theory.

The "Precocious Gully Theory" (or "Stream Capture Theory") proposes the Upper Colorado River flowed in a southeasterly direction along the course of the current Little Colorado River and, inexplicably, the river cut across the Kaibab Upwarp and formed the enormous gully called the Grand Canyon. This theory would have to reconcile an older, ancestral Upper Colorado River many millions of years old and a much younger Grand Canyon that was eroded in just a few million years. These theories create more problems than solutions and have been essentially discarded.

Colorado River, Dead Horse Point State Park near Canyonlands National Park, Utah
Similar to the Grand Canyon, this landform is the result of rapid drainage of floodwaters over a short period of time. Today, the Colorado River gradually erodes these canyons. *Photo by Roger Gallop*

The "Breached Dam Theory," or the catastrophic drainage of lakes, provides the most reasonable explanation and is remarkably similar to the Lake Missoula flood and Bonneville flood (explained in the following sections). This theory proposes that the Grand Canyon was eroded largely by torrential drainage of a large lake system (30,000 square miles, or over three times the size of Lake Michigan). These ancient drainage lakes, whose dams or barriers failed during or some time following the flood, are named Vernal Lake, Canyonlands Lake, and Hopi Lake.

Evidences for this theory include:

- sediment strata indicative of ancient drainage lakes east of the Kaibab Upwarp,
- unusual deltaic deposits near the Gulf of California,
- geomorphic evidence of accelerated drainage, and
- relict landforms.

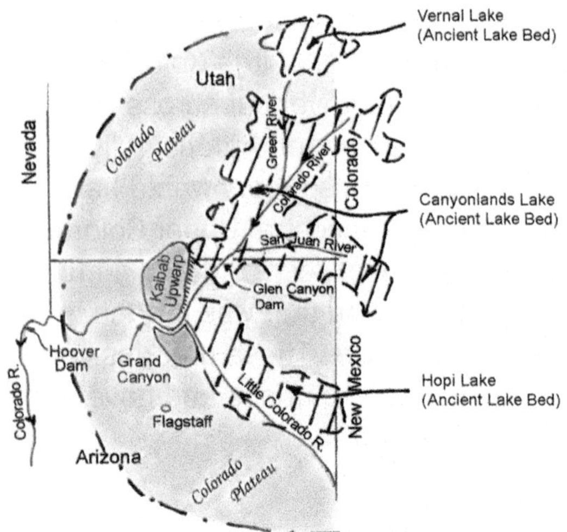

Earle E. Spamer summarized the situation:[33]

> The greatest of Grand Canyon's enigmas is the problem of how it was made. This is the most volatile aspect of Grand Canyon geological studies ... Grand Canyon has held tight to her secrets of origin and age. Every approach to this problem has been cloaked in hypothesis, drawing upon the incomplete empirical evidence of stratigraphy, sedimentology, and radiometric dating.

For decades secular geologists have considered the Grand Canyon to be a showcase of uniformitarian processes—the belief that the Colorado River eroded the canyon over a span of 50 to 70 million years following the Laramide uplift. Today, most geologists concede that this explanation is wrong and the canyon was created by a cataclysmic flood in a short period of time. This is due to overwhelming visual and empirical evidence in support of such an event.

Relict landforms are geomorphic features that have remained essentially unchanged after surrounding landforms have changed or disappeared. Such features have survived decay and disintegration. Plateaus, mesas, buttes, spires, and coulees are examples of relict landforms.

Erosional Features of the Colorado Plateau

While there has been significant vertical erosion of the Grand Canyon, the Great Goosenecks, and other drainageways for supposedly 70 million years, much of the top of the Colorado Plateau remains untouched—virtually flat and featureless except for isolated erosional remnants. A "flat" plateau is contrary to uniformity theory over tens of millions of years. So what really happened?

As flat-lying strata were inundated by floodwaters (described in chapter 4, phase two of the flood) and then uplifted 5,000 feet (phase three), retreating floodwaters stripped away upper sedimentary layers except for resistant erosional remnants. Drainage waters receded into areas of least resistance creating deeply entrenched meanders, canyons, and gorges, all in a matter of months, not millions of years.

Confluence of Green and Colorado Rivers, looking south, southeast Utah

Layers such as Moenkopi and Chinle formations, commonly found in remaining erosional remnants such as mesas, buttes, spires, and hoodoos, comprise a wide assortment of marine fossils deposited during the flood. These layers were *rapidly swept away* "by broad, sheetlike erosion" in a matter of weeks or months.[34]

Great Goosenecks of the San Juan River, Gooseneck State Park, SR 261 at SR 163, Mexican Hat, Utah. The rocks exposed in the gorge are fossiliferous marine limestone, sandstone, and shale--named Honaker Trail and Paradox Formations. Secular geologists mistakenly believe that a river flowing on a flat plain was "entrenched" as the Colorado Plateau slowly uplifted over millions of years. Similar to the Grand Canyon, this geologic landform was actually the result of rapid drainage of floodwaters over a short period of time.

Grand Canyon, Arizona. As the plateau was uplifted, floodwaters receded within drainageways of least resistance resulting in vertical erosion and creating canyons and narrow gorges throughout the plateau. *Photos by Roger Gallop*

Once the water was gone, isolated relict landforms or erosional remnants remained. These landforms, which today undergo much-reduced wind and rain erosion, are monuments to a worldwide flood.

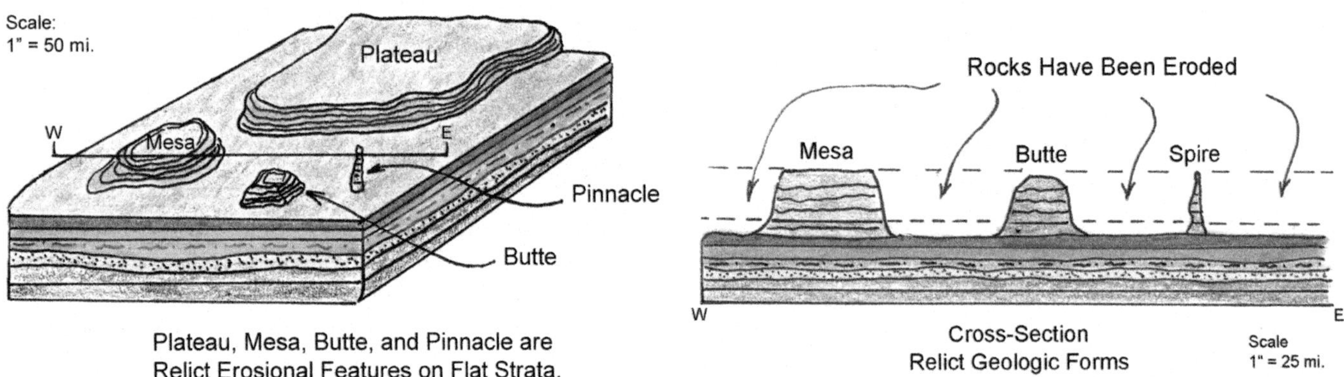

Plateau, Mesa, Butte, and Pinnacle are Relict Erosional Features on Flat Strata.

Cross-Section Relict Geologic Forms

A **mesa** is a broad, flat-topped hill rising above the surrounding lowlands; a **butte** has a smaller, narrower top than a mesa, with steep-sided edges; and a **spire or pinnacle** is a tall totem pole feature that tapers from the ground upward. **Hoodoos** are similar to spires except they have variable thickness. *Sketches by Roger Gallop*

Red Butte, SR 180, Arizona
Photo by Roger Gallop

Erosional Features, Grand Canyon, Arizona. *Photos by Roger Gallop*

Balanced Rocks, SR 89A, AZ

Hoodoos and an Arch at Bryce Canyon National Park, Utah *Photos by Roger Gallop*

Goblin Valley State Park, Utah Entrada Sandstone

Balanced Rock Arches National Park, Utah Entrada Sandstone, 128 feet high

Delicate Arch, Arches National Park, Utah *Photos by Roger Gallop*

Such features are the result of continental drainage (tidal channels that directed currents to deepening ocean basins) and post-flood wind, rain, and frost. Followed by erosive action of moving floodwaters, gradual erosion by wind, rain, and frost has worked to enlarge natural arches and bridges.

Young Earth Interpretation

Tapeats Sandstone was deposited early in the flood, and all subsequent formations including the Coconino Sandstone and Kaibab Limestone were deposited later in the flood. These sediment layers were deposited horizontally underwater and remained horizontal during continental and regional uplift (except the Kaibab Upwarp that deformed sediment layers in a vertical orientation). The Precambrian Cardenas Basalt (base of these sediment layers) existed before the flood as part of the supercontinent, Pangaea.

During the third phase of the flood (ocean basin subsidence, continental uplift, and drainage; see chapter 4), floodwaters rapidly swept away sediments and, in the process, floodwaters became trapped as gigantic lake systems (Hopi Lake, Canyonlands Lake, and Vernal Lake). Sometime following the flood (likely during the Great Ice Age; see chapter 4), these lake systems broke through plateau barriers and eroded the Grand Canyon in a matter of days. Today, the Grand Canyon has become a monument to the power of the Creator God and Noah's Flood described in Genesis 6-8.

The Missoula Flood – Channeled Scablands

The Channeled Scablands are a unique geologic erosional feature that was created across eastern Washington and much of the Columbia Plateau. This 16,000-square-mile drainage pattern, which begins in the northeastern portion of the state and exits at the Pacific Ocean, has a braided, gorge-like appearance with immense potholes, ripple marks, and hundreds of small lakes surrounded by flat-top mountains. The landscape also comprises dry, braided canyons known as "coulees"—ancient ravines, basins, and dry waterfalls. Grand Coulee Canyon, for example, is about 50 miles in length and one to five miles across (see photograph). All these relict erosional features are found several hundred feet above the present course of the Columbia River.

Cordilleran Ice Sheet
Canada
Washington
Montana
Pacific Ocean
Oregon
Idaho
Portland
Columbia R.
Snake R.

1. Glacial Lake Missoula
2. Dry Falls (Grand Coulee Canyon)
3. Channeled Scablands
4. Wallula Gap
5. Columbia River Gorge
6. Mount St. Helens

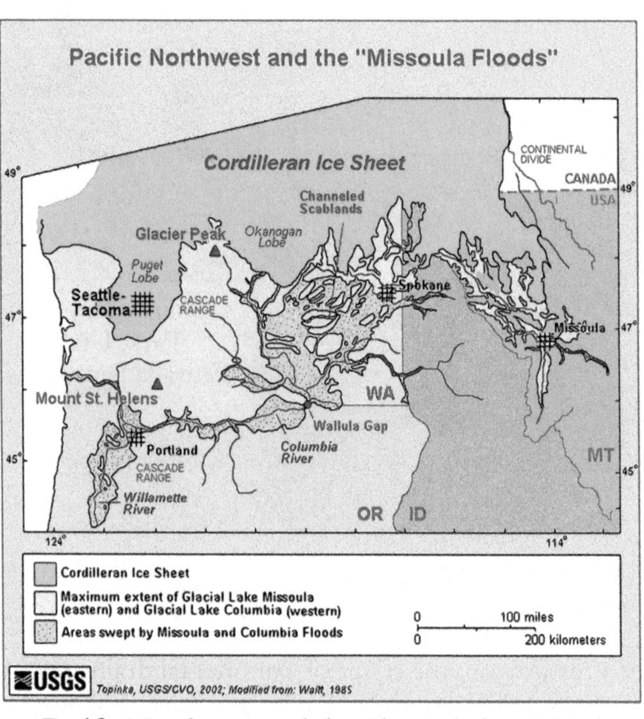

Pacific Northwest and the Channeled Scablands

At the end of the last Ice Age, Dry Falls, in eastern Washington's Lower Grand Coulee Canyon, is believed to have been the world's largest waterfall. As the name suggests, Dry Falls no longer carries water, but is a relic of what was the largest waterfall known to have existed on earth, with 3.5 miles of vertical cliffs that dropped 400 feet. By comparison, Niagara Falls, is about one mile wide with a drop of 165 feet.

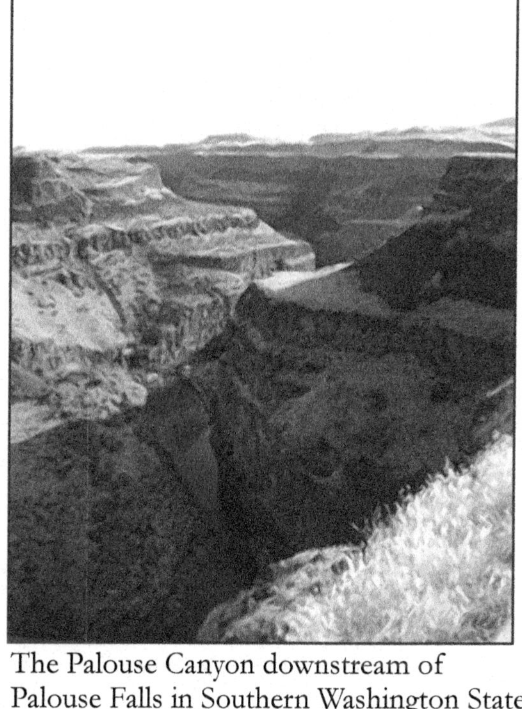

The Palouse Canyon downstream of Palouse Falls in Southern Washington State

In the early 1920s, Dr. J. Harlen Bretz (1882–1981), American geologist, maintained the Scablands were created by a cataclysmic flood that swept across the panhandle of Idaho, eastern Washington, and down the Columbia River Gorge and into Oregon. The idea that land features such as the Channeled Scablands, Dry Falls, and Palouse Falls Gorge were the result of floodwaters was considered "outrageous" and "lunacy" by secular geologists. After great opposition for more than 50 years, these same geologists finally realized the Scablands were formed by a catastrophic flood, and Dr. Bretz was eventually recognized for his research and was awarded the Penrose Medal in 1979—the most prestigious honor in geology.

> Like most people, scientists become entrenched in their beliefs, whether right or wrong, and "do not like to admit a long-held error even in the face of otherwise over-whelming evidence."[35]

Most geologists now believe the Channeled Scablands were created by thawing of the Cordilleran Ice Sheet and a catastrophic collapse of a massive ice dam holding back waters of "Glacial Lake Missoula." The rising waters of Lake Missoula, covering over 3,000 square miles of northwest Montana (and estimated to contain half the volume of Lake Michigan), lifted a massive ice dam and allowed waters 400 to 600 feet deep to rush out with incredible force. At over 50 miles per hour, floodwaters carved braided gorges and ravines in just a matter of a few days.

Although geologists now recognize that the Channeled Scablands were created by a cataclysmic flood, they nevertheless maintain a "tight grip" on old age doctrine and believe this event occurred during the Pleistocene epoch at the end of the Wisconsin Ice Age, about 10,000 to 15,000 years ago. Remarkably similar to the breached dam theory of the Grand Canyon, creation scientists maintain the scablands were created during the Great Ice Age approximately 4,000 years ago following a catastrophic worldwide flood.

The Bonneville Flood – Snake River Plain

Lake Bonneville was an ancient lake system that covered much of North America's Great Basin region (also known as Bonneville Basin)—today, a large, arid region located between the Wasatch Mountains of Utah and the Sierra Nevada mountains of California. The lake was almost 1,000 feet (305 m) deep and more than 20,000 square miles (51,800 km^2), nearly the size of Lake Michigan.

With the collapse of rock barriers between elevations 5,220 feet and 4,840 feet, the lake emptied about 380 feet, or almost 850 cubic miles of water through Red Rock Pass in southern Idaho.[36] (See section, Dry Lake Basins, for additional information.) These floodwaters swept into what is now the Snake River Plain, a semi-arid valley of the Snake River, and a major tributary of the Columbia River in Washington. The drainage flow that carved out this plain was far greater than the flow of water today.

Similar to events of the Channeled Scablands of Washington and the Grand Canyon of Arizona, an enclosed lake system was formed during the worldwide flood followed by collapse of physical barriers that occurred during the Great Ice Age. See chapter 4. Relict erosional landforms (e.g., great dry lake basins, canyons and ravines, and overfit valleys) associated with such events are common throughout our world today. Secular geologists are quick to interpret such geologic land features as the result of slow and gradual processes because they cannot accept the biblical worldwide flood and a Divine Creator.

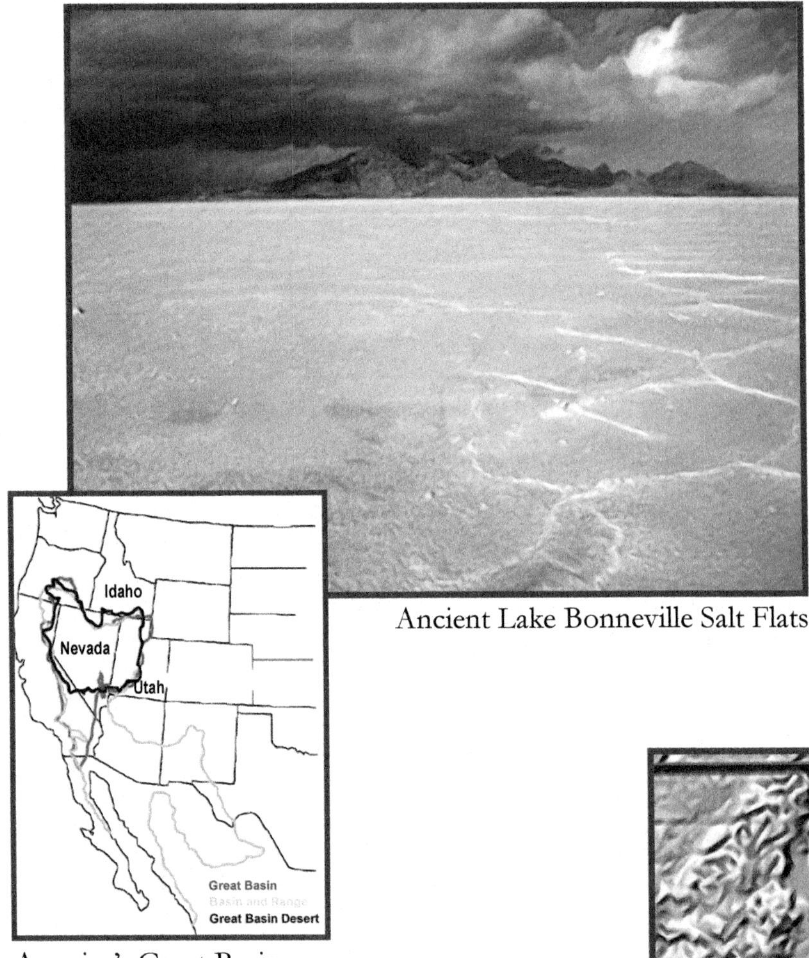

Ancient Lake Bonneville Salt Flats

America's Great Basin Desert (black)

Ancient lakes were once filled with seawater during and following the worldwide flood, but remained enclosed only for a short time. Soon after the flood, waters broke through physical barriers, eroding great canyons and basins while flowing to newly formed ocean basins. See section, Erosional Theories of the Grand Canyon.

Snake River Plain across Southern Idaho

The Snake River near Oxbow, Oregon

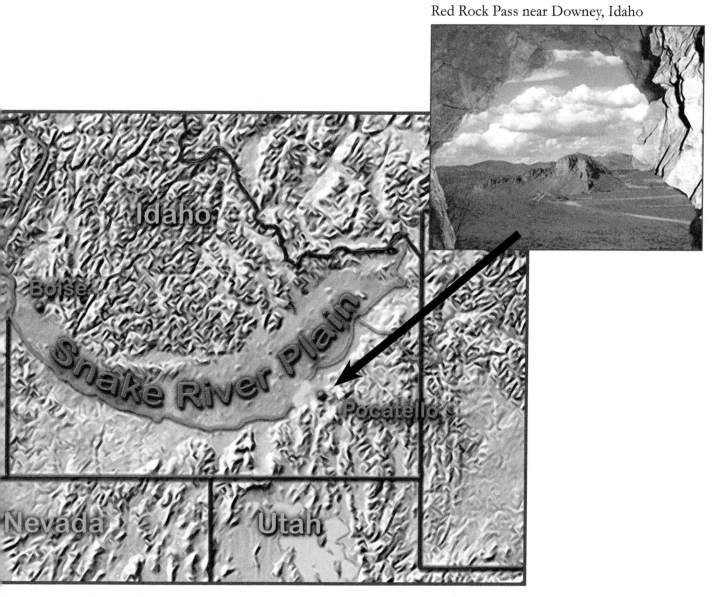

Red Rock Pass near Downey, Idaho

Rapid Continental Drainage – Relict Landforms

As newly formed mountain ranges and continental landmasses appeared from the waters of the flood, suspended sediments of receding waters formed vast alluvial plains that today extend from the Rocky Mountains to the Mississippi, and from Canada to Mexico, and throughout most of the world. Relict landforms of receding floodwaters include alluvial plains, underfit rivers and overfit valleys, raised river terraces, great monoliths (monadnocks) and peneplains, and dry lake basins.

Alluvial Plains, Underfit Rivers and Overfit Valleys, and Raised Terraces

Alluvial plains found throughout the world (including the fertile agricultural breadbasket of the Midwest) comprise enormous volumes of sedimentary rock and soil hundreds (sometimes thousands) of feet thick. Alluvial describes rich fertile soils deposited underwater. Except for mountain ranges characterized by igneous intrusion and massive tectonic uplift, the entire earth is covered by hundreds of feet of massive alluvial layers.

Most geologists agree that alluvium was deposited by floodwaters. Today, great alluvial plains have no stream or river drainage or have slow-moving streams. Although some of these plains receive periodic flooding and deposits of silt and clay by present-day processes, such stream flows are inadequate to produce the vast alluvial plains we see now. Most alluvial plains (and alluvial deserts) were created by the deposition of suspended sediments and then shaped by receding floodwaters in a relatively short period of time. Alluvial plains are relict landforms of a worldwide flood.

> Alluvial plains (or alluvial deserts) are relatively flat landforms created by the deposition of sediment by rivers and floodwaters. According to secular geologists, the alluvial plain is a region in which the river floods and the floodplain has shifted over geologic time.
>
> In reality, however, alluvial plains are geologic landforms created by the deposition of suspended sediments (as water velocity diminished prior to continental uplift) during the worldwide flood and then shaped by receding floodwaters.

Similarly, many river valleys throughout much of the world have slow-moving streams in broad floodplains that are strewn with boulders, cobbles, gravel, and sand typical of enormous flood drainage. Also, raised river terraces extending from the Mississippi River to the Rocky Mountains to the west and Appalachian Mountains to the east suggest that rivers in North America, such as the Ohio, Missouri, and Arkansas rivers that feed into the Mississippi River, "once carried much larger volumes of water than do their present remnants."[37] The drainage flow that carved out such river valleys was far greater than flow of water today.

During the global flood, the Mississippi River discharged enormous quantities of sediment into the Gulf of Mexico, forming vast marine plains or deltas. This is consistent with torrential floodwaters draining from rising land surfaces at the close of Noah's Flood and flowing into rapidly sinking, deeper ocean basins.

> An **underfit river** exhibits evidence of a much greater discharge in the past, but now has a stream of greatly reduced size—that is, the present stream is too small to have eroded a U-shaped valley. An **overfit valley** is much too large for modern day rivers or river meanders.

> The erosion caused by receding floodwaters is the reason that U-shaped river valleys are much larger than current river channels and their historic floodplains. Most secular geologists reject a worldwide flood because it implies Divine creation and a young earth.

Great Peneplains and Monoliths

Great peneplains are low, flat featureless plains found throughout the world. Rather than depositional features like alluvial plains, peneplains are formed by "broad, sheetlike erosion" as discussed in section, Erosional Features of the Colorado Plateau. The predominance of flat, featureless surfaces is the primary characteristic surrounding monoliths or monadnocks (such as Red Butte of Arizona).

A **peneplain** is a nearly featureless plain formed by sheetlike erosion (as water velocity varied, often exceeding 60 mph). Common erosional features of peneplains are buttes, monadnocks or monoliths.

A **monadnock** is an isolated hill or small mountain that rises abruptly from a virtually level surrounding peneplain.

A **monolith** is a single massive mountain or rock feature, often comprising very hard metamorphic or sedimentary rock, or igneous rock.

Pilot Mountain, a quartzite monadnock ("Big Pinnacle" and "Little Pinnacle"), rises to a peak 2,421 feet (738 m) and is considered one of the most distinctive natural features in North Carolina. Mount Airy and the town of Pilot Mountain, North Carolina, are situated near the mountain. According to secular geologists, Big Pinnacle and Little Pinnacle and the surrounding peneplain represent an advanced stage of erosion over many millions of years. Such an explanation is based on the assumption of uniformity theory and old earth doctrine.

Creation geologists argue the peneplains were formed by sheetlike erosion as floodwaters receded from the continent and that such features are more consistent with a catastrophic flood event than with millions of years of erosion.

Another prominent example is Uluru (also known as Ayers Rock), a sandstone monolith formation in the Northern Territory of Australia about 217 miles (350 km) southwest of Alice Springs. The formation, located in Uluru-Kata Tjuta National Park, is more than 986 feet (301 m) high and 5 miles (8 km) around. It also extends 1.5 miles (2.41 km) into the ground.

Relict landforms such as sandstone monoliths remain after surrounding rock formations have disappeared. Such features have survived disintegration.

Mount Airy, NC: This is known as the fictional town of Mayberry on the TV classic, The Andy Griffith Show. Mount Airy is known for its rock quarries and being the birthplace of actor Andy Griffith.

Photo by Roger Gallop

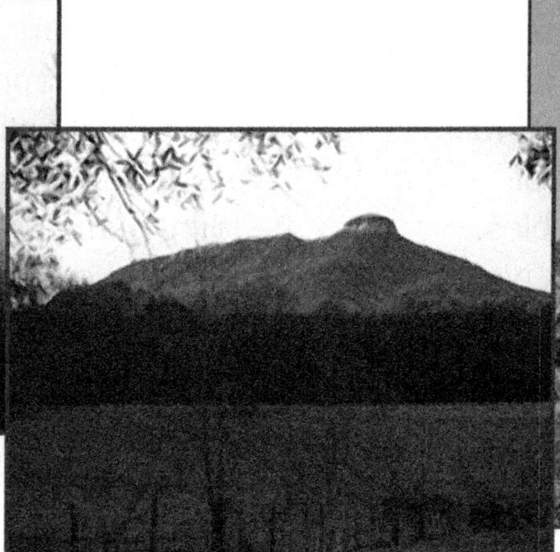

Big Pinnacle of Pilot Mountain, NC

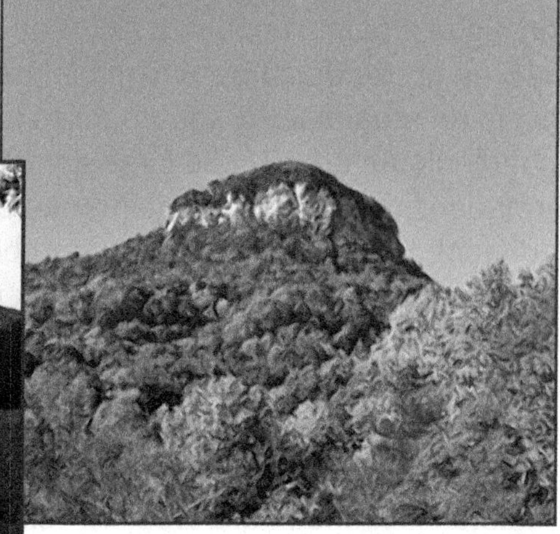

Second and Third Photos by Cindy McCrary, www.cindymccraryphotography.com. Photos used by permission.

Map of Australia

Ayers Rock, Australia, also known by its Aboriginal name, "Uluru," with surrounding peneplain

This monolith is the result of underwater deposition of horizontal sediment layers (turbidites), followed by uplift and tilting, and broad, sheetlike erosion of surrounding lands as floodwaters receded. "The feldspar-rich sand that makes up Uluru must have been deposited very quickly and recently. Long-distance transport of the sand would have caused the grains to be rounded and sorted, whereas they are jagged and unsorted. If they had sat accumulating slowly in a lake bed drying in the sun over eons of time… the feldspar would have weathered into clay."[38] Evidence indicates that sediment deposition, uplift and tilting and erosion all took place in a relatively short period of time, not millions of years as maintained by secular geologists.

Similarly, the nearby Kata Tjuta (also known as Mount Olga) comprises 36 domes. The highest point is about 1,791 feet (546 m) above the surrounding plain and covers an area of 8.33 square miles (21.57 square kilometers). These large domed rock formations comprise a conglomerate of sediment consisting of boulders, cobbles, gravel, sand, and mud of varying rock types, all cemented together. Similar to Uluru, this unsorted sediment mixture indicates rapid transport and deposition.[39]

Another monolith known as Mount Augustus is found in the Western Territory about 529 miles north of Perth. Located in Mount Augustus National Park, it is widely claimed to be the "world's largest monolith," more than 2,822 feet (860 meters) high and covers an area of 18.51 square miles (47.95 square kilometers). Similar to Kata Tjuta, this monolith is also composed of an unsorted mixture of boulders, cobbles, gravel, sand, and mud, which overlies older granite near its northern end. Again, the unsorted sediment mixture suggests the material was transported and deposited very rapidly.

Geologic monoliths and accompanying peneplains are found throughout the world—Africa, Antarctica, Asia, Europe, and North and South America.[40] These monoliths and domed rock formations are surrounded by the world's best developed peneplains that have eroded to almost perfect flatness except for isolated rocky remnants. This is further evidence that sedimentary landforms were deposited in a short period of time by a catastrophic worldwide flood.

Today's world has countless examples of horizontal deposition that has been subsequently uplifted, folded, tilted, and inverted with cross-bedding and sharp, angular bedding planes typical of a cataclysmic flood. Big and Little Pinnacles of North Carolina, Uluru-Kata Tjuta (Ayers Rock and Mount Olga) and Mount Augustus of Australia, and many other monoliths and peneplains found worldwide including buttes, mesas, spires, and pinnacles of the Colorado Plateau all stand as monuments to the cataclysmic flood and drainage described in Genesis.

Dry Lake Basins and Salt Lakes

Dry lake basins, also known as enclosed or endorheic basins, are found throughout the world in semi-arid and arid climates. They have no outflow to rivers or oceans—water enters the basin by rainfall and limited surface drainage, and exits by evaporation to the atmosphere and ground seepage. These basins often contain extensive salt pans (also called salt flats, or alkali flats). Evolutionists believe that salt pans were formed by evaporation of seawater trapped in enclosed basins by alternating periods of tranquil seas and dry conditions over a span of many millions of years. But as a general rule, these salt pans are relatively thin and contaminated with wind-blown sediment, dust, and debris[41]—typical of a one-time occurrence, not multiple cycles over a vast time period.

The Lake Bonneville Basin, for example, is a major topographic depression which encompasses more than 20,000 square miles over much of western Utah and parts of Nevada and Idaho. It is the largest endorheic watershed of North America. As discussed in a previous section, The Bonneville Flood - Snake River Plain, the basin retained a large portion of its seawater following the collapse of several hundred feet of rock barriers at Red Rock Pass, Idaho, near elevation 5,220 feet.

As climatic conditions changed, the basin began to dry up—water evaporated and became hypersaline, and salts or evaporites, primarily sodium chloride, accumulated as crystalline deposits on the basin bottom. What remains today is Great Salt Lake, the fourth largest salt lake in the world covering 1,700 square miles. The salt pan or crust in the Bonneville Basin is less than an inch thick near its fringes to 5 feet thick at Great Salt Lake, the lowest elevation at less than 4,200 feet.

The three largest salt lakes in the world are the Caspian Sea, Aral Sea, and Lake Balkhash.[42] The salt lake with the highest elevation is Namtso in Tibet (15,479 ft.), and the one with the lowest elevation is the Dead Sea in the Middle East (1,385 ft. below sea level). The salt lake with the highest salinity is the Don Juan Pond in Antarctica.

Shoreline terraces or shelves of the original Lake Bonneville can be observed within surrounding mountains (as high as 5,220 feet at the Bonneville Bench), approximately 1,000 feet above the basin floor. Dry lake basins, salt lakes, and elevated shoreline terraces are a testament to a one-time worldwide flood, not slow and gradual cycles over millions of years.

Three major shorelines or terraces in the Lake Bonneville Basin: Bonneville at elevation ~5,220 feet following the worldwide flood; Provo at elevation ~4,840 feet following the Bonneville flood; and Gilbert at elevation ~4,275 feet at Great Salt Lake. Salt Lake is presently at 4,200 feet. How did seawater find its way into an enclosed basin reaching an elevation of 5,220 feet?

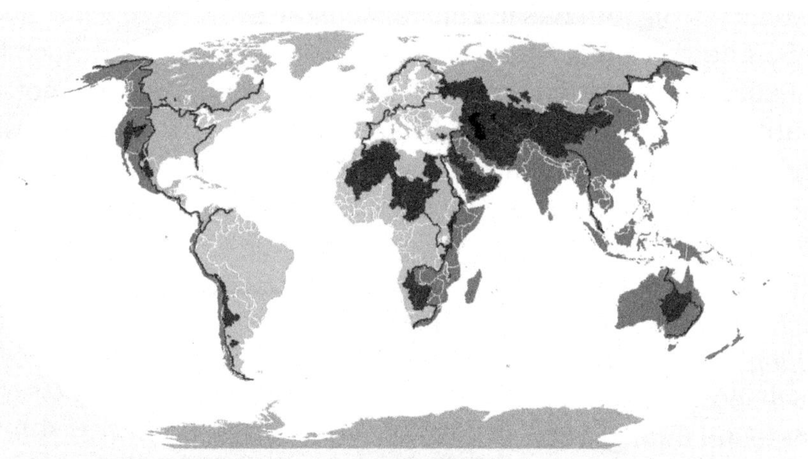

Major Endorheic Basins of the World. Endorheic basins are shown in dark-gray and enclosed lakes are in black. Black linear lines are general drainage basin boundaries.

Summary

Mountain ranges throughout the world provide powerful evidence for catastrophic mountain building resulting from runaway subduction toward the end of a worldwide flood. Uniformity theory cannot explain mountain uplift with no denudation (erosion) and thrusting, tilting, and folding of rock formations *AFTER* the rocks have cooled and become brittle over many millions of years.

It is clear to most unbiased researchers (setting aside the presumption of uniformity theory) that deformation of rock layers *occurred over a short period of time when the rock was still soft (either hot igneous rock or wet sedimentary rock) and pliable*. The presence of enormous rock formations throughout the world with no denudation; massive folding, tilting, and thrusting; and igneous intrusion into layers of water-deposited sedimentary rock remains *immensely problematic* for secular geologists.

> Uniformity theory cannot explain massive igneous structures and most igneous formations associated with mountain building.

Other examples of complete failure of uniformity theory are the Lewis Thrust Fault in northwestern Montana; Heart Mountain in Wyoming's Bighorn Basin; Empire Mountains in Southern Arizona; Franklin Mountains near El Paso, Texas; Matterhorn of the Pennine Alps; Glarus Alps near Schwanden, Switzerland; and other examples found worldwide. The thrust fault hypothesis is a futile attempt by *secular geologists trying to explain how older rock formations were deposited over much younger formations.* Such inconsistencies are found throughout the world.

Although there are two uniformity theories regarding the formation of the Grand Canyon, these have been discarded and are simply wrong. Breached Dam Theory, or the catastrophic drainage of lakes, provides the most reasonable explanation based on observational and empirical data. This theory proposes that the Grand Canyon was eroded largely by accelerated drainage of a large lake system sometime following the worldwide flood. Most secular geologists continue to reject this theory because it implies massive catastrophic processes and a young earth.

When flat-lying strata such as those of the Colorado Plateau are inundated by floodwaters and then uplifted more than 5,000 feet, sheetlike erosion will quickly sweep away the uppermost sediment layers. Once the water is gone, isolated relict landforms—buttes, mesas, pinnacles, and great monoliths—would be surrounded by peneplains that have eroded to almost perfect flatness.

Alluvial plains (and alluvial deserts) were created by the deposition of suspended sediments (as water velocity diminished at the end of Phase 2; see chapter 4) and then shaped by receding floodwaters in a relatively short period of time. Other remnants of receding floodwaters include underfit rivers and overfit valleys, raised river or shoreline terraces, dry lake basins, and salt lakes.

Such features are the result of a worldwide flood, not erosion over millions of years while the land was exposed to dry conditions. For many esteemed geologists who have researched geologic landforms and catastrophic processes, evidence of a global flood is **indisputable**.

> For more information about flood geomorphology, refer to the article, "Geomorphology Provides Multiple Evidences for the Global Flood," by Michael J. Oard (2015, Creation magazine, vol. 7, no. 1, 47-49). Oard explains the problems of 'slow and gradual' geomorphology and provides compelling evidence that landforms found worldwide confirm a worldwide flood. www.Creation.com

notes: Chapter 5

1. Baumgardner, J. (March 2005). Recent Rapid Uplift of Today's Mountains. Institute for Creation Research, ICR *Impact*, Article 381; and Whitcomb, J.C., and Morris, H.M. (1961). *The Genesis Flood*. Phillipsburg, NJ: The Presbyterian and Reformed Publishing Company, 127-128.

2. Eardley, A.J. (June 1957). The cause of mountain building – An enigma. *American Scientist*, 189; as cited in Whitcomb, J.C. (1988). *The World That Perished*. Grand Rapids, MI: Baker Book House, 84.

3. Thompson, G.R., and Turk, J. (1991). *Modern Physical Geology*. Philadelphia, PA: Saunders College Publishing, 370.

4. Whitcomb, J.C. (1988). *The World That Perished*. Grand Rapids, MI: Baker Book House, 86; and Whitcomb and Morris (1961), 139.

5. Whitcomb (1988), op. cit., 86-87.

6. Whitcomb and Morris (1961), op. cit., 191; and Whitcomb (1988), op. cit., 87.

7. Hubbert, M.K., and Rubey, W.W. (February 1959). Role of fluid pressure in mechanics of overthrust faulting. *Bulletin of Geological Society of America,* 70: 122, 127. Also cited in Whitcomb, 1988, op. cit., 87.

8. Whitcomb, J.C., and Morris, H.M. (1961), op. cit., 198-200, and Morris, J.D. (August 2013). The upside-down mountain. *Acts & Facts*, 42 (8), Dallas, TX: Institute for Creation Research, 13.

9. Slusher, H.S. (May 1966). Supposed overthrust in Franklin Mountains, El Paso, Texas. *Creation Research Society Annual*, p. 59; Burdick, C.L. (December 1975). Geological formations near Loch Assynt compared with Glarus formation. *Creation Research Society Quarterly*, 12 (3); and Burdick, C.L. and Slusher, H.S. (June 1969). The Empire Mountains – A thrust fault? *Creation Research Society Annual*, p. 40; as cited in Sharp, D. (1986). *The Revolution Against Evolution*. Douglas Bruce Sharp, 9. Also see von Fange, E. (1981). *Time Upside Down*, Chapter 4. Retrieved November 2013, from www.creationism.org/vonfange/vonFangeTimeUpDownChap03.htm.

10. Morris, J.D. (1994). *The Young Earth*. Green Forest, AR: Master Books, 107.

11. Morris (1994), op. cit., 108. Also see Austin, S., and Morris, J. (1986). Tight folds and clastic dikes as evidence for rapid deposition of two very thick stratigraphic sequences. *Proceedings of the First International Conference on Creationism*, Pittsburg, PA, 3-15; and Humphreys, R. (June 2005). Evidence for a young earth. Institute for Creation Research, ICR *Impact*, Article 384.

12. Ibid.

13. Morris, J.D. (2000). *The Geology Book*. Green Forest, AR: Master Books, 35.

14. Morris, J.D. (1994), op. cit., 99-100. Also Austin, S.A. (Ed.). (1994). *Grand Canyon: Monument to Catastrophe*. Institute for Creation Research, El Cajon, CA, 74-75.

15. Snelling, A.A., and Austin, S.A. (December 1992). Startling evidence for Noah's flood. *Creation,* 15 (1): 46-50. Retrieved December 2007, from http://creation.com/contents-all-creation-magazines. Permission received August 20 and 27, 2007.

16. Ibid. Also see Middleton, L.T., Elliott, D.K., and Morales, M. (1990). Coconino Sandstone. In: *Grand Canyon Geology*. (S.S. Beus and M. Morales, Eds.). New York: Oxford University Press and Museum of Northern Arizona Press, chapter 10, 183-202.

17. Ibid. Also see McKee, E.D. (1947). Experiments on the development of tracks in fine cross-bedded sand. *Journal of Sedimentary Petrology*, 17: 23-28.

18. Ibid. Also see Brand, L.R. (1979). Field and laboratory studies on the Coconino Sandstone (Permian) vertebrate footprints and their palaeoecological implications. *Palaeogeography, Palaeoclimatology, Palaeoecology*, 28: 25-38.

19. Ibid. Also see Brand, L.R., and Tang, T. (1991). Fossil vertebrate footprints in the Coconino Sandstone (Permian) of northern Arizona: Evidence for underwater origin. *Geology*, 19: 1201-1204.

20. Ibid. Also see Visher, G.S. (1990). *Exploration Stratigraphy.* (2nd edition). Tulsa, OK: Penn Well Publishing Co., 211-213.

21. Ibid. Also see Kuenen, P.H., and Perdok, W.G. (1962). Experimental abrasion – frosting and defrosting of quartz grains. *Journal of Geology*, 70: 648-658.

22. Ibid. Also see Amos, C.L., and King, E.L. (1984). Bedforms of the Canadian eastern seaboard: a comparison with global occurrences. *Marine Geology*, 57: 167-208.

23. Ibid. Also see Allen, J.R.L. (1970). *Physical Processes of Sedimentation*, George Allen and Unwin Ltd, London, 78.

24. Ibid. Also see Beus, S.S. (1979). Trail log third day: South Kaibab Trail, Grand Canyon, Arizona. In: *Carboniferous Stratigraphy in the Grand Canyon Country*, Northern Arizona and Southern Nevada, (S.S. Beus and R.R. Rawson, Editors), Falls Church, VA: American Geological Institute, 16.

25. Ibid.

26. Ibid. Also see Blakey, R.C. (1990). Supai Group and Hermit Formation. In: *Grand Canyon Geology* (S.S. Beus and M. Morales, Eds.). New York: Oxford University Press and Museum of Northern Arizona Press, chapter 9, 147-202.

27. Ibid. Also see Turner, C.E. (1990). Toroweap Formation. In: *Grand Canyon Geology.* (S.S. Beus and M. Morales, Editors). New York: Oxford University Press and Museum of Northern Arizona Press, chapter 11, 203-223.

28. Morris (1994), op. cit., 99.

29. Austin, S.A. (Ed.). (1994). *Grand Canyon: Monument to Catastrophe*. Institute for Creation Research, El Cajon, CA, 26. Also see Shinn, E.A., Steinen, R.P., Lidz, B.H., and Swart, P.K. (1989). Whitings, a sedimentologic dilemma. *Journal of Sedimentary Petrology,* 59: 147-161; Dill, R.F., Kendall, C.G.S., and Shinn, E.A. (1989). Giant subtidal stromatolites and related sedimentary features. *American Geophysical Union, Field Trip Guidebook* T373, 33; and Dill R.F. and Steinen, R.P. (1988). Deposition of carbonate mud beds within high-energy subtidal sand dunes, Bahamas. *American Association of Petroleum Geologists Bulletin,*72: 178-179.

30. Ibid., 26–28.

31. Lagerstätte (May 2009, last modified). In Wikipedia, the free encyclopedia. Retrieved May 2009, from http://en.wikipedia.org/wiki/Lagerst%C3%A4tten.

32. Austin (1994), op. cit., 83-107 (chapter 5).

33. Spamer, E.E. (1989). The development of geological studies in the Grand Canyon. *Tryonia,* 17: 39; as cited in Austin (1994), op. cit., 107.

34. Austin (1994), op. cit., 84; and Hoesch, W.A. (March 2008). Red Butte: Remnant of the flood. *Acts & Facts*, 37 (3), Dallas, TX: Institute for Creation Research, 14. Also see Doelling, H. et al. (2000). Geology of Grand Staircase-Escalante National Monument, Utah. In Sprinkel, D.A. et al. (Eds.). *Geology of Utah's Parks and Monuments, Utah Geological Association Publication 28,* Salt Lake City: Utah Geological Association, 189-231.

35. Sean W. Pitman, M.D. (April 2004). J. Harlen Bretz and the great scabland debate. Retrieved December 2007, from http://www.detectingdesign.com/harlenbretz.html.

36. Austin, S.A. (July 2008). Red rock pass: Spillway of the Bonneville flood. *Acts & Facts*, 37 (7), Dallas, TX: Institute for Creation Research, 10-12. Also see O'Connor, J.E. (1993). Hydrology, hydraulics, and geomorphology of the Bonneville flood. *Geological Society of America,* Special Paper 274, 83; and Lake Bonneville, Wikipedia, the free encyclopedia. Retrieved July 2010, from http://en.wikipedia.org/wiki/Lake_Bonneville.

37. Whitcomb and Morris (1961), op. cit., 318 and Whitcomb (1988), op. cit., 84. Also see Garner, H.F. (1974). *The Origin of Landscapes*. New York: Oxford, 734, as cited in Austin, S.A. (April 1983). Did landscapes evolve? Institute for Creation Research, ICR *Impact,* Article 118.

38. Catchpoole, D., Sarfati, J., and Wieland, C. (2008). *The Creation Answers Book*. (D. Batten, Ed.). Atlanta, GA: Creation Book Publishers, 180. Also see Snelling, A.A. (1998). Uluru and Kata Tjuta: testimony to the flood. *Creation*, 20 (2): 36-40.

39. Ibid.

40. Monolith (January 3, 2009). In Wikipedia, the free encyclopedia. Retrieved September 2008, from http://en.wikipedia.org/wiki/Monolith.

41. Morris, J.D. (June 2010). Evaporites and the flood. *Acts & Facts*, 39 (6), Dallas, TX: Institute for Creation Research, 17.

42. Caspian Sea lies between northern Iran, southern Russia, western Kazakhstan and Turkmenistan, and eastern Azerbaijan; Aral Sea lies between Kazakhstan in the north and Karakalpakstan, an autonomous region of Uzbekistan, in the south; and Lake Balkhash is in Central Asia.

Chapter 6

Sedimentary Features of a Worldwide Flood

"By the first day of the first month of Noah's six hundred and first year, the water had dried up from the earth. Noah then removed the covering from the ark and saw that the surface of the ground was dry. By the twenty-seventh day of the second month the earth was completely dry."—Genesis 8:13-14, NIV

Evidence of Flood Deposition

There is plenty of geologic evidence that a worldwide flood over a relatively short period of time is the best explanation for the types of rock layers and formations found on earth.

Presence of Tightly Bent Strata Without Cracking

In mountain regions throughout the world, sediment strata just a few feet to hundreds of feet thick are tightly bent and folded. According to conventional geologic time scale (that is, old earth scenario), these sediment layers are separated in time by *many millions of years*—and they were deeply buried and solidified *hundreds of millions of years before* they were subjected to deformation (bending, folding, tilting). In most cases, visual and empirical evidence indicates that folding occurred *without cracking*—which, in turn, implies that folding occurred shortly after deposition while the sediment was still wet and pliable.[1]

As explained in most geology textbooks, every rock layer has a limit beyond which it cannot deform without fracture—this point is called the "elastic limit." When a rock layer is deformed beyond this limit, it will rupture or behave in what hydraulic engineers call a hard "brittle" fashion. As described in most structural geology textbooks, "when a rock undergoes brittle fracture, it breaks sharply and the fracture becomes a permanent feature of the rock."[2] If, on the other hand, sediment layers are submerged and then subjected to uplift and deformation within a year, these sediments will remain soft and pliable and will bend without rupture.

Folded Rock Strata, Moruya, New South Wales, Australia. Note person at bottom right for scale.

The Kaibab Upwarp of the Grand Canyon is an excellent example of tightly bent strata without cracking (as described in chapter 5). Dr. John D. Morris points out that such layers "deformed as would an unconsolidated mud"—and he

Anticline near Bcharre, Lebanon

Rainbow Basin Syncline, Barstow, California
See other photos of folded strata in chapter 5, Mountain Chains.

further explains that "the situation at the Grand Canyon is far from unique. There are many, many other places where rocks have deformed while in a soft, unconsolidated condition... One such occurrence might be passed off as an anomaly, but the world is full of examples of soft sediment deformation..."[3]—such features indicate a worldwide flood.

According to the young earth scenario, great thicknesses of sediment were deposited during the year of the flood as described in Genesis. Subsidence of ocean basins, uplift of mountains and plateaus, and continental drainage would have taken place late in the flood. In most cases, evidence indicates that *uplifts and deformation occurred when sediments were still wet—soft and pliable, less than a year old—not hard and brittle as they are today*. Such geologic features are difficult for the secular geologist and evolutionist to explain away—so they have simply been ignored and rejected.

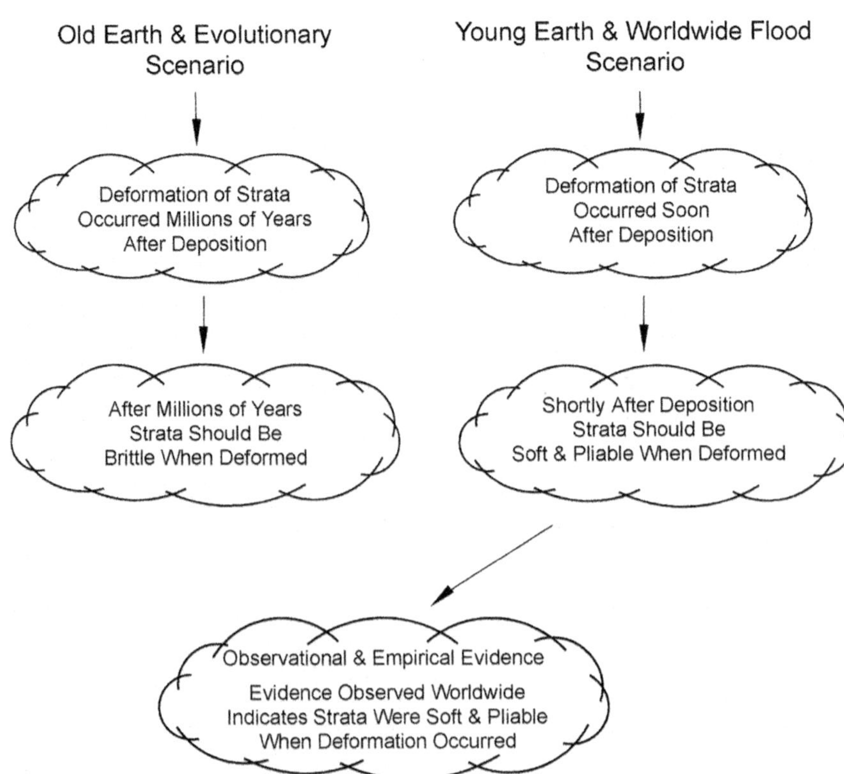
Old Earth & Evolutionary Scenario
Young Earth & Worldwide Flood Scenario
Deformation of Strata Occurred Millions of Years After Deposition
Deformation of Strata Occurred Soon After Deposition
After Millions of Years Strata Should Be Brittle When Deformed
Shortly After Deposition Strata Should Be Soft & Pliable When Deformed
Observational & Empirical Evidence
Evidence Observed Worldwide Indicates Strata Were Soft & Pliable When Deformation Occurred

Presence of Sharp Contact Bedding Planes

According to old earth scenario, most sediment layers and formations throughout the world were formed by slow and gradual deposition during alternating periods of inundation in a calm sea and dry conditions over a span of many millions of years. Most geologists agree that sediment rock layers, in almost all cases, were deposited by water. When a sediment formation is exposed as dry land, erosion takes place—not deposition. Most secular geologists maintain that erosional episodes lasting many millions of years occurred between the uplift of one sedimentary rock formation (group of layers) and the deposition of another—so, if dry conditions lasted many millions of years, significant erosional features would be expected between layers and formations.

> A **formation** is a distinct body of sedimentary, igneous, or metamorphic rock with similar characteristics, and can be visually recognized in the field. A formation usually comprises a group of similar **layers**. Examples of sedimentary rock formations are Tapeats Sandstone, Redwall Limestone, Hermit Shale, Coconino Sandstone, Toroweap Limestone, and Kaibab Limestone of the Grand Canyon.

Typically, the geologist will find two formations lying one on top of the other, with a "knife edge" bedding plane (sharp contact) between them. The existence of sharp contacts is completely at odds with the passage of long periods of time. Sharp contacts are characteristic of most geologic layers and formations throughout the world, and provide unequivocal support for a worldwide flood.

> A **bedding plane** is a surface which separates layers of sedimentary rocks. Most geologists agree that sediment layers are deposited while underwater.

As described in chapter 5, rock formations in the Grand Canyon indicate that they were rapidly deposited *without* time breaks between the deposition of each layer and formation. A good example is the abrupt, sharp contact between Coconino Sandstone and Toroweap Limestone above and the sharp contact between Coconino Sandstone and Hermit Shale below. Obviously, the sharp, undisturbed bedding planes indicate little or no time between the deposition of these rock formations.

South Rim, Grand Canyon, Arizona. The sharp contact shows no time breaks between the deposition of the sediment formations.
Photo by Roger Gallop

Coconino Sandstone and Hermit Shale,
Grand Canyon, Arizona. *Photos by Roger Gallop*

Capitol Reef National Park, Utah
Navajo Sandstone and Kayenta Formation

The lack of erosion along bedding planes of rock formations supposedly separated by many millions of years provides strong evidence of rapid deposition of sediment strata. These sharp contacts are typical of most geologic formations throughout the world.

Lack of Soil Layers and Bioturbation

The same reasoning can be applied to other land features common to upland areas found worldwide. If sediment layers were deposited with periodic exposure to dry land lasting many millions of years, as maintained by uniformity theory, sediment layers would show evidence of soil accumulation similar to today's environment. Soil is the natural occurrence of intermixed fine sand, clay, and decaying organic matter (humus) on the surface of the earth that supports plant and animal life.

Life is impossible without a soil layer and we know from the fossil record that abundant populations of plants and burrowing animals existed throughout much of earth's history. However, there is an absence of soil layers anywhere in the geologic column. So where are the soils after millions of years of exposure as dry land? Obviously, there was not enough time to produce soils. Again, one can reasonably conclude that sediment layers were not exposed to an environment of soil accumulation.

The same logic can be used for the absence of physical disturbance by living organisms such as plants and animals, commonly referred to as bioturbation. The earth's surface at any given time is abundant with life forms. Organisms such as clams, sea urchins, starfish, worms, and fish are continually disturbing the surface of aquatic sediments, and roots of grasses, shrubs, and trees, and worms and moles, and small animals grubbing for food are continually disturbing the surface of upland sediments. Animals like clams will burrow through sediment to hide from predators swimming or crawling along the ocean floor. Burrowing animals and insects, and plant roots will quickly change soil morphology.

According to secular geologists and old earth advocates, the vast majority of sediment layers represents millions of years, but such layers show *no evidence* of soil layers or bioturbation. Again, one can logically conclude that sediment layers were never exposed to an environment of plant and animal activity before they were rapidly buried.[4]

Tapeats Sandstone, Little Colorado Gorge, Arizona, is typical of many sediment layers found throughout the world—yet, there is no evidence of soil accumulation and biological disturbance. *Photos by Roger Gallop*

Presence of Cross-Bedding and Fragile Surface Features

Another common feature found worldwide is cross-bedding. For many years secular geologists have compared cross-bedding of the Coconino Sandstone and other similar formations to eolian processes—that is, formation of sand dunes in deserts.

Contrary to evolutionary, old earth doctrine, evidence indicates that *cross-bedding* typical of Coconino Sandstone is *not indicative of desert sand dunes, but rather, "underwater" sand dunes.* Sand dunes in deserts rarely, if ever, result in cross-bedding that hardens or becomes "frozen" into rock. Why?

Desert dune features are fragile and will soon erode and disappear because of wind erosion. Preservation of delicate features *requires immediate burial, moisture, and cement* (dissolved calcium carbonate, silica, and iron). Such necessities are *not available* in desert environments. Refer to chapter 5, Coconino Sandstone, for more information.

Cross-bedding is an arrangement of beds at an angle to the main sedimentary layer. See photos in this section.

Similar to Coconino Sandstone, the checkerboard Navajo Sandstone at Zion National Park is the result of water-laid deposits requiring vast ebb and flow currents—not eolian processes (migrating desert dunes) over many millions of years as maintained by secular geologists. As floodwaters receded, these water-laid sand deposits, including mudflows (which are commonly observed throughout the southwest), dried and solidified, and vertical cracks formed as a result of sediment composition and the drying process over hundreds of years. Cracks were further accentuated by contraction and expansion due to temperature changes. Any other interpretation is an attempt by secular geologists to hold onto uniformity doctrine.

Most geologists agree that almost every sediment layer we observe today was deposited by water and gravity.

Lithification is the process in which sediments gradually become solid rock.

Cross-bedding in sandstone along South Kaibab Trail, Grand Canyon, Arizona. Cross-bedding is the result of water-laid deposits requiring vast currents—not migrating desert dunes over many millions of years.

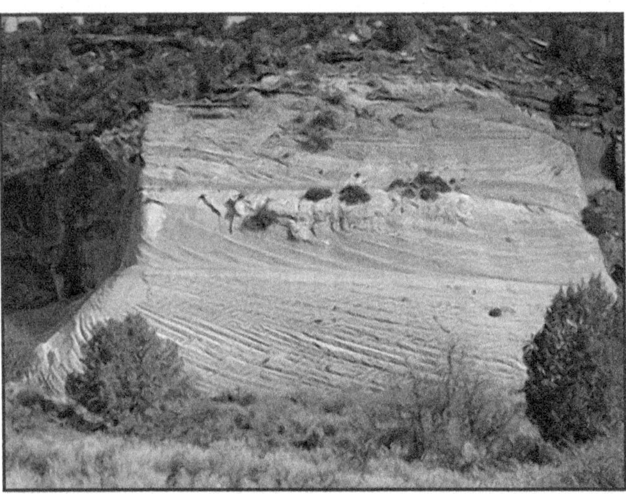

Cross-Bedding in sandstone, Highway 89, Utah
Photos by Roger Gallop

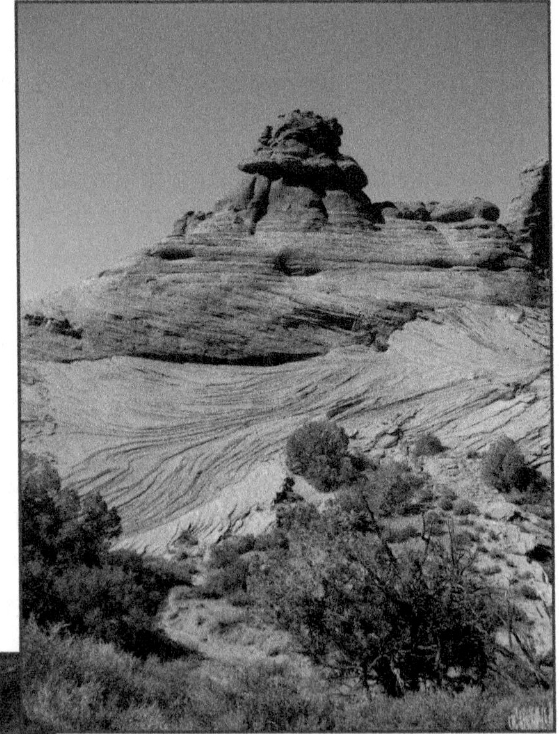

Cross-Bedding
Arches National Park, Utah

Cross-Bedding in sandstone, Sedona, Arizona

Cross-Bedding, Aztec Sandstone, Zion National Park, Utah. The checkerboard Navajo Sandstone at Zion National Park was the result of water-laid deposits—not migrating desert dunes. *Photos by Roger Gallop*

Zion National Park, Utah. Mudflows found throughout the plateau are another example of recent water-laid deposits.
Photos by Roger Gallop

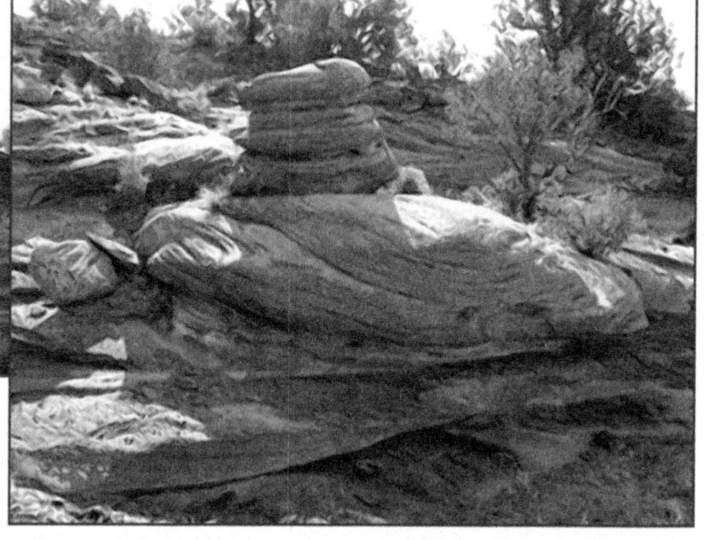

Mudflows near Zion National Park, Utah
Photos by Roger Gallop

Cross-Bedding with mudflow, Zion National Park, Utah

Cross-Bedding, Mudflow (ebb and flow tidal currents) SR12 east of Escalante, Utah

Highway 89, Arizona. Note the formation of the checkerboard texture in the mudflow.
Photos by Roger Gallop

Another common feature found in many sedimentary rock strata is the presence of surface features (sometimes referred to as "sedimentary structure") such as ripple marks, animal tracks, and fossil impressions which have been "frozen" into solid rock. Again the same question is asked, how can such surface features be preserved? As stated by Dr. J. D. Morris, "Sedimentary structure is fragile and short-lived, yet such features abound in nearly every sedimentary rock layer."[5] If such features are exposed to dry conditions on any surface (i.e., dry, windy desert conditions), they will quickly erode and disappear.[6]

Even on a hard rock surface exposed to wind and rain, surface features will quickly erode in just a few decades. There is no possibility that delicate surface features will last for days, not to mention millions of years, without immediate preservation. Such surface features can be preserved only if they are quickly buried in the presence of moisture and cement (dissolved calcium carbonate, silica, and iron). If fragile surface features are found between two layers, one can reasonably conclude that little time, if any, has passed between the two layers or strata.

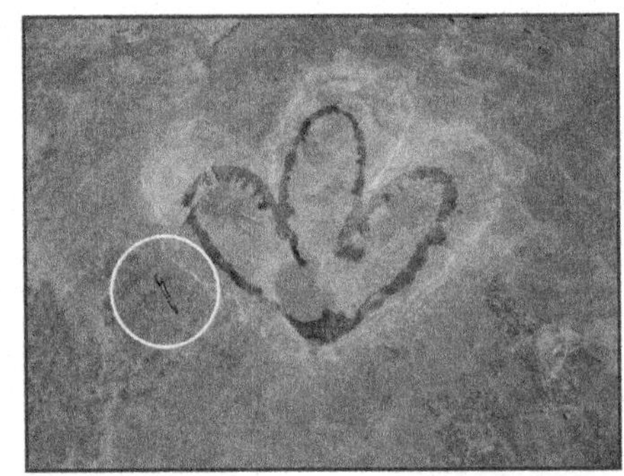

Fossil Footprint in an Ancient Shallow Stream—Dilophosaurus tracks (dinosaur), Kayenta Mudstone, Tuba City, Arizona. Note 5 inch green pen for scale.
Photos by Roger Gallop

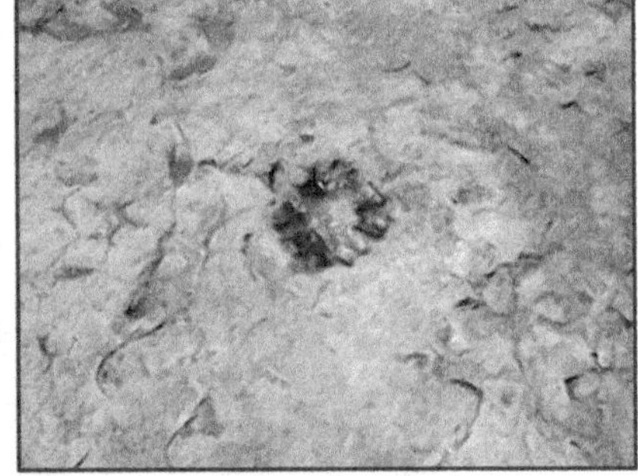

Ammonite Sea Fossil, Kayenta Mudstone near Tuba City, Arizona, Hwy 89 off US 160

Ancient Fossil Plant, Museum of Natural History, San Diego, California

Presence of Polystrate Fossils

Polystrate fossil trees are trees that extend through multiple sediment layers where supposedly, according to standard geologic time scale, each layer is separated in time by many millions of years. Entire "fossil forests" have been found throughout the world—in the eastern United States, eastern Canada, England, France, Germany, and Australia. The world contains many thousands of examples of polystrate trees and fossils which tie sediment rock layers together into a relatively short time frame.

If a tree grew in place and eventually died, how long would it take the trunk to decay—and could such a tree remain standing for millions of years as sediment slowly accumulated? Most foresters agree that dead trees (snags) will not remain upright for more than a few decades, but secular geologists who maintain uniformity doctrine would have you believe otherwise. Wood decays in just a few years, whether in an aquatic environment or exposed to open air.

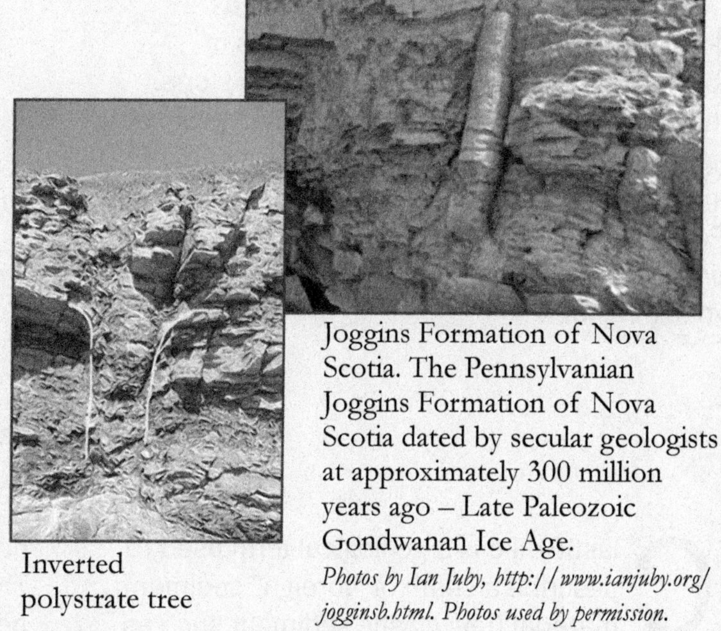

Inverted polystrate tree

Joggins Formation of Nova Scotia. The Pennsylvanian Joggins Formation of Nova Scotia dated by secular geologists at approximately 300 million years ago – Late Paleozoic Gondwanan Ice Age.

Photos by Ian Juby, http://www.ianjuby.org/jogginsb.html. Photos used by permission.

According to author, Ian Juby, "… the entire fossil story embedded in these cliffs is clearly one of chaos: The plants were broken and crushed. Many were strewn all over the place and lapped all over each other when found horizontal."

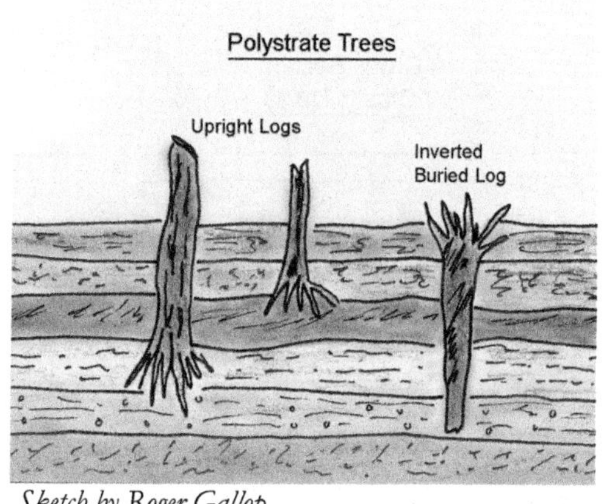

Sketch by Roger Gallop

Other types of fossils give evidence to the same conclusion. Obviously, sediment could not have slowly accumulated around a dead animal that would have decayed in just a few weeks. In nature, land animals have a "low-fossilization potential"—that is, when animals die, they are quickly devoured by other animals or quickly decomposed by bacteria in the presence of oxygen. And when land animals die in water, they will either float to the surface or sometimes sink to the bottom, and disintegrate in a matter of days or weeks.

Yet, what is found today are billions of well-preserved fish with scales, fins, and other soft parts, including shellfish, plant fossils, and insects, in *massive sediment deposits* throughout the world. See Appendix E, Lagerstätten. Occasionally, fossilized creatures have been found actually giving birth or eating.[7] The Green River Formation in Wyoming, for example, comprises an abundance of fossilized catfish with soft tissue, reptiles, birds, and other fossils. These fossils are found in different orientations, overlapping thin layers of sediment and "obviously buried quickly."[8]

Fossilization (petrification) is the process by which organic material is converted into stone. It is essentially the replacement of organic material with silica (silicon dioxide, SiO_2) and other minerals such as calcite and pyrite.

Once the animal was buried, and as floodwaters drained away, the process of fossilization converted soft tissue into stone. Instead of uniform, gradual deposition over vast ages, fossils throughout the geologic column are evidence of rapid burial while underwater. *Polystrate plant and animal fossils are classic proof of a rapid worldwide flood event and young earth.*

> Plants and animals can be preserved only if they are quickly buried in the presence of moisture and cement (dissolved calcium carbonate, silica, and iron), and without the presence of oxygen; otherwise, decay will occur.

Presence of Clastic Dikes

Clastic dike is a geological term used to describe a seam of "foreign" sediment material (i.e., clastic sediment such as sandstone) that fills vertical cracks in horizontal sediment strata. These dikes are "stone walls" that can be seen at the surface (as shown in the photos) or sometimes hidden just below the ground surface. Clastic dikes may vary in thickness from a few inches to several feet or more; their height may vary from 10 to 100 feet; and their length may be a few miles. Smaller dikes have been found to branch and rejoin a larger one. Clastic dikes are found worldwide and, according to standard geologic time scale, are dated between 50 and 250 million years old. So what is the significance of dikes?

> There are many igneous dikes (solidified form of magma) but our focus is with "clastic dikes."

Visual and microscopic examination of sand grains of these clastic dikes suggests the dike sand is homologous with the source sandstone bed buried beneath the dikes—that is, there is little or no change in grain size and chemistry.[9] In his book, *The Young Earth,* Morris explains, "the only difference between the dikes and the mother sandstone bed is that the sand grains in the dike appear to be similarly oriented, with their long axes tending to point in the same direction. This would result if the material were squeezed upward from below... No deformed sand grains are seen, and there is no hint of broken and recrystallized cement."[10] This hydraulic mechanism is similar to wet sand being squeezed through the fingers of the hand.

When squeezing occurred, according to old earth advocates and standard geologic time scale, the source sandstone was already many millions of years old. In some cases, the source bed was deposited 150 to 250 million years *before* injection of clastic sediment into the younger overlying sediment. But evidence indicates that clastic sediment in the dikes was *saturated mud* when it was squeezed up through cracks in the overlying strata.[11]

Cross Section
Clastic Dikes

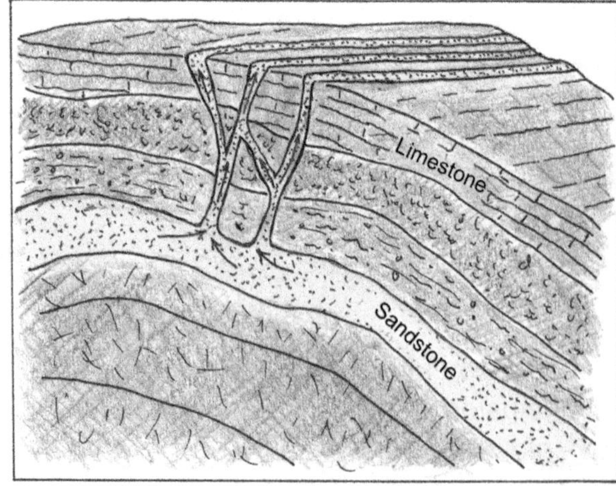

Sketch by Roger Gallop

Sandstone dikes in the Moreno Formation in the Panoche Hills, looking north across Moreno Gulch, Fresno County, California

Obviously, there is a 150 to 250 million year time discrepancy between the source sandstone and overlying sediments—that is, *the source sandstone could not have remained uncemented due to the immense pressure of overlying rocks accumulating over this vast span of time* as suggested by old-earth advocates.

A fascinating example of classic dikes (i.e., vertical pillars or "pipes") is located at Kodachrome State Park Basin in Utah. This amphitheater-like basin is surrounded on three sides by 600- to 700-foot steep elevations (escarpments) with pancake-like strata. Within the basin are numerous relict sand and cobble pillars or pipes about 50 feet in diameter and 100 to 170 feet high. These relict landforms were once fluid unconsolidated mud that was squeezed into the overlying sandstone strata from a depth of several hundred feet. But according to secular geologists and standard geologic time scale, the source bed was more than 150 million years old at the time of injection.[12]

So how long does it take sediments (mineralized fragments such as quartz sand or silt) to turn into solid rock? It may take just a few years but no longer than a few hundred years depending on pressure, temperature, and presence of dissolved minerals.

It is extremely unlikely the source sand would remain uncemented for many millions of years, particularly with the pressure of overburden hundreds of feet thick. One can reasonably conclude that *the intruded, overlying sediments and source sandstone (from below) formed at approximately the same time*—that is, the source bed did not have time to harden prior to squeezing—thus, such an event likely occurred during tectonic activity (such as continental uplift) in the latter phase of the worldwide flood. Surrounding sediment was subsequently eroded away by rapid, sheetlike erosion consistent with drainage of floodwaters.

Clastic Dikes, Kodachrome State Park, Utah. Vertical dikes remain while surrounding sediments have washed away.
Photos by Roger Gallop

Origin of Immense Limestone Formations and Chalk Beds

Limestone Formations

Limestone is a sedimentary rock composed largely of the mineral calcite (calcium carbonate, $CaCO_3$). Sometimes limestone is almost pure calcite, but most limestone formations are intermixed with sand, silt, and shell debris. In today's world, limestone deposits are derived from the remains of marine organisms such as corals, clams, mussels, and snails. Such organisms are able to extract calcium from seawater to build their skeletal structure. When they die, their shells disintegrate and intermingle with bottom sediment and collect as calcareous ooze (lime muds).

According to most geology textbooks, accumulation rates of lime mud in shallow seas are about one foot per 1,000 years. Because evolutionists believe that "the present is the key to the past"—that is, physical processes have remained constant throughout earth's history, they believe it is impossible to explain the formation of massive limestone by a worldwide flood. However, there is substantial visual and empirical evidence for rapid and massive accumulation of limestone—evidence that has been ignored by secular scientists because it implies a young earth and Divine Creation.

Massive limestone formations are found in all continents (except Antarctica) within low elevations and high mountain ranges. Mount Everest (Himalayas) and other mountain ranges throughout the world are *composed of fossil-bearing, marine limestone* which had its origin beneath water. Marine fossils such as shells and coral are common on earth's high mountains. (See chapter 4, Phase Three.)

Some of the most compelling evidence for rapid formation of lime muds is the extraordinary fossilization of animals that were rapidly covered with lime mud of fine texture. For example, some of "the 'world's most perfect fossils'... come from the fine-grained limestones of the Santana Formation of northeast Brazil."[13] These formations are found worldwide and earn the label of *Lagerstätten* because fossil assemblies are diverse and extremely well-preserved. While describing fossil fish of the Santana Formation, it was concluded that except "for a few notable exceptions, lithification was instantaneous and fossilization may even have been the cause of death."[14]

Lagerstätten (German; literally *places of storage*) are sedimentary deposits that exhibit extraordinary fossil richness. See Appendix E.

According to Dr. Steven Austin, "These limestones appeared to have formed as animals were smothered in lime mud."[15]

Some of the world's major **Lagerstätten** are found in limestone (carbonate muds), volcanic ash, shale, slate, chert, and chalk, and a mixture of limestone, sandstone, and shale.

Limestone Formations. First Photo: Kaibab Limestone. Note bulging, pancake layers. Oak Creek Canyon between Sedona and Flagstaff, Arizona. Second Photo: Kaibab Limestone. Yavapai Point, South Rim, Grand Canyon, Arizona. *Photos by Roger Gallop.* Third Photo: Pancake Limestone Layers, Northwest Coastline of New Zealand. *Photo by David A. Howard, December 21, 2009. Photo used by permission.*

Primary mechanisms for rapid limestone formation are: 1) aggregation or flocculation of fine-grain crystals of aragonite or calcite in high energy environments (turbulent seawater), 2) warming of seawater and direct precipitation of dissolved calcium carbonate, and 3) precipitation of dissolved calcium carbonate due to the depletion of carbon dioxide in the atmosphere, all as a result of the worldwide flood.

- Microscopic examination of modern lime muds shows that fine grain crystal particles aggregate into "pelletoids [which]...exhibit hydraulic characteristics of sand" in high energy environments and quickly settle out.[16] In tidal channels or during storms, fine grain aragonite particulates clump together as a "creamy white lime mud" and quickly settle to the bottom.

  Massive ancient limestone formations found throughout the world range from fine-textured limestone to pancake-like formations with mixtures of sand and shell fragments. The latter was formed by underwater flow of lime mud, sand, and skeletal shell fragments forming turbidites. Bulging lime mud strata (see photos) indicate soft mudflow and turbidite formation.

- During the flood, great tectonic events would have included extensive volcanic activity, earthquakes, hydrothermal vents, hot springs, and superheated geysers on a massive scale—and consequently, floodwaters would have become very warm and extremely hot in some locations. See phase diagram in chapter 4.

  Gases like CO_2 and O_2 are less soluble in warm water and more soluble in cold water.

  Carbon dioxide is less soluble in warm water and more soluble in cold water. As seawater is heated, carbonate (CO_3^{-2}) comes out of solution, combines with calcium (Ca^{+2}), flocculates or clumps together, and quickly settles to the ocean bottom. Today, $CaCO_3$ is often found precipitating near hot springs. In contrast, $CaCO_3$ is not found in sediments in the colder, deep marine environment. Warming of seawater would have precipitated massive quantities of lime mud deposits resulting in enormous fine-textured limestone formations we observe today (e.g., Redwall Limestone, Grand Canyon).

- The flood buried massive amounts of carbon in the form of plants that became coal, oil, and gas, thus depleting and "locking away" the total carbon available in the biosphere (earth's plant and animal communities and atmosphere). The normal production of carbon dioxide by decaying plants would have been greatly curtailed. This would have caused a *major shift in carbon dioxide equilibrium* (between the oceans and the atmosphere) resulting in massive precipitation of calcium carbonate in sea water—and ultimately, the immense limestone formations.

  Plants absorb carbon dioxide and use it to grow, but when plants decay or burn, carbon dioxide is released back into the atmosphere.

These fossil evidences and mechanisms are discarded by secular geologists because they have *bound themselves to one interpretation*—uniformity doctrine and evolution. A detailed discussion of the effects of a worldwide flood and mechanisms for rapid accumulation of lime muds is presented in Appendices E and F; also, see chapter 4, Limestone - Rapid Formation of Mud Layers, and chapter 5, Limestone Formations.

Chalk Beds

Chalk is a form of limestone composed of the mineral calcite, a fine grain calcium carbonate. It is a soft, compact white rock that is a very pure type of limestone. Chalk is found worldwide—the most notable site is the White Cliffs of Dover in southern England. The cliffs are 350 feet high (105 m) at the Strait of Dover, English Channel, and the beds are 1,300 feet thick (400 m) across much of England. These beds can be traced to Northern Ireland, and across France and other parts of Europe, and to Turkey and the Middle East. These same beds are also found throughout North America (Nebraska, Kansas, Texas, Tennessee, Mississippi, and Alabama) and Perth basin of Western Australia. Approximately 25 percent of the earth's surface is covered with deposits of pure chalk.

> **Foraminifera** are single celled protists with shells. There are about 275,000 species, both living and fossil. They are usually less than 1 mm in size, but some are much larger, and the largest recorded specimen reached 19 cm.
>
> **Algae** are simple plants lacking roots, leaves, and a vascular system. The term calcareous algae is used for algae capable of forming minerals of calcium carbonate ($CaCO_3$).

Unlike other forms of limestone, chalk consists of tiny shells called tests, which comprise billions of microorganisms called foraminifera and calcareous algae known as cocoliths and rhabdoliths. These planktonic (floating) microorganisms live in the upper 100 to 200 meters of the open seas. Similar to limestone formations, evolutionists believe that chalk beds accumulated at a very slow rate (less than 1 inch per 1,000 years) over a span of 30 to 35 million years during the Cretaceous Period about 100 million years ago.

According to Dr. Andrew A. Snelling, in his book *Earth's Catastrophic Past*, the coccolith content on the floor of the Atlantic and Indian Oceans is "regarded as the modern version of the chalk beds found in the geologic record..."[17] These modern ocean sediments have a coccolith content (by weight) of 5 to 33 percent and 4 to 71 percent, respectively[18] whereas the ancient French and Kansas chalk beds have a coccolith content of 90 to 98 percent and 88 to 98 percent, respectively.[19]

The extreme purity of ancient chalk beds found in England, France, and throughout Europe and North America is very problematic for evolutionists. How can pristine chalk beds be deposited over a span of tens of millions of years without contamination by sediment and debris from other events? Common sense would dictate that this is not possible.

White Cliffs of Dover, England
Photo by Roger Gallop

For chalk beds to have reached a thickness of 1,300 feet such as the White Cliffs of Dover, there would have been a continuous cycle of massive "white water" blooms and rapid deposition of these planktonic organisms. During the flood, warming seawater combined with massive nutrient loading as the continents were reshaped (see chapter 4) would have caused a continual explosive proliferation of planktonic algae and foraminifera. As the population of these microorganisms increased to exceedingly high concentrations,[20] warm, turbulent seawaters would have caused these fine-grained calcite particles to flocculate or aggregate and rapidly settle to the bottom as pure chalk—a continuous cycle lasting until floodwaters were drained from the continents. See Appendix F.

Bloom is a rapid increase in population density of planktonic organisms because of the input of nutrients, increase in temperature, and other unknown factors.

Chalk beds of the White Cliffs give evidence for three explosive blooms during the flood. Although there are variations in consistency, there is no variation in purity. Dr. Snelling concludes that "it is the extreme purity of the chalk beds that argues for their rapid deposition and formation" during the worldwide flood as described in the Book of Genesis.[21]

Primary reasons for blooms were availability of nutrients and carbon dioxide, warming temperatures, and turbulence. An explosive bloom would not be dependent on sunlight, but rather a sufficient amount of nutrients, especially nitrogen.[22] Nutrients (nitrogen, phosphorus, iron, and sulfur, and other trace elements) and carbon dioxide were continually available in the water column from dead fish, upwelling of nutrient-laden waters, and volcanic eruptions.

Ammonite Fossils, Museum of Natural History, San Diego, California. *Photo by Roger Gallop*

Kingdom: Animalia
Phylum: Mollusca
Class: Cephalopoda
Subclass: Ammonoidea (extinct)

Animal fossils are commonly found within chalk beds. For example, ammonites, some as large as 3 feet across, were free-swimming, deep sea animals and could not have been buried slowly—fossilization requires rapid burial.

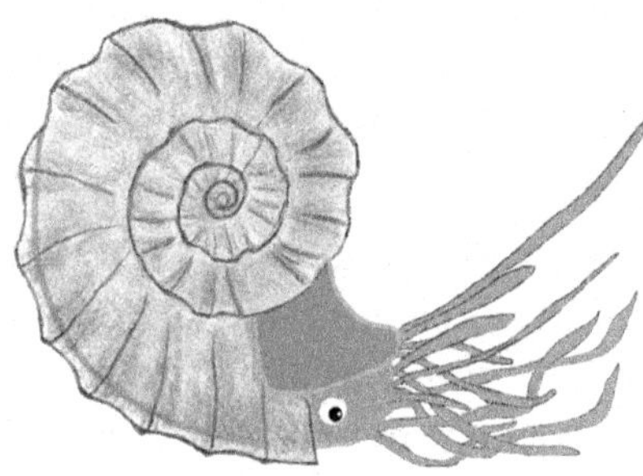

Extinct Ammonite
Sketch by Roger Gallop

The ammonites are extinct cousins of the nautiloids—free-swimming, deep-sea mollusks. Similar to its cousin, it used jet propulsion as it drew water into and out of the chamber through a hyponome—a small, flexible propulsion tube. See Appendix E.

Origin of Immense Underground Salt Beds

Immense underground salt beds, hundreds of feet thick (in some cases, 1,000 feet thick; 300 meters) are found worldwide. For example, an immense salt bed extends from the Appalachian basin of western New York through parts of Ontario and under much of the Michigan basin, an area of more than 600,000 square miles (155,400 km^2). Other deposits are located in Texas, Ohio, Kansas, New Mexico, Nova Scotia, and throughout much of the world. Similar to chalk deposits, these immense underground salt beds are relatively pristine—that is, they are "free of contaminants...[and] easily cleanable and appropriate for consumption."[23]

Rock salt or halite (NaCl), gypsum ($CaSO_4 - 2H_2O$), and anhydrite ($CaSO_4$) are water soluble mineral sediments that result from 1) the evaporation of seawater; that is, as water evaporates to the atmosphere, the concentration of salts becomes high enough so that they fall out of solution, and 2) rapid temperature (and pressure) changes. Warmer water will hold more salts in solution than cooler waters.

Can such immense salt beds be reasonably explained within the constraints of an evaporation lake model proposed by evolutionists—that is, evaporation of seawater trapped in enclosed basins by alternating periods of tranquil seas and dry conditions over a span of many millions of years? How many times must an enclosed basin be inundated by seawater and then exposed to dry land in an arid climate to produce such huge volumes of pristine salt over thousands of square miles—and *without contamination* by sediment and debris? As with chalk beds, these underground salt beds hundreds of feet thick are problematic for evolutionists. A much more plausible explanation is rapid deposition of salts during the worldwide flood.

Events associated with the flood included catastrophic rifting and subduction, uplift of oceans basins, flooding of the low-lying continents, sedimentary deposition, and mountain and continental uplift. See chapter 4. Waters spewed by hydrothermal vents would have been superhot.

During the year-long flood, hydrothermal vents, hot springs, superheated geysers, and super volcanoes would have spewed magma and enormous volumes of volcanically heated seawater. Hot temperatures enable seawater to become supersaturated—hypersaline. As saturated hot waters encountered cooler thermal gradients, salts quickly fell out of solution and deposited as massive, pristine salt beds.

Another explanation offered by Dr. M. Hovland is the Hydrothermal Salt Theory[24]—rapid salt production under high heat and pressure in deep oceanic hydrothermal systems. Examples of continuous salt production today are the Red Sea, Danakil depression in Ethiopia, and mid-oceanic ridge or plate boundary in the Arctic Ocean. Whenever seawater becomes super-critical, "temperatures above 380°C, combined with pressures above 230 bars, [water] turns into a non-polar fluid"—consequently, it will rapidly deposit its salt load with little chance of contamination. At the time of the flood, salt deposits were rapidly covered by clastic sediment preventing salts from redissolving as temperature and pressure changed.

Water is a polar molecule—that is, it has electrical poles, more negative charge on the oxygen and more positive charge on the hydrogen. Polar molecules can only dissolve with polar molecules (such as NaCl), and non-polar molecules can only dissolve with non-polar molecules, but polar NaCl will not dissolve in non-polar water to form a solution.

Origin of Coal and Oil

As routinely taught in classrooms and presented in evolutionary-based science textbooks (from elementary school through college), coal and oil formed about 300 million years ago during the Pennsylvanian Period (see chapter 7, Standard Geologic Column and Time Scale) in a swamp ecosystem where plant remains were trapped and preserved from biodegradation by water and mud. Assuming uniformity theory, it is explained that uncompacted plant matter known as peat collected in enclosed inland bogs and swamps—and, as wetland trees and shrubs lived and died, decaying peat was eventually buried by lime muds and silty sand over vast periods of time. As the land was slowly uplifted (or the seas receded), the overlying mud turned into rock (that is, limestone or shale) while the peat gradually changed into coal (bituminous and anthracite), and oil and natural gas by heat and pressure.

> Keep in mind that no one has ever viewed the supposed gradual change of peat into coal and oil under these conditions—theorists have merely *assumed* the process based on the old earth model.

> What is the basic difference between coal and oil formation? We know that coal and oil are derived primarily from decaying vegetation (terrestrial plants and planktonic sea organisms). The process of organic decomposition produces heat that begins to dehydrate the organic matter. Dehydration results in more heating and water loss—ultimately, the cycle produces a dried, charred material known as coal. If, on the other hand, water is confined within geological rock strata and cannot escape, the end product is oil.

> The softer bituminous coal is regarded as a sedimentary rock and the harder forms, such as anthracite coal, are regarded as metamorphic rock because of exposure to increased temperature and pressure.

Research has shown the creation of coal and oil (and natural gas) does not take millions of years of heat and pressure but, as demonstrated in the laboratory, it can develop in a matter of days with just a few ingredients.[25] As stated by Dr. Carl Wieland in his book, *Stones and Bones*, researchers at Argonne National Laboratory have produced high-grade black coal from the following laboratory procedure:[26]

> Take lignin (the main component of wood), mix it with some acid-activated clay [catalyst] and water, and heat all this at only 150°C with no added pressure in an air-free sealed quartz tube. Geologically, this is not very hot at all; in fact, there is nothing exceptional or 'unnatural' about any ingredient. The process does not need millions of years, either, just 4–36 weeks!

The process doesn't require pressure but rather, hot water and a catalyst—a clay-like material (kaolin clays) that causes the reaction to occur rapidly. This catalytic material is a derivative of volcanic ash, commonly known as "partings"—and such material is found in coal beds throughout the world.[27]

> Oil and gas are rapidly produced after thermal maturation of organic matter by hot hydrothermal fluids (generated by volcanic heat below the ocean basin).

Research has also shown that rapid production of oil and gas is not only feasible under controlled laboratory conditions, but rapid formation has been shown to occur naturally.[28] An example is the discovery of accelerated production of petroleum in the Guaymas Basin of the Gulf of California—that is, the hydrothermally produced, naturally occurring "very young" oil and gas (less than a few thousand years old).

Such an on-going hydrothermal phenomenon has provided geologists with the opportunity to actually observe petroleum forming under natural conditions. In his article, "How Fast Can Oil Form?", Dr. Andrew Snelling concluded that, "...hydrothermal generation of petroleum is more than a feasible process for the generation of today's oil and gas deposits in the time-scale subsequent to Noah's Flood as suggested by creation scientists."[29]

> Additionally, a worldwide flood would have altered the carbon balance of the world resulting in the rapid accumulation of vast deposits of limestone. See section, Limestone Formations, and Appendix F.

The conclusion by many is vast deposits of coal and oil found throughout the world, including widespread discoveries of surface coal in today's polar regions,[30] are the result of a worldwide flood—a flood which would have buried and sealed huge amounts of carbon in the form of plants that became coal and oil. Tectonic events produced superheated water (similar to the hydrothermal fluids found in the Guaymas Basin) and enormous quantities of heated sediment. Volcanic ash fell on billions of tons of tree logs and vegetative debris that were buried between heated sediment layers, forming coal and oil in a short period of time.

> See chapter 8, Carbon-14 in Ancient Fossils and Rocks, for strong evidence of young coal formations and a young earth. Also see effects of a global flood in chapters 4, 6, and Appendices F and H.

A recent example of this phenomenon is found with the 1980 eruption of Mount St. Helens, which destroyed (crushed) more than 160 square miles of forest. Within a matter of minutes massive quantities of tree logs and enormous volumes of volcanic ash were floating on Spirit Lake. Many of the floating trees became waterlogged and sank to the bottom—and they were often buried upright.

According to Dr. John Morris in *The Young Earth*, "within just a few years" an organic deposit of tree bark, together with a mixture of vegetative debris and volcanic ash, had accumulated in the lake.[31] This black shiny peaty material has the same characteristics as ancient coal beds—including partings (volcanic ash) found throughout the deposit. Also, this organic material has collected around many of the upright trees in a "polystrate" position.[32]

It is interesting to note that coal regions of the Eastern U. S. (northeastern Pennsylvania in the central Appalachian Mountains, West Virginia and Kentucky) are commonly characterized by 50 to 60 coal seams separated by layers of limestone and shale rock. The coal seams range in thickness from a foot to hundreds of feet—and according to uniformity theory, this represents millions of years of swamp environment.

One of the most common and dangerous facets of coal mining is the presence of polystrate trees or kettles—commonly seen as circular shapes in the mine's roof that occasionally detach and fall on the rail cars or the miners. Polystrate trees, similar to those found at Mount St. Helens, tie two or more coal seams and intervening shale or limestone layers into a very short period of time. This, of course, is incompatible with old earth doctrine.

An upright tree preserved in the cliffs at Joggins, Nova Scotia, Canada (printed 1868)

Before Photo: A pre-eruption view of Mount St. Helens from Spirit Lake shows the smooth, conical slopes of a very young and potentially explosive volcano.

Before Photo: Prior to the eruption, high mountain lakes like Obscurity Lake, 10 miles north of Mount St. Helens, were characteristically clear due to extremely low levels of dissolved nutrients.

May 18, 1980, eruption of Mount St. Helens

1992, view of the volcano and Spirit Lake from the Boundary Trail at Norway Pass

1980, the same view (see page 125) after the eruption shows the extent of damage to the forest surrounding Obscurity Lake. Note the large quantity of volcanic ash that was eroded from adjacent hillsides and deposited on large deltas at the mouth of inlet streams.

1980, the same view (see page 125) shows the profound change in the volcano and Spirit Lake. The formerly clear mountain lake was completely displaced by the massive landslide and choked with ash and organic debris.

Summary

There is abundant evidence that most rock layers and formations found throughout the world were deposited quickly, one after the other without significant time breaks. The folding or bending of layers without cracking is convincing evidence that layers were wet (soft and pliable) when they were folded or bent. Sharp bedding planes separating layers and formations that are supposedly many millions of years old with no evidence of erosion, soil layers, or animal activity (bioturbation), all testify to rapid continual deposition. Other evidence of rapid deposition is the preservation of delicate features, polystrate fossils, and clastic dikes.

Uniformity theory assumes that limestone, chalk, and salt formations were formed by slow deposition over tens of millions of years. Visual and empirical evidence indicates otherwise. The most persuasive evidence for rapid formation is the "extreme purity" of salt and chalk beds—and such purity is not possible with slow deposition over vast time periods. Animal fossils found within limestone and chalk formations require rapid burial. Also, there are compelling geochemical mechanisms for rapid deposition of limestone, chalk, and salt associated with a worldwide flood.

Secular scientists assume that coal and oil were formed by heat and pressure of organic matter in inland swamp environments with periodic burial by water deposited sediment over many millions of years. No one has ever observed the supposed gradual change of peat into coal and oil under such conditions—theorists merely assume the process. Observational and empirical evidence, however, points to rapid formation of coal and oil as vast forests were uprooted and buried beneath massive amounts of heated sediment during a catastrophic worldwide flood.

Evolutionary doctrine does not take into account a worldwide flood because it implies Divine creation and a young earth. Secular geologists believe that existing physical processes are sufficient to account for all past changes and have tightly bound themselves to this single system of interpretation, despite compelling evidence to the contrary.

notes: Chapter 6

1. Morris, J.D. (1994). *The Young Earth*. Green Forest, AR: Master Books, 106-109; Humphreys, D.R. (June 2005). Evidence for a young world. Institute for Creation Research, ICR *Impact,* Article 384 (Item #7); and Austin, S.A. and J.D. Morris (1986). Tight folds and clastic dikes as evidence for rapid deposition and deformation of two very thick stratigraphic sequences. *Proceedings of the First International Conference on Creationism*, vol. II, Creation Science Fellowship, Pittsburg, PA, 113-126.

2. Thompson, G.R., and Turk, J. (1991). *Modern Physical Geology*. Philadelphia, PA: Saunders College Publishing, 229.

3. Morris (1994), op. cit., 107, 109.

4. Ibid., 95-97.

5. Morris, J.D. (July 2009). Sedimentary structure shows a young earth. *Acts & Facts*, 38 (7), Dallas, TX; Institute for Creation Research, 15.

6. Ibid. Also see Hayes, M.O. (1961). Hurricanes as geological agents: Case studies of hurricanes Carla, 1961, and Cindy, 1963. *University of Texas, Bureau of Economic Geology, Report of Investigation,* No. 61, 56.

7. Morris, J.D. (2000). *The Geology Book*. Green Forest, AR: Master Books, 38.

8. Morris (1994), op. cit., 102; Buchheim, H.P., and Surdam, R.C. (April 1979). Fossil catfish and the depositional environment of the Green River formation, Wyoming. *Geology*, 5: 196; and Green River Formation. Retrieved 2009, from http://en.wikipedia.org/wiki/Green_River_Formation.

9. Ibid., 111. Also see Kelsey, M., and Denton, H. (1932). Sandstone dikes near Rockwall, Texas. *University of Texas Bulletin*, No. 3201, 138-148.

10. Ibid.

11. Roth, Ariel A. (1977). Clastic dikes. *Origins*, 4 (1), Geoscience Research Institute, 53-55. Retrieved April 2008, from http://www.grisda.org/orgins/Q4053.htm.

12. Roth, Ariel A. (1992). Clastic pipes and dikes in Kodachrome Basin. *Origins,* 19 (1): 44-48; and Hoesch (May 2008). Sand injectites. *Acts & Facts*, 37 (5), Dallas, TX: Institute for Creation Research, 14.

13. Austin, S.A. (Ed.). (1994). *Grand Canyon: Monument to Catastrophe.* Institute for Creation Research, El Cajon, CA, 26.

14. Martill, D.M. (1989). The medusa effect: Instantaneous fossilization. *Geology Today*, 5: 201. Also cited in Austin (1994), 26. Also, see Martill, D.M. (1989). A new 'solenhofen' in Mexico. *Geology Today*, 5: 25-28.

15. Austin (1994), op. cit., 26. Also see Brett, C.E., and Seilacher, A. (1991). *Fossil Lagerstatten: A Taphonomic Consequence of Event Sedimentation*. (G. Einsele, W. Ricken, and S. Seilacher, Eds.). *Cycles and Events in Stratigraphy.* New York, NY: Springer-Verlag, 296.

16. Ibid., 26.

17. Snelling, A.A. (2009). *Earth's Catastrophic Past* (Geology, Creation & The Flood, Volume 2). Institute for Creation Research, Dallas, TX, 925.

18. Jeans, C.V. and P.F. Rawson (Ed.). (1980). *Andros Island, Chalk and Oceanic Oozes*. UK: Yorkshire Geological Society, Leeds, Occasional Publication No. 5; as cited in Snelling (2009), 925.

19. Pettijohn, F.J. (1957). *Sedimentary Rocks*. New York: Harper and Rowe, 400-401; as cited in Snelling (2009), 925.

20. Roth, A.A. (1985). Are millions of years required to produce biogenic sediments in the deep oceans? *Origins*, 12 (1): 48-56, and Woodmorappee, J. (1986). The antediluvian biosphere and its capability of supplying the entire fossil record. *Proceedings of the First International Conference on Creationism.* (R.E. Walsh, C.L. Brooks, and R.S. Crowell, Eds.). Creation Science Fellowship, Pittsburg, PA, 2: 205-218; as cited in Snelling, A.A. (1994). Can flood geology explain thick chalk beds? *Journal of Creation*, 8 (1): 11-15.

21. Snelling (2009), op. cit., 928.

22. Tappan, H. (1982). Extinction or survival: Selectivity and causes of Phanerozoic crises. *Geological Society of America,* Special Paper 190, 270; as cited in Snelling (1994) and Snelling (2009), 928. Also see Worrest, R.C. (1983). Impact of solar ultraviolet-B radiation (290-320nm) upon marine microalgae. *Physiologica Plantarum*, 58 (3): 432; as cited in Snelling (1994).

23. Morris, J.D. (June 2010). Evaporites and the flood. *Acts & Facts*, 38 (7), Dallas, TX: Institute for Creation Research, 17.

24. Hovland, M. The hydrothermal salt theory. Marine Geological Discoveries. Retrieved July 2010, from http://www.martinhovland.com/new_salt_theory.htm.

25. Major, T. (1990). *Genesis and The Origin of Coal and Oil*. Montgomery, AL: Apologetics Press, 8, 4-8; Snelling, A.A. (March 1990). How fast can oil form? *Creation,* 12 (2): 30-34. Retrieved October 2009, from http://creation.com/contents-all-creation-magazines; and Major, T. (1990), 10-14; Hayatsu, R. et al. (1984). Artificial coalification study: Preparation and characterization of synthetic macerals. *Organic Geochemistry*, 6: 463-471; and Morris (1994), 102. Also see Davis, A. and Spackman, W. (1964). The role of cellulosic and lignitic components in articulate coalification. *Fuel*, 43: 215-224; and Larson, John (March 1985). From lignin to coal in a year. *Nature,* 31: 16.

26. Wieland, C. (2001). *Stones and Bones*. Green Forest, AR: Master Books, 12-13. Also see Major, T. (1990); and Hayatsu, R. et al. (1984).

27. Morris (1994), op. cit., 102. Also see Pennisi, E. (February 1993). Water, water everywhere: Surreptitiously converting dead matter into oil and coal. *Science News*, 121-125.

28. Snelling, A.A. (March 1990), op. cit.; and Major, T. (1990), 10–14. Also see Didyk, B.M. and Simoneit, B.R.T. (1989). Hydrothermal oil of Guaymas Basin and implications for petroleum formation. *Nature*, 342, 65-69; and Clarey, T. (March 2014). Rapidly forming oil supports flood timeframe. *Acts & Facts*, 43 (3), Dallas, TX; Institute for Creation Research, 14-15.

29. Snelling (1990), op. cit., 34.

30. *National Geographic* magazine (February 1963). Antarctica: New look at a continent (Researched by George W. Beatty), 123 (2): 288, 296; cf. Nov. 1971, 653. Also cited in Whitcomb, J.C. (1988). *The World That Perished*. Grand Rapids, MI: Baker Book House, 80, 83; and Dillow, Joseph C. (1982). *The Waters Above*. Chicago, IL: Moody Press, 141, 335-353.

31. Morris (1994), op. cit., 103.

32. Ibid. Also see Austin, Steven A. (1986). Mount St. Helens and catastrophism. Institute for Creation Research, ICR *Impact,* Article 175, 103.

Chapter 7

Fossil Dating - Evidence for a Young Earth

"The Lord saw how great man's wickedness on the earth had become, and that every inclination of the thoughts of his heart was only evil all the time. The Lord was grieved that he had made man on the earth, and his heart was filled with pain. So the Lord said, 'I will wipe [blot out] *mankind, whom I have created, from the face of the earth—men and animals, and creatures that move along the ground, and birds of the air—for I am grieved that I have made them.'"*—Genesis 6:5–7, NIV

Dating sedimentary rock layers is limited to one relative dating technique: index fossils with visual stratigraphic review. '*Relative*' dating of sediment layers '*estimates*' the order of deposition by observing fossils assigned to the standard geologic column and by using a few basic stratigraphic rules. The word *relative* is used because there is *no credible technique* for establishing the absolute age of any sedimentary rock formation—the *baseline assumption of this dating technique is evolution and uniformitarianism*.

Uniformitarian doctrine assumes that physical and meteorological processes have remained constant–that is, fossils were buried by uniform rates of erosion and deposition consistent with today's rates. In such a 'slow and gradual' environment, land animals have a low fossilization potential–when animals die they are quickly devoured by other animals or rapidly decompose by bacteria in the presence of oxygen. See fossilization in Appendix A, Glossary.

Yet what is found today are billions of well-preserved fish and land fossils. Plants and animals can only be preserved if they are quickly buried in the presence of moisture and cement (calcium carbonate, silica, and iron) without the presence of oxygen. Fossils found throughout the geologic column are *evidence for rapid burial while underwater.* As floodwaters eventually drained away, the process of fossilization converted soft tissue into "mineralized fossils"—or **stone.**

The word '*relative*' is used because there is no credible technique for establishing the absolute age of sedimentary rock and embedded mineralized fossils. The word '*estimates*' is used because there is a significant amount of uncertainty in this dating technique.

Stratigraphic rules include: superposition, original horizontality, lateral continuity, inclusions, cross-cutting, and biostatic fossil correlations. Superposition, for example, assumes the youngest rock is on the top and oldest rock is on the bottom. See glossary.

Sedimentary rocks cannot be dated using radiometric techniques because these rocks comprise clastic fragments and minerals from preexisting rocks that were eroded and redeposited. In "mineralized fossils" organic materials have been replaced with silica, calcite, pyrite, and other minerals from preexisting rocks and cannot be dated for the same reason.

Although there are a few "proposed schemes" to potentially determine the age of certain minerals in sedimentary rocks, such minerals cannot be dated for reasons cited above–also, dating methods are unreliable and "are seldom considered helpful."[1]

Dating organic material, igneous rock, and metamorphic rock is limited to radiometric dating–and such techniques are based on unprovable assumptions that are fatally flawed. (See chapter 8.)

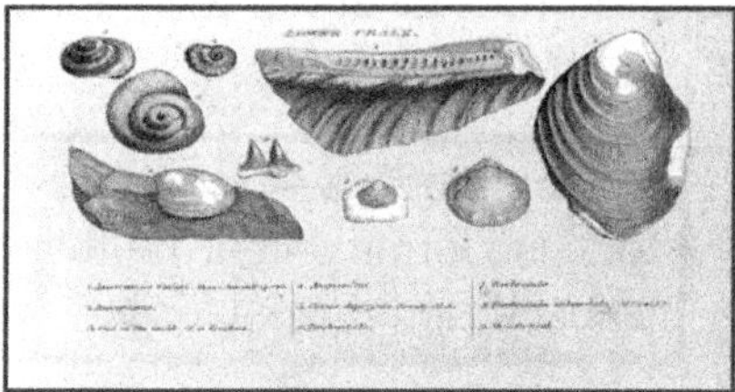

Monograph identifying geological strata using fossils by William Smith (1769 – 1839). He created the first geologic map of England using fossil markers to correlate rock strata.

The Fossil Record

Uniformitarian Approach and Evolution

Paleontologists assume uniformity theory, evolution and an old earth as their initial premise and use animal or plant fossils to establish relative ages of specific rock layers and formations. Sedimentary rocks contain billions of fossils and some of these remains are unique to certain layers or formations. Fossils are recorded and arranged in a specific order that reflects macroevolution; that is, the assumption that all life descended from a common ancestor. Based on the rule of superposition—that is, the lower the strata, the older the fossil—creatures thought to have evolved first are considered to be the oldest and are placed at the bottom of the geologic column and creatures thought to have evolved later are placed higher up on the geologic column.

Uniformity theory is the belief that natural physical processes have remained constant ('slow and gradual') during earth's history; there have been no worldwide cataclysmic geologic events; and sediment strata throughout the world exhibit the same order. See beginning of chapter 4.

Evolution has been declared a fact in many science textbooks and evolutionary biologists and school teachers have publicly stated that evolution has been "tested and proved by scientists."[2] *No one bothers to mention the fact that evolution is an assumption that has never been proven by empirical science nor observed within living populations—it is merely assumed to be true.*

Thus, various fossil assemblages known as "index fossils" have been specifically chosen by paleontologists to represent a particular sedimentary rock layer and a place in the standard geologic time scale. Millions or billions of years are hypothetically assigned to fossils based on the assumption of evolution. Paleontologists simply ***guess*** how long it would take for one fossil kind to evolve into another and then date the fossils and rock layers accordingly. As long as the layers remain within the framework of various stratigraphic rules, then index fossils are believed by evolutionists to document the age of sediment layers, and sediment layers in turn document the age of index fossils.

Paleontologists study prehistoric life forms on earth—that is, plant and animal fossils.

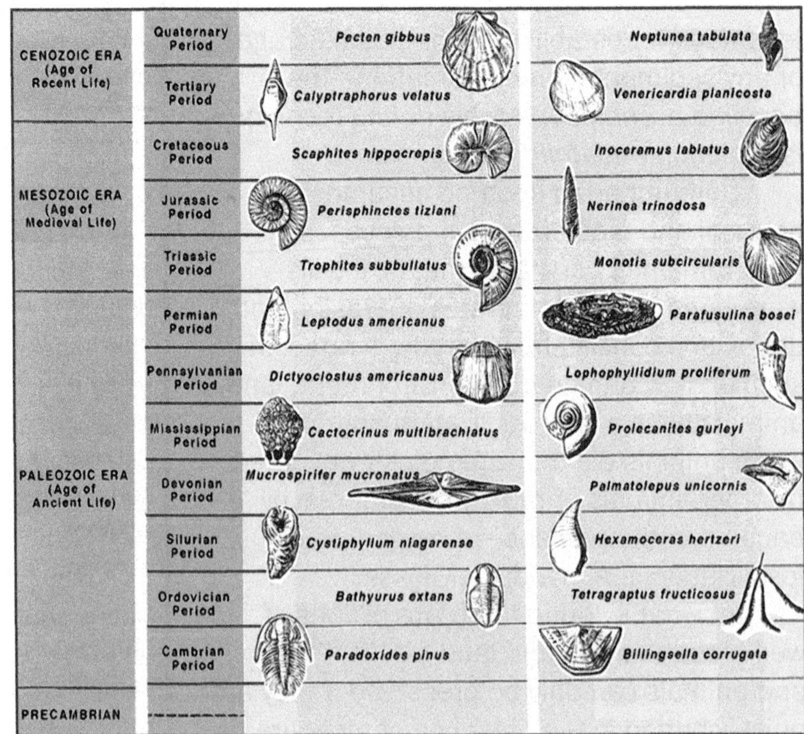

Examples of Index Fossils

Brachiopods (marine invertebrates; lamp shells) appeared in the Cambrian period; ammonites (another marine invertebrate; ribbed-spiral form shell) appeared in the Triassic and Jurassic periods; and nanofossils (microscopic fossils) are widely distributed and are considered time-specific.

In other words, paleontologists assume vast ages for index fossils based on 'slow and gradual' natural physical processes, evolution and the superposition of geologic strata—an older age is assigned to fossils in the lower strata. In turn, the geologist assumes the age of rock strata based on index fossils. Isn't this circular reasoning based on the presumption of evolution and uniformity theory?[3] J. E. O'Rourke comments, "The intelligent layman has long suspected circular reasoning in the use of rocks to date fossils and fossils to date rocks."[4]

So, does the assumption of evolution and uniformity theory, and the geologic column and fossil record as depicted in most textbooks today, actually represent the real world?

> A circular argument assumes as fact what one is trying to prove...*x* is used to prove *y* and *y* is used to prove *x*. For example, secular scientists may assert "evolution is true ***because*** living organisms created themselves, organisms tend to move from simple to more complex forms, and natural processes have remained constant during the world's history." The reasons offered after the word ***because*** provide ***no evidence***—these reasons are assumptions and offer no proof of the claim. This is called blind trust, or circular reasoning.

> Superposition assumes the youngest rock layer is on the top and oldest rock layer is on the bottom—but in reality, this is **not the case throughout much of the world**. *Most often, rock formations are inverted, folded, inserted, repeated, or missing, and fossils are in reverse order contrary to uniformity theory.* Such formations are consistent with catastrophic processes occurring within a short period of time.

Standard Geologic Column and Time Scale

Probably every school textbook in use in America dealing with evolutionary subjects (that is, biology, geology, paleontology, physical science, and anthropology) includes some type of presentation of the geologic column and fossil record. This display of the various geologic eras, periods, and epochs supposedly illustrates the natural history of the geologic strata and fossil record, with recent fossils on the top and older fossils on the bottom.

As indicated in the Standard Geologic Column, Time Scale, and Fossil Record (see table on the next page), one can "visualize" evolution: how single-celled organisms changed into marine invertebrates which, in turn, changed into fish, and fish eventually changed into amphibians and then reptiles, which finally gave way to birds and mammals, and, eventually, man. As shown on the column, modern geologists and geophysicists consider the age of the earth to be 3 to 5 billion years.

As stated by Drs. J. C. Whitcomb and H. M. Morris, "Such a presentation obviously indicates a gradual progression of life from the simple to the complex, from lower to higher, and therefore implies organic evolution. This is considered by geologists [and evolutionists] to be a tremendously important key to the interpretation of geologic history. Modern biologists in turn regard the geologic record as the *cornerstone* of their hypothesis of organic evolution."[5] [Italics added]

In reviewing diagrams and pictures as depicted in most science textbooks, with transitional creatures evolving toward more complex forms, one would certainly "see and believe" that evolution is true but, in fact, such depictions are *misleading and fictitious* and do not represent the real world.

Eon	Era	Period		Epoch		Ma*	Events
Phanerozoic	Cenozoic	Quaternary		Holocene		0.01	evolution of humans
				Pleistocene	Late		
					Early	1.8	
		Tertiary	Neogene	Pliocene	Late		
					Early	5.3	
				Miocene	Late		mammals diversify
					Middle		
					Early	23.7	
			Paleogene	Oligocene	Late		
					Early	33.7	
				Eocene	Late		
					Middle		
					Early	54.8	
				Paleocene	Late		
					Early	65.0	
	Mesozoic	Cretaceous		Late			extinction of dinosours first primates
				Early		144	
		Jurassic		Late			first birds
				Early		206	
		Triassic		Late			
				Early		248	dinosaurs diversify
	Paleozoic	Permian		Late			first reptiles
				Early		290	
		Pennsylvanian					
		Mississippian				354	first trees
		Devonian		Late			
				Early		417	first amphibians
		Silurian		Late			first vascular land plants
				Early		443	
		Ordovician		Late			sudden diversification of metazoan families
				Early		490	
		Cambrian				543	first fishes first chordates
Precambrian	Proterozoic	Late				900	first soft-bodied metazoans
		Middle				1600	
		Early				2500	first animal traces
	Archean	Late				3000	
		Middle				3400	
		Early				3800?	

Standard Geologic Column, Time Scale, and Fossil Record

*Million Years Old

Closer Look at the Fossil Record

Let's take a closer look at the fossil record. The following are interesting statistics provided by Dr. John D. Morris in his books, *Is the Big Bang Biblical?* and *The Young Earth.*[6]

- Although the standard geologic column and fossil record found in textbooks suggests that vertebrate fossils are most abundant, in reality, "95 percent of all fossils are shallow marine invertebrates, mostly shellfish."
- Of the remaining 5 percent, about 95 percent are algae and plant fossils (0.05 x 0.95, or 4.75 percent of the total) and 5 percent (0.05 x 0.05, or 0.25 percent of the total) are insects and other non-marine invertebrates and vertebrates.
- Of the remaining 0.25 percent of the total, 95 percent are insects and other non-marine invertebrates and only 5 percent (0.05 x 0.0025, or 0.0125 percent of the total) are vertebrate fossils (mostly fish; and finally, amphibians, reptiles, birds, and mammals).
- And of the vertebrates, only one percent consist of a single bone (0.01 x 0.000125, or 0.000125 percent of the total).

> The standard geologic column, time scale, and fossil record as depicted in most textbooks **does not represent the real world.**

As Dr. Morris explains, "As it turns out, in the real world of fossils, 95 percent of all fossils are shallow marine invertebrates, mostly shellfish. For instance, clams are found in the bottom layer, the top layer, and every layer in between. There are many different varieties of clams, but clams are in every layer and are still alive today. There is no evolution, just clams! The same could be said for corals, jellyfish, and many others. The fossil record documents primarily marine organisms buried in marine sediments, which were catastrophically deposited." Further, Morris notes that "each basic body style (phylum) has been present from the start. In the lowest level of abundant multi-celled organisms, the Cambrian Period, fossils of each phylum have been found, including vertebrates!"

In Cambrian rocks, supposedly 500 million years old, invertebrates such as sponges, clams, trilobites, sea urchins, and starfish are found with *no evolutionary ancestors*. Also, marine invertebrates and vertebrate fish that have been placed in the upper Cambrian by evolutionists, also *just suddenly appeared*. If evolution were true, then the ancestors of trilobites, fish, and other marine life would have been found by now. This is immensely problematic for evolutionists and secular geologists.

As explained by Dr. Steven Austin, "All the animal phyla, including chordate fish, are known as fossils in the Cambrian System. No ancestral forms can be found for the protozoans, arthropods, brachiopods, mollusks, bryozoans, coelenterates, sponges, annelids, echinoderms or chordates. These phyla appear in the fossil record fully formed and distinct, in better agreement with the concept of 'multiple, abrupt beginnings' (creation) than with the notion of 'descent from a common ancestor' (evolution)."[7]

> Noah's Flood occurred over a span of just one year, and the geologic column and fossil record reflect the order of burial (see section, Rapid Depositional Fossil Pattern).
>
> Creation and pre-flood geology are associated with the Precambrian Eon; and early-flood deposition is associated with the Paleozoic Era (including the Cambrian period with trillions of immobile and mobile marine life); mid- and late-flood with the Mesozoic and Cenozoic Era.
>
> Cambrian Explosion: Marine life that existed in the oceans of Noah's day was carried by great floodwaters (tsunamis) over the splitting super-continent (see chapter 4), and then deposited and buried in a short period of time.

Lack of In-Between Types

If evolution were true, there should be billions of transitional intermediate forms. Or to put it another way, if there were a slow evolutionary transition from amoeba to higher animal to man over billions of years, one would expect to find billions of examples of transitional fossils and living in-between forms. For instance, if the limb of an amphibian had slowly transitioned into the wing of a bird, one would expect to find a long succession of fossil forms with multiple stages of transition over these billions of years—one stage gradually transitioning to another. So where are all the in-between forms?

> *Not a single unequivocal transitional form with transitional structures in the process of evolving has ever been observed within the billions of known fossils.*

Even Darwin admitted the absence of in-between stages was "the most obvious and gravest objection which can be urged against my theory."[8] Evolutionist Dr. Colin Patterson of the British Museum of Natural History responded to a written question asking why he failed to include illustrations of transitional forms in a book he wrote on evolution. "You say that I should at least show a photo of the fossil from which each type of organism was derived. I will lay it on the line—there is not one such fossil for which one could make a watertight argument."[9]

Where is the chain of evidence? Why isn't every geologic stratum full of intermediate forms? In fact, *there are none*. Some evolutionists claim there may be a few transitional types, but other leading experts have rejected all of these! "It is easy enough to *make up stories* of how one form gave rise to another... But such stories are not part of science, for there is no way of putting them to the test."[10] [Italics added]

Some evolutionists claim that evolution proceeds "too slowly" and the real evidence of evolution is found in the fossil record. If so, there should be millions, if not billions, of true transitional structures—after all, there are billions of non-transitional fossil structures! Some evolutionists insist that evolution proceeds "too fast" and transitional structures would not be observed, but most evolutionists agree that many generations would be required for one distinct "kind" of animal to supposedly evolve into a different, more complex kind. If evolution were true, there ought to be a considerable number of transitional structures preserved in the fossil record but *none have been found*.

> Each basic kind of animal is distinct in our modern world and in the fossil record, although there is much variation within these basic groups or kinds. Variation (speciation and natural selection) within animal kinds is not evolution. See chapter 3.

To overcome this problem, some evolutionists insist that evolution occurred in small groups of genetically isolated individuals while other groups remained the same, but such claims are fictitious story-telling. Fossil evidence ***exactly fits*** the creation model, for creation insists that each animal "kind" was created fully formed, with no evolutionary transition.

> As asserted by evolutionists, if a bird evolved from a reptile, in-between forms would have neither functional wings nor functional legs–just protrusions or stubs. Certainly, such halfway limbs would be an impediment to survival. So where is the impulse to "naturally select" such appendages or protrusions 'best suited for a specific environment'?
>
> It is also impossible for vital features of one kind (e.g., reptile's two-way lung) to transform into another kind (e.g., bird's one-way lung) without ending the so-called evolutionary process.[11]
>
> And if whales evolved from a four legged cougar-like animal, where is the impulse of the animal to leave the land to which it was "highly adaptive" and where there was abundant food and water, to venture into the ocean and evolve into a whale?[12]
>
> A few very doubtful examples of transitional forms provided by evolutionists have been thoroughly refuted by Drs. Jonathan Sarfati and Duane T. Gish.[13]

Consider the following quotes concerning fossil evidence:[14]

> "The absence of fossil evidence for intermediary stages between major transitions in organic design, indeed our ability, even in our imagination, to construct functional intermediates in many cases, has been a persistent and nagging problem for gradualistic accounts of evolution." —Stephen Jay Gould (deceased professor of geology and paleontology, Harvard University), "Is a new and general theory of evolution emerging?" *Paleobiology*, vol. 6 (1), January 1980, p. 127.
>
> "The more one studies paleontology, the more certain one becomes that evolution is based on faith alone; exactly the same sort of faith which it is necessary to have when one encounters the great mysteries of religion." —Louis T. More, *The Dogma of Evolution*, (Princeton, NJ: Princeton University Press, 1925, Second Printing), p. 160.
>
> "In fact, the fossil record does not convincingly document a single transition from one species to another." —S. M. Stanley, *The New Evolutionary Timetable: Fossils, Genes, and the Origin of Species*, 1981, p. 95.
>
> "The fossil record had caused Darwin more grief than joy. Nothing distressed him more than the Cambrian explosion, the coincident appearance of almost all complex organic designs ..." —Stephen J. Gould, *The Panda's Thumb*, 1980, pp. 238-239.
>
> "The majority of major groups appear suddenly in the rocks, with virtually no evidence of transition from their ancestors." —D. Futuyma, *Science on Trial: The Case for Evolution*, 1983, p. 82.
>
> "Instead of finding the gradual unfolding of life, what geologists of Darwin's time, and geologists of the present day actually find is a highly uneven or jerky record; that is, species appear in the sequence very suddenly, show little or no change during their existence in the record, then abruptly go out of the record and it is not always clear, in fact it's rarely clear, that the descendants were actually better adapted than their predecessors. In other words, biological improvement is hard to find." —David M. Raup, "Conflicts Between Darwin and Paleontology," Bulletin, Field Museum of Natural History, vol. 50, 1979, p. 23.
>
> "But fossil species remain unchanged throughout most of their history and the record fails to contain a single example of a significant transition." —D. S. Woodroff, Science, vol. 208, 1980, p. 716.
>
> "Stasis, or nonchange, of most fossil species during their lengthy geological lifespans was tacitly acknowledged by all paleontologists, but almost never studied explicitly because prevailing theory treated stasis as uninteresting nonevidence for nonevolution. ... The overwhelming prevalence of stasis became an embarrassing feature of the fossil record, best left ignored as a manifestation of nothing (that is, nonevolution)." —Stephen J. Gould, "Cordelia's Dilemma," *Natural History*, 1993, p. 15.
>
> "Fossil lizards perfectly preserved in amber from the Caribbean are said to be 20 million years old but show absolutely no difference from their modern-day counterparts, even down to the finest details such as their tiny scales." — *PNAS Journal* (2015). 112 (32):9961-9966; as cited in *Creation* magazine (2016). Lizards identical after '20 million years', 38 (1):9.
>
> "The geological record has provided no evidence as to the origin of the fishes." —J. Norman, *A History of Fishes*, 1963, p. 298.

Contradictions in Uniformity Theory

If the uniformitarian process over many millions of years were true, rock layers and formations would be uniformly deposited regionally and globally. Uniformity theory is based on the "supposed fact that the strata everywhere exhibit the same order, thus permitting the development of a worldwide system of identification and correlation."[15]

In reality, *this is not the case*. The typical situation in geology is that *"only a few rock formations are ever superposed* [that is, younger rock layers placed over older rock layers] in any one locality and that it is very difficult or impossible to correlate strata in different localities by this principle of superposition."[16] *Most often, rock formations are inverted, folded, inserted, repeated, or missing, and fossils are in reverse order from that required by evolution.*

> Geologists rely on fossils to date sedimentary rock layers, and the fossil sequence is assumed implicitly (without question) to agree with uniformitarian processes and evolutionary doctrine.
>
> So when sedimentary rock layers or fossils are found out of order or missing, standard explanations are maintained by secular geologists by assuming reverse thrusting or stagnation with no erosion or deposition over millions of years. But such occurrences are nonexistent in the real world.
>
> See chapter 5, Mountain Overthrust, and Appendix G for more information.

Here are standard explanations used by uniformitarians to resolve discrepancies regardless of the physical evidence.[17]

- When fossils are found in reverse order, this difficulty is resolved by simply assuming the strata were inverted by reverse thrust faulting or folding.
- When rock strata are found with intervening layers missing, this is explained by assuming a period of erosion or dormancy rather than deposition.
- When the fossil is older than the rock layer, it is assumed the fossil was redeposited from a previous deposit or the fossil animal survived longer than assumed.
- When the fossil is younger than the rock layer (for example, human artifacts found in Cambrian strata), it is assumed the fossil was reworked and mixed, or the fossil animal lived earlier than assumed.

With these tidy explanations for reconciling inconsistencies, obviously all but the most difficult cases can be easily explained away. In cases that cannot be explained, then it is assumed "that there must have been some mistake in the field evidence or its description."[18] The fact is, these explanations are *FALSE.*

Such inconsistencies are found throughout the world and are *completely consistent* with catastrophic geologic processes (that is, earthquakes, volcanism, flood, sedimentation, orogeny, and erosion) occurring within a short period of time. One prominent geologist states:

> "Current methods of delimiting intervals of time, which are the fundamental units of historical geology, and of establishing chronology are of dubious validity....The findings of historical geology are suspect because the principles upon which they are based are either inadequate, in which case they should be reformulated, or false, in which case they should be discarded. *Most of us refuse to discard or to reformulate*, and the result is the present deplorable state of our discipline."[19] [Italics added]

Rapid Burial versus Slow Deposition

The entire earth is covered by hundreds of feet of sediment layers containing fossil remains of drowned animals and plants (except for mountain ranges and massive igneous intrusion). Many sites throughout the world have billions of well-preserved mineralized creatures—such as shellfish, coral, sea fans, algae and plant fossils, insects, and fish with scales, fins, and other soft parts, and polystrate trees—violently mixed together.

"Formations have been discovered containing hundreds of billions of fossils and our museums now are filled with over 100 million fossils of 250,000 different species."[20] See Appendix E, Lagerstätten. Obviously these fossil remains, many of them almost perfectly preserved in huge deposits, could not have been buried by gradual depositional processes described by secular geologists!

> Plants and animals can only be preserved if they are quickly buried in the presence of moisture and cement (calcium carbonate, silica, and iron) without the presence of oxygen. Fossils found throughout the geologic column are *evidence for rapid burial while underwater.*

As observed today, dead fish and land animals are quickly devoured by scavengers. For example, buffalo carcasses were once scattered over America's Western plains by the tens of thousands just a few generations ago, but now there is scarcely a trace. Also, the Middle East was overrun with lions (described in 1 Kings 13:24, 20:36; 2 Kings 17:25; Job 38:39; and Proverbs 22:13), but fossils of lions have never been found in that region.

In nature, flesh is devoured by wolves, vultures, or other predators within hours or days after death, and within just a few years, bones decay and quickly vanish due to weather and bacterial decomposition. When animals die in water, they will either float or sometimes sink to the bottom and disintegrate in a matter of a few days or weeks.

Unless fish and land animals, including insects and other fossil remains, are quickly buried in the presence of moisture and cement (i.e., dissolved calcium carbonate, silica, and iron) and without the presence of oxygen, soft tissues will promptly decay. *Preservation of fossils suggests rapid burial,* not slow and gradual processes as proposed by secular geologists and paleontologists.

Another interesting example is the fossil remains of woolly mammoths (genera: *Mammuthus*) and other large animals (such as the saber-toothed tiger, woolly rhinocerous, bear, reindeer, musk ox, antelope, bison, and horse) that were found buried along the coastline of northern Siberia and Alaska and islands of the Arctic Ocean. An estimated 5 million mammoths and other animals lived perhaps no longer than 4,000 years ago near the end of the Ice Age. The instant death of these animals is indicated by their remarkable state of preservation. Many mammoths were found in a standing position with the presence of "unchewed bean pods still containing the beans that were found between its teeth" and the "well-preserved state of their stomach contents."[21]

Saber-Toothed Tiger and Woolly Mammoth
Drawing by Patrick Gerrity

The mammoths were possibly overcome by a regional flood similar to the Missoula Flood (see chapter 5)—thawing of a great ice sheet and collapse of a massive ice dam holding back floodwaters. Because many are buried in wind-blown silt (loess), some scientists postulate the mammoths may have suffocated during "*gigantic* dust storms."[22] However these animals met their demise, they were *buried rapidly* and uniformitarian processes did not apply.

And what about those 80 baleen whales mysteriously buried in a Chilean desert?[23] Some 20 whale specimens were found completely intact. Paleontologists used 'index fossils' (see section, Uniformitarian Approach and Evolution) in nearby sediment strata to date these bones. They assigned ages between 2 and 7 million years based on the assumption of evolution and then assumed the whales were stranded in a lagoon and poisoned by algal toxins. Preservation of complete fossil skeletons indicates rapid burial, not slow and gradual processes. Such findings ***fit perfectly*** within Phases 2 and 3 of the worldwide flood as described in chapter 4—but secular scientists reject a worldwide flood because it implies Divine creation and a young earth.

Rapid Depositional Fossil Pattern

The fossil record as we know it today has been *grossly misinterpreted* by a secular scientific community that assumes slow depositional processes and evolutionary doctrine and ignores overwhelming evidence of a worldwide flood. The fossil record does not represent a history of life over millions of years, but rather the *order of burial during a one-time cataclysmic worldwide flood* that entombed plants and animals in massive tons of sediment.

Immobile, bottom-dwelling marine creatures include sponges, sea fans, coral, sea urchins, trilobites, and lampshells.

A global flood (beginning with the breaking up of the foundations of the deep, Genesis 7:11) would bury plants and animals in the following sequence:

- Bottom-dwelling sea creatures (i.e., marine invertebrates such as shellfish) would not escape massive sediment deposition—and they tend to be found mainly in bottom layers, but they are also found in all layers. Shellfish are relatively light and may be picked up by strong ocean currents and redeposited.
- Vertebrate fish and similar creatures would be the next to go because of continuing sediment deposition, turbulence, and temperature changes—and they tend to be found in intermediate layers.
- Land animals, such as mammals and especially birds, would have instinctively escaped to higher ground and be the last to succumb to the onslaught of floodwaters. Although animals tend to be found in the upper layers, most were destroyed and were not preserved by burial.
- Marine plants would be buried before freshwater swamp vegetation, and swamp vegetation would be buried before upland trees and shrubs.

The immobility of plants and marine invertebrates accounts for over 99.75 percent of fossils found in the geologic column, and the mobility of animals and humans accounts for less than 5 percent of the remaining 0.25 percent—reflecting the total destruction and general absence of animals and human fossils (including birds).

Ancestry of Mankind

Human Fossils and World Population

According to evolutionists and the standard geologic time scale, man became human (*Homo erectus;* homo is Latin for "human") about 1.8 million years ago and became "fully" human (*Homo sapiens*) about 250,000 to 185,000 years ago during the Pleistocene epoch.[24] Supposing evolution were true and humans lived on earth about 250,000 years ago without a worldwide flood, then many billions, if not trillions, of humans would have lived and died. Remember that only 0.0125 percent of the total number of fossils represents vertebrates (fish, amphibians, reptiles, birds, and mammals including humans)—and only one percent of these consists of a single bone! So, where are the human fossils?

One could reasonably assume that family groups would have buried or would have attempted to preserve human remains and artifacts when man supposedly became *Homo sapiens*. With trillions of humans having lived and died, assuming uniformitarian processes similar to today (without a worldwide flood), where are the human fossils, artifacts, and cultural evidence? This lack of evidence is very difficult for uniformitarians to overcome. *Evidence of early man is extremely rare.*

We know the current population growth, for example, is between 1 and 2 percent (allowing for disease and plagues).[25] Beginning with Noah's family (eight people; Noah and his wife, his three sons and their wives) with a conservative growth rate of 0.5 percent since the Great Flood (2385 BC), and allowing for disease (and everyday ailments and accidents), famine, pestilence, wars, and so forth, it would take about 4,000 to 5,000 years to reach today's population which has exceeded 7.0 billion.[26]

The worldwide flood would account for so few human fossil remains found today. What did humanity do when tsunamis began to come ashore during the time of the flood? As the most intelligent and mobile of all animals, many thousands of humans would have retreated to higher ground and been the last to surrender to the onrushing floodwaters. People would cling to fishing boats, rafts, trees, and logs until the end. According to the Bible, the purpose of Noah's Flood was to destroy the human race, not preserve it.

In Genesis 6:5-7 (NIV): "*The Lord saw how great man's wickedness on the earth had become and that every inclination of the thoughts of his heart was only evil all the time* [continually]. *The Lord was grieved that he had made man on the earth, and his heart was filled with pain. So the Lord said, 'I will wipe* [blot out] *mankind, whom I have created, from the face of the earth—men and animals, and creatures that move along the ground, and birds of the air—for I am grieved that I have made them.'*"

The fact of the matter is, evidence of early man with written records cannot be documented more than 4,000 years ago (time of the Mesopotamia civilization and early Egyptian empire). The evolutionary picture has men surviving as "hunters and gatherers for 185,000 years during the Stone Age" before discovering language, the written record, and agriculture less than 4,000 years ago—it is *extremely unlikely, or absurd* to think that it took human man 181,000 years to make such discoveries.[27]

The Biblical time scale is much more likely—that is, man was without agriculture for five to seven hundred years following the worldwide flood during the Great Ice Age. Contrary to evolutionary belief, evidence indicates that early man descended from Noah and his family approximately 4,400 years ago—and that mankind was *intelligent, highly skilled, possibly believed in an afterlife, and had an advanced social structure.*[28]

Upside Down Evolutionary Vision of Ancient Man

Although it is impossible to keep track of the many changing claims of anthropologists (see glossary) over the past century, celebrated "ancestors" of mankind and "missing links" have been silently discarded. In an attempt by anthropologists to further their careers and advance their professional self-esteem, and in their effort to certify their claims of evolution and uniformity theory, some have deceptively misled the world by rebuilding fossil animals from a few bone fragments or planting fake bones. The following are a few examples.

Neanderthal Man (*Homo sapiens neanderthalensis)* fossils were found in France in 1908. Reconstruction by secular anthropologists pictures a "brutish, stooped, knuckle-dragging ape-man"—but in fact, many competent scientists admit the so-called Neanderthals were fully human just like modern-day man. These people likely lived in the immediate post-flood period (Ice Age) when man was beginning to colonize the world. Many now recognize the stooped posture was because of the disease, rickets[29] (see glossary)—worsened by the post-flood, Ice Age environment. It is worth repeating—these people were intelligent and highly skilled with an advanced social structure. There is also evidence suggesting their belief in the existence of an afterlife.

Neanderthals

Ramapithecus (*Ramapithecus*) fossils were discovered in northern India and in East Africa in 1932. Although this was an apelike creature, *Ramapithecus* was considered a possible human ancestor based on a reconstructed jaw and dental characteristics of fragmentary fossils. A complete jaw discovered in 1976 was clearly nonhuman and the fossils are now recognized as an *extinct orangutan*.

Piltdown Man (*Eoanthropus)* fossils were found in a gravel pit in Sussex, England, in 1912 and were considered to be the second most important fossil proving the evolution of man. This was a hoax based on a human skull cap of modern age and an *orangutan's jaw*. The fragments had been chemically stained to appear old and the teeth had been filed down. For 40 years it was widely publicized as the "missing link."

Nebraska Man (*Hesperopithecus)* fossil was based on a single tooth in 1922. Nevertheless, this fossil tooth grew to be considered an evolutionary link between man and monkey. An identical tooth was found in a *wild pig* now living only in Paraguay.

Java Man (*Pithecanthropus)* and **Peking Man** (*Sinanthropus*) fossil remains were once promoted as "missing links," but morphology, and archaeological and cultural findings suggest that ***Homo erectus*** was fully human (*Homo sapiens)*. As with the Neanderthal Man, these people lived in the immediate post-flood period of the Great Ice Age (chapter 4). As suggested by Dr. Alan Thorne, an evolutionist, in 1993, "They're not *Homo erectus*, they're people."[30]

Orce Man, a fossil discovered in 1982 near the Spanish town of Orce, was claimed to be a human cranial fragment. It was hailed as the oldest fossilized human remains found in Europe. Scientists said the skull belonged to a 17-year-old man who lived 900,000 to 1.6 million years ago, and even depicted detailed drawings to represent what he would have looked like. It was later learned that the skull fragment was not human but actually an *infant donkey*.

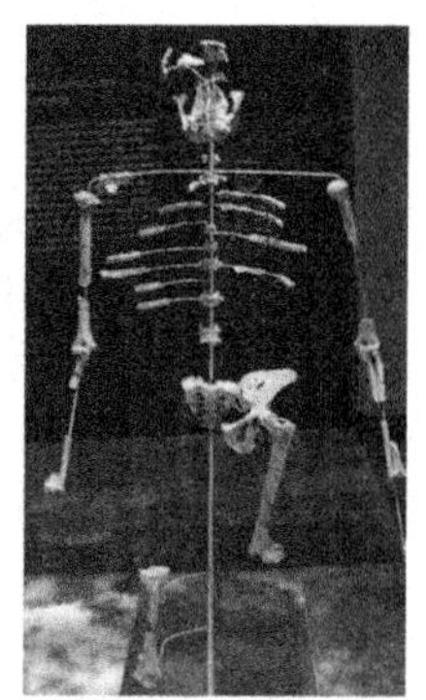
Full replica of Lucy (*Australopithecus*

Lucy (*Australopithecus afarensis)* was once promoted as the missing link. It is ape-like, and many evolutionists no longer consider it to be transitional to humans.

Ida (*Darwinius masillae;* name in honor of Darwin), recently promoted by the media as the "missing link," is nothing more than a fossilized *lemur*. Another false claim that has been quietly discarded.

Homo naledi is the most recent claim (2013-2015) of a humanlike ancestor to man. Researchers discovered 1,550 bone fragments representing at least 15 individuals, likely a family of apes. Geologists believe the limestone cave developed during the Pliocene-Pleistocene Epoch—arbitrarily assigning ages of 1 to 3 million years. Jaws did not fit the craniums with "tiny little brains stuck on bodies [bones] that weren't tiny..."—essentially, the bones did not match. It appeared that the cave had previously been entered and bones rearranged. Most pieces were found in "uncemented, loose sediment" on the cave floor and "partially unmineralized" implying the bones may be quite young, possibly the post-Flood Ice Age. (Hardened sedimentary rock normally encases these types of fossils to allow preservation.)[31]

Like variations or races of mankind (Caucasoid, Mongoloid, Native American, Ethiopian, Australoid, and Malayan), there are today's variations of apes (gorillas, orangutans, chimpanzees, and gibbons with many more variations now extinct) due to natural selection or speciation—a mechanism for change in animal populations which selects favorable gene traits from an "existing gene pool" best suited for a given environment. This is also known as 'survival of the fittest' which is ***NOT evolution***. (See chapter 3, Natural Selection and Extinction for more information.) ***Homo naledi*** is likely an extinct species of ape with curved hands for vertical climbing and flat feet for wading in swamps adapted to a specific environment. This species is ***not*** an ancestor of mankind.

If evolution were true, there would be billions of transitional types throughout the world—but there are none, yet secular paleontologists insist on concocting imaginary links between man and apes, and the scientific secular community clamors and exclaims over the find. It is the same one to two year repetition of evolutionists emerging with a fossil find trying to make "a name for themselves." Also, it is worth noting that the media frequently sensationalizes these so-called evidentiary "claims" but later, repudiation is barely worth mentioning.

Suffice it to say there is no fossil evidence that man ever evolved from apes. Evolutionary scientists will go to any length to certify evolutionary doctrine and advance their professional image, even in the face of *overwhelming evidence* to the contrary. But the truth is, no ancestor of man has ever been successfully documented—the so-called "missing link" is still missing *because* ***it does not exist***. Considering the history of invented, artificially fabricated ape-men, all claims should be treated with immense skepticism.

"Evolutionists appear to largely live in what could be described as a will-driven reality—i.e., they see what they want to see; they see a past they believe has happened [evolution], and that desire drives their vision." The science of origins cannot be observed in the past or tested using "empirical analysis"—evolution is just assumed as a basis for interpretation. (See chapter 1, Empirical Science and the Past.) Secular paleontologists try to fit the fossil into an evolutionary worldview—an "imaginary extrapolation that fosters visualizations and mental constructs that do not exist in a reality outside of their minds."[32]

So-Called Vestigial Organs and Recapitulation

According to evolutionists, vestigial organs are those that supposedly are no longer needed because we have evolved past their usefulness. But hardly anyone uses the "leftover" argument for evolution any more—probably because it has caused too much embarrassment in the past. In 1895, Dr. Robert Wiedersheim in his book, *The Structure of Man: An Index to His Past History,* confidently declared that 186 organs of the human body were useless, leftover (vestigial or retrogressive) relics of our evolutionary ancestors and these organs have been bypassed by the progress of evolution.

Just a few of these examples claimed by Wiedersheim were the appendix, the coccyx, the little toes, the parathyroid, the thymus, the pituitary gland, the pineal gland, and wisdom teeth. Also, Darwin claimed in his book, *The Origin of Species*, that vestigial organs were essential to the proof of evolution. With medical advances and research, one by one, organ functions were discovered and today, all are known to have one or more specific physiologic purpose.

> **Ontogeny** is the growth (size) and development (shape) of an individual organism; **phylogeny** is the so-called evolutionary history of a species as claimed by evolutionists. **Recapitulate** means to repeat the principal stages.

Those who studied science several decades ago may also remember the theory of recapitulation or the adage commonly taught in college classrooms, "Ontogeny Recapitulates Phylogeny." This theory asserts the human fetus goes through stages of evolutionary history as it develops. Ernst Haeckel proposed this theory in the late 1860s in his effort to promote Darwin's so-called theory of evolution. He made detailed drawings of embryonic development of eight different embryos in three stages of development—and such drawings were depicted in science textbooks and encyclopedias. A few years later his drawings and data were shown to have been fabricated.

This is one of many examples of various deceptions used by evolutionists over the years to validate their so-called theory. Yet, unbelievably, science textbooks today continue to promote vestigial structures and embryonic development as evidence for evolution.

Dinosaur Fossils

According to evolutionary scientists, dinosaurs evolved about 248 million years ago during the early Mesozoic Era. Evolutionists surmise that dinosaurs lived and ruled the earth for about 180 million years during the Triassic, Jurassic, and Cretaceous periods, long before man walked the earth. Most evolutionists believe that a natural event destroyed the dinosaur (for instance, meteorite, glaciation and marine regression, predation by man, and disease), or dinosaurs evolved into another type of animal such as birds (*Archaeopteryx*).[33]

There are numerous problems with the bizarre and absurd notion that birds evolved from dinosaurs, but the most prevalent is the lack of fossil evidence—transitional types have never been found because they do not exist. As with other animal kinds, evolutionists cannot point to any transitional (in-between) forms over this vast period of time to verify evolution from one dinosaur to another. Although secular books about dinosaurs show distinct kinds, only imaginary lines join dinosaurs to some common ancestor.

In his book, *Evolution: The Fossils Still Say No!*, Dr. Duane Gish states, "It is only because of their desperate lack of transitional forms that evolutionists have trumpeted so loudly about *Archaeopteryx*. *Archaeopteryx* appears abruptly in the fossil record, a powered flyer with wings of the basic pattern and proportions of the modern avian wing, and feathers identical to those of modern flying birds, an undoubted true bird without a single structure in a transitional state."[34]

Although fossil bones of dinosaurs are found throughout the world, most remain entombed under tons of flood sediment. Discoveries usually consist of just a few bones but, in some cases, nearly complete skeletons have been recovered. With only present day bones to evaluate, dinosaur paleontologists try to answer some basic questions: what did they look like, what did they eat, and how did they die? Rebuilding the dinosaur (and other fossilized animals) is an inexact art—and restructuring muscles and skin is, at best, guesswork.

> Preservation of complete fossil skeletons suggests rapid burial, not slow and gradual processes as suggested by uniformity theory.

Dating Dinosaur Fossils

Although evolutionists claim dinosaurs lived millions of years ago, it is important to realize that there is no gauge to indicate the age of mineralized fossilized bones. Evolutionists obtain their dates by indirect or relative dating methods as mentioned in the first section—that is, the use of index fossils and sediment strata with the presumption of evolution and an old earth. But there is considerable evidence that dinosaurs roamed the earth within the last 4,000 years, not millions of years as maintained by evolutionists and old earth advocates.

> In addition to bone and tissue of dinosaurs, DNA strands have been recovered from insects, supposedly 120 to 135 million years old, and from bacteria, supposedly 250 million years old.[37]

Soft tissue has been recovered from dinosaurs that supposedly walked the earth 68 to 248 million years ago. Scientists from the University of Montana found frozen, unfossilized *Tyrannosaurus rex* bones in a sandstone formation of eastern Montana that contained blood vessel channels, red blood cells, and hemoglobin.[35]

Also found were fragile proteins and DNA—yet paleontologists remain entrenched in their "long-age paradigm."[36] Although they have tried to explain away the evidence, it is highly unlikely that such tissues (including the presence of carbon-14; see chapter 8) could have survived for more than a few thousand years under optimum real world conditions.[38] Unfossilized dinosaur bones have also been found in Alaska.[39] Such findings have amazed researchers and have confounded those who believe that dinosaurs lived millions of years ago.

Also, rock formations containing dinosaur fossils often contain little or no evidence of fossilized plant material, soil layers, or bioturbation—a good example is the Morrison Formation in North America.[40] If the strata represent an age of dinosaurs, what did they eat? A large dinosaur, such as *Apatosaurus* (see chapter 10, Dinosaurs in the Bible), would need several tons of vegetation per day, yet there is no significant evidence of vegetation in these dinosaur-bearing strata.

Dinosaur fossils are usually found in sedimentary rock and graveyards with vast numbers of creatures violently mixed together.[41] Great fossil graveyards, which exist in just about every major country of the world, could not have been buried by gradual sediment deposition as described by secular geologists and paleontologists. Although there is much we don't know about dinosaurs, the evidence is consistent with Noah's Flood and a young earth.

Theropods are a diverse group of bipedal dinosaurs (animal with two legs) as shown below. Sauropods had long necks, long tails, small heads, and thick legs—see chapter 10, Dinosaurs in the Bible.

Allosaurus fraglis, bipedal predator 28 feet in length, Natural Museum of History, San Diego, CA. *Photo by Roger Gallop*

Tyrannosaurus rex

Illustration of a *Dilophosaurus*

Dilophosaurus sinensis

Dilophosaurus tracks. Kayenta Mudstone[42] (limestone and sandstone) near Tuba City, Arizona, Highway 89 on US 160.

Dilophosaurus tracks. Note 5-inch green pen in photos. Kayenta Mudstone (limestone and sandstone) near Tuba City, Arizona, Highway 89 on US 160. Such tracks are relatively fresh, certainly incompatible with 250 million years. *Photos by Roger Gallop*

Triceratops fossil. Note 5-inch green pen in photo. Kayenta Mudstone (limestone and sandstone) near Tuba City, Arizona, Hwy 89 on U.S. 160.
Photo by Roger Gallop

1904 Illustration by Charles R. Knight

Dinosaur Extinction

All humans and animals, including dinosaurs, were destroyed by a worldwide flood, except those that were taken on Noah's ark (described in chapter 10). After the flood (2385 BC), "kinds" of land animals, including all dinosaur kinds, came off the ark to face a cold, bleak devastated earth and the beginning of the Great Ice Age. Dinosaurs and other animals such as the woolly mammoth were overwhelmed with post-flood ice age conditions, lack of food, and natural calamities such as the collapse of massive ice dams and drainage of enormous lake systems (i.e., Missoula and Bonneville Floods). It is likely that some of these animals survived and acclimated, and lived for some time in the postdiluvian world. For more information, see chapter 10, Dinosaurs on the Ark.

As we know, animal species are prone to complete extinction if environmental conditions change.[43] Evolutionists are unable to explain the demise of dinosaurs because of their assumption of evolution and uniformitarian doctrine.

Summary

Paleontologists assume evolution and an old earth as their initial premise, and use fossils to establish relative ages of rock layers and formations. They simply *guess* how long it would take for one fossil animal to evolve into another, and then date the fossils and rocks accordingly. Many scientists object to this circular reasoning; that is, the age of index fossils is validated by the superposition of sediment layers, and the age of sediment layers is validated by index fossils—all based on the presumption of evolution, uniformity theory, and an old earth.

The standard geologic column and time scale depicted in most science textbooks, with transitional creatures evolving toward more complex forms, are utterly *misleading and fictitiou*s and *do not represent the real world.* If the process occurred over many millions of years, rock layers and formations would be uniformly deposited regionally and globally. But *this is not the case.*

Typically, only a few rock formations are superposed—most often, rock formations are inverted, folded, inserted, repeated, or missing, and fossils are in reverse order from that demanded by evolution. This is completely consistent with catastrophic processes (that is, earthquakes, volcanism, flood, sedimentation, orogeny, and erosion) occurring within a short period of time.

If evolution were true, there should be billions of transitional forms. If there were a slow evolutionary transition from amoeba to animal to man over billions of years, one would expect to find billions of examples of transitional fossils and living types. The fact is, *none have been found.* Fossil evidence *exactly fits* the creation model, for creation insists that each animal "kind" was created fully formed. Evolutionists look for "the" missing link—but there should be billions of examples if evolution were true, and yet there are none. This alone should convince most rational people that evolution is a false doctrine.

Most of the earth is covered by hundreds of feet of sediment layers containing fossil remains of drowned animals and plants. Many sites throughout the world reveal billions of creatures and polystrate trees violently mixed together. During gradual erosional and depositional processes as we observe today, dead fish and land animals are quickly devoured by scavengers. Unless creatures are rapidly buried underwater without the presence of oxygen, soft tissues will quickly decay. *Preservation of fossils suggests rapid burial, not slow and gradual processes as proposed by secular geologists and paleontologists.*

The standard geologic column, time scale, and fossil record has been ***grossly misinterpreted*** by a secular scientific community that assumes slow depositional processes and ignores overwhelming evidence of a worldwide flood. The standard fossil record depicts creatures evolving from a single-celled organism toward more complex forms over many millions of years. But such a representation is fictitious—it does not represent the real world. In reality, shell fish and plants account for 99.75 percent of fossils found in the geologic column—and mobile animals account for less than 5 percent of the remaining 0.25 percent. In fact, the fossil record represents the *order of burial during a cataclysmic worldwide flood* that entombed plants and animals in massive tons of sediment!

In uniformity doctrine, trillions of humans would have lived and died over a long period—250,000 to 185,000 years. But where are the human fossils, artifacts, and cultural evidence? *Evidence of early man is extremely rare* and cannot be documented more than 4,000 years ago. Contrary to evolutionary belief, evidence indicates that early man descended from Noah and his family at that time—and that mankind was *intelligent, highly skilled, possibly believed in an afterlife, and had an advanced social structure.*

Over the past century, celebrated "ancestors" of mankind and "missing links" have been silently discarded. Some scientists have deceptively misled the world by rebuilding fossil animals from a few bone fragments or planting fake bones.

What about dinosaur fossils? Although evolutionists claim dinosaurs lived millions of years ago, there is no test or gauge to date mineralized bones—fossil bones are simply assumed to be millions of years old based on a fictitious geologic column. But there is considerable evidence in favor of less than a few thousand years: soft tissue including red blood cells and hemoglobin have been found in unmineralized dinosaur fossils supposedly millions of years old—evidence that has been dismissed or suppressed by evolutionists in favor of false assumptions and poor science.[44]

Additionally, the assumption of evolution and uniformity doctrine leaves paleontologists unable to explain the demise of dinosaurs. However, a worldwide flood and the Great Ice Age provide a reasonable explanation.

***notes:* Chapter 7**

1. Morris, J.D. (1994). *The Young Earth*. Green Forest, AR: Master Books, 51; and Morris, J.D. (2007), 48.

2. The Florida Times-Union, Jacksonville, Florida. Tia Mitchell's article titled "Evolution Under the Microscope," appeared on January 8, 2008.

3. Geologic Time Scale – The Misconceptions. (2002). All About Science. Retrieved December 2007, from http://www.allaboutcreation.org/geologic-time-scale.htm (AllAboutScience.org). Also see Morris, Henry (June 1977). Circular reasoning in evolutionary biology. Institute for Creation Research, ICR *Impact,* Article 94; O'Rourke, J.E. (1976). Pragmatism versus materialism in stratigraphy. *American Journal of Science*, 276: 51, 47-55; and Azar, Larry (November 1978). Biologists, help! *Bioscience*, 28: 174.

4. O'Rourke (1976), op. cit., 47. Also cited in Morris (1977), op. cit.

5. Whitcomb, J.C., and Morris, H.M. (1961). *The Genesis Flood*. Phillipsburg, NJ: The Presbyterian and Reformed Publishing Company, 132-134.

6. Morris, J.D. (2003). *Is the Big Bang Biblical?* Green Forest, AR: Master Books, 108-109; and Morris, J.D. (1994). *The Young Earth*. Green Forest, AR: Master Books, 70. Statistics provided by paleontologist Kurt P. Wise, Ph.D. Geology (Paleontology). Also see interview (February 1998) by Dr. C. Werner with Dr. D. Gish, author of the book *Evolution: The Fossils Still Say No!* (1995); as cited in Werner, C. (2007). *Evolution: The Grand Experiment Vol. 1.* Green Forest, AR: New Leaf Press, 163.

7. Austin, Steven A. (November 1984). Ten misconceptions about the geologic column. El Cajon, CA: Institute for Creation Research, ICR *Impact,* Article 137 (Item #10). Retrieved from http://www.cnt.ru/users/chas/imp-137.htm. Austin, S.A. (1984). Ten misconceptions about the geologic column. *Acts & Facts,* 13 (11). Copyright © 1984 Institute for Creation Research, used by permission.

8. Darwin, Charles (1859). *On the Origin of Species by Means of Natural Selection, or the Preservation of Favoured Races in the Struggle for Life* (1st ed.). London: John Murray, 279-280.

9. Personal letter (written April 10, 1979) from Dr. Colin Patterson, then senior paleontologist at the British Museum of Natural History in London, to Luther D. Sunderland; as cited in Sunderland, L.D. (1998). *Darwin's Enigma*, rev. ed. (2002). Green Forest, AR: Master Books, 102.

10. Ibid.

11. Thomas, B. (May 2012). Four scientific reasons that refute evolution. *Acts & Facts*, 41 (5), Dallas, TX: Institute for Creation Research, 17.

12. Interview (February 1998) by Dr. C. Werner with Dr. D. Gish, author of the book *Evolution: The Fossils Still Say No!* (1995); as cited in Werner, C. (2007). *Evolution: The Grand Experiment Vol. 1.* Green Forest, AR: New Leaf Press, 132.

13. Sarfati, J. (2011). *Refuting Evolution 2.* Atlanta, GA: Creation Book Publishers, 129-153; Sarfati, J. (2007). *Refuting Evolution.* Atlanta, GA: Creation Book Publishers, 47-78; and Gish, D.T. (1995). *Evolution: The Fossils Still Say No!* El Cajon, CA: Institute for Creation Research, 305.

14. Truth and Science Ministries, In Search of the Truth. Retrieved April 2008, from http://www.truthandscience.net/quotes.htm. Also see Morris, H.M. (1997). *That Their Words May Be Used Against Them.* Green Forest, AR: Master Books, 142, 181; and *PNAS Journal* (2015). 112(32):9961-9966; as cited in *Creation* magazine (2016). Lizards identical after '20 million years', 38 (1):9.

15. Whitcomb and Morris (1961), op. cit., 169.

16. Ibid., 135. Also see Austin, Steven A. (1984), op. cit.; Woodmorappe, J. (1981). The essential nonexistence of the evolutionary-uniformitarian geologic column: a quantitative assessment. *Creation Research Society Quarterly*, 18: 46-71; and Woodmorappe, J. (1999). The geologic column: Does it exist? *Journal of Creation*, 13 (2): 77-82. Retrieved from http://creation.com/journal-of-creation-archive-index.

17. Ibid., 171.

18. Whitcomb and Morris, op. cit., 171. Also see O'Brien, J. (2015). Radiometric backflip, bird footprints overturn 'dating certainty'. *Creation*, 37 (1): 26-28.

19. Allan, Robin S. (January 1948). Geological correlation and paleoecology. *Bulletin of the Geological Society of America*, 59: 2; as cited in Whitcomb and Morris (1961), 170.

20. Sunderland, L.D. (1998, rev. ed. 2002). *Darwin's Enigma.* Green Forest, AR: Master Books, 11. Also cited in Whitcomb, J.C. (1988). *The World That Perished*. Grand Rapids, MI: Baker Book House, 76.

21. Dillow, Joseph C. (1982). *The Waters Above*. Chicago, IL: Moody Press, 377. Also cited in Whitcomb, 80.

22. Oard, M. (December 2000). The extinction of the woolly mammoth: Was it a quick freeze? *Journal of Creation*, 14 (3), 24-34. Retrieved December 2011, from http://creation.com/the-extinction-of-the-woolly-mammoth-was-it-a-quick-freeze.

23. Walker, T. (2011). 80 whales buried mysteriously in Chilean desert. Retrieved February 2014, from http://creation.com/chile-desert-whale-fossils. Speculation of algal toxicity was reported February 2014 by numerous news outlets including Chile (AP), The Washington Post, National Geographic, and others. Sherwin, Frank (May 2014). Ancient Whale Graveyard Points to Genesis Flood. *Acts & Facts*, 43 (5), Dallas, TX: Institute for Creation Research, 16.

24. The word *homo* is Latin for "human." There is no real distinction between *H. erectus* and *H. sapiens*–in modern taxonomy, *Homo sapiens* is the only extant (actually existing) species of its genus, *Homo*.

25. U.S. Census Bureau. Retrieved April 2008, from http://www.census.gov/ and Population growth. Retrieved from http://en.wikipedia.org/wiki/Population_growth.

26. Wieland, C. (2001). *Stones and Bones.* Green Forest, AR: Master Books, 32. Also see Rosenburg, M. (January 2006). Population growth. Retrieved December 2007, from http://geography.about.com/od/populationgeography/a/populationgrow.htm and World Population (May 29, 2013, last modified). In Wikipedia, the free encyclopedia. Retrieved January 2011, and May 2013, from http://en.wikipedia.org/wiki/world_population.

The world population is the total number of humans on earth at a given time. As of May 31, 2009, the world's population was over 6,792,467,727 (6.7 billion) and the world's growth rate was about 1.14 percent, representing doubling of the population in 61 years. In modern times, human populations have been increasing consistently at more than one percent per year. An automatically daily calculation by the United States Census Bureau estimates the current figure to be approximately 7.088 billion (May 29, 2013).

27. Humphreys, R. (June 2005). Evidence for a young world. Institute for Creation Research, ICR *Impact,* Article 384 (Item #13). Also see Deevey, E.S. (September 1960). The human population. *Scientific American*, 203: 194-204.

28. Catchpoole, D., Sarfati, J., and Wieland, C. (2008). *The Creation Answers Book*. (D. Batten, Ed.). Atlanta, GA: Creation Book Publishers, 123; Lubenow, M.L. (1998). Recovery of Neanderthal mtDNA: An evaluation. *Journal of Creation*, 12 (1): 87-97; Humphreys, R. (Item #13), op. cit.; and Marshack, A. (January 1976). Exploring the mind of ice age man. *National Geographic*, 147: 64-69.

29. Ibid., 205; and Gish (1995), op. cit., 305. See glossary for rickets.

30. *The Australian*, August 19, 1993. Dr. Thorne was then a paleoanthropologist at the Australian National Laboratory; as cited in Wieland, C. (2001). *Stones and Bones*. Green Forest, AR: Master Books, 20. Also see Mehlert, A.W. (1994). *Homo erectus* 'to' modern man: evolution or human variability. *Journal of Creation,* 8 (1): 105-116; and Bowden, M. (1988). *Homo Erectus:* A Fabricated Class of 'Ape-Men', *EN Tech.J.*, vol. 3, 152-153; as cited in creation.com/images/pdfs/tj/j03_1/j03_1_152-153.pdf.

31. Clarey, T. (December 2015). *Homo naledi*: New Claims of a Missing Link. *Acts & Facts*, 44 (12), Dallas, TX: Institute for Creation Research, 17. See Ambler, Marc (September 2015). What to make of *Homo naledi*? Creation.com/homo-naledi.

32. Guliuzza, R.J. (December 2015). The Imaginary Piltdown Man. *Acts & Facts*, 44 (12), Dallas, TX: Institute for Creation Research, 13-14.

33. Gish (1995), op. cit., 129-141; Tomkins, J. (October 2009). Dinosaur protein sequences and the dino-to-bird model. *Acts & Facts*, 38 (10), Dallas, TX: Institute for Creation Research, 12-14; and Clarey, T.L. (September 2006). Dinosaurs vs. birds: The fossils don't' lie. ICR *Impact,* Article 399. Also see Catchpoole, Sarfati, and Wieland (2008), op. cit., 252; and Werner, C. (2007). op. cit., 162-164 (145-164).

34. Ibid., 141

35. Schweitzer, M.H., Wittmeyer, J.L., Horner, J.R., and Toporski, J.K. (March 2005). Soft tissue vessels and cellular preservation in *Tyrannosaurus rex*. *Science,* 307 (5757): 1952-1955; Schweitzer, M.H., and Staedter, T. (June 1997). The real Jurassic Park. *Earth*, 6 (3): 55-57; Scientists recover *T. rex* soft tissue. The Associated Press (2008). Retrieved April 2008, from http://www.msnbc.msn.com/id/7285683/; Fields, H. (May 2006). Dinosaur shocker. *Smithsonian Magazine;* Wieland, C. (September 1997). Sensational dinosaur blood report! *Creation*, 19 (4): 42-43; Wieland, C. (1999). Dinosaur bones: Just how old are they really. *Creation*, 21 (1): 54-55; Thomas, B. (October 2008). Dinosaur soft tissue: Biofilm or blood vessels? *Acts & Facts*, 37 (10), Dallas, TX: Institute for Creation Research, 14. Also see Catchpoole, Sarfati, and Wieland (2008), op. cit., 255; and Cupps, V.R. (December 2015). Soft-Tissue Time Paradox. *Acts & Facts*, 44 (12), Dallas, TX: Institute for Creation Research, 16.

36. Catchpoole, D. (2014). Double-decade dinosaur disquiet. *Creation*, 36 (1): 12-14.

37. Humphreys, R. (Item #8), op. cit. Also see Cherfas, J. (September 20, 1991), op. cit., 1354-1356; Cano, R.J., Poinar, H.N., Pieniazek, N.J., Acra A., and Poinar, Jr., G.O. (June 10, 1993). Amplification and sequencing of DNA from a 120-135-million-year-old weevil. *Nature,* 363: 536–538; and Vreeland, R.H., Rosenzweig, W.D., and Powers, D.W. (October 19, 2000). Isolation of a 250-million-year-old halotolerant bacterium from a primary salt crystal. *Nature*, 407: 897-900.

Greenwalt, D.E. et al. (October 14, 2013). Hemoglobin-derived porphyrins in a Middle Eocene blood-engorged mosquito. *Proceedings of the National Academy of Sciences*; and Haley, G. (October 21, 2013). Unprecedented blood-filled mosquito fossil raises questions over evolutionary dating methods. *Christian News*, Christian News Network. Also see Thomas. B. and Clarey, T. (January 2014). Bloody mosquito pierces standard fossil dating procedure. *Acts & Facts*, 43 (1), Dallas, TX: Institute for Creation Research, 13-15.

38. Wile, J.L. (December 2013). An explanation that is not exactly iron-clad. *Proslogion*. Retrieved December 2013, from http://blog.drwile.com/?p=11753.

Schweitzer, M.H. et al. (2013). A role for iron and oxygen chemistry in preserving soft tissues, cells and molecules from deep time. *Proceedings of the Royal Society B,* 281:20132741. In a two-year lab experiment conducted at room temperature, it was found that iron in blood may help preserve soft tissue. The study concluded that soft tissue was preserved 'over deep time'—an assumption apparently made in an effort to uphold evolutionary doctrine—but the study fails to solve the mystery of 'age' and 'real world conditions.' Fragile tissue would certainly deteriorate after undergoing a freeze/thaw cycle each year, and it would be extremely hard to comprehend how fragile cell tissue could remain intact for 68 to 248 million years (see section, Standard Geologic Column and Time Scale). In the real world, blood and soft tissue could not persist for this vast period of time. The bottom line: The presence of unmineralized bones and soft tissues attests to a relatively short time span.

39. Batten, D. (1997). Buddy Davis - The creation music man (who makes dinosaurs). *Creation,* 19 (3): 49-51; and Davies, K. (1987). Duckbill dinosaurs (Hadrosauridae, Ornithishia) from the north slope of Alaska. *Journal of Paleontology*, 61 (1): 198-200.

40. The Morrison Formation is a distinctive sequence of Late Jurassic sedimentary rock that is found in the western United States and Canada, which has been the most fertile source of dinosaur fossils in North America. It is composed of mudstone, sandstone, siltstone, and limestone. Most of the fossils occur in the green siltstone beds and lower sandstones, relics of the rivers and floodplains. Wikipedia, the free encyclopedia.

The formation is centered in Wyoming and Colorado, with outcrops in Montana, Saskatchewan, Alberta, North Dakota, South Dakota, Nebraska, Kansas, the panhandles of Oklahoma and Texas, New Mexico, Arizona, Utah, and Idaho. It covers an area of 600,000 square miles (1.5 million square km), although only a small fraction is exposed and accessible to geologists and paleontologists. Morrison Formation (January 3, 2009, last modified). In Wikipedia, a free encyclopedia. Retrieved from http://en.wikipedia.org/wiki/Morrison_Formation.

41. Oard, Michael J. (August 1997). The extinction of the dinosaurs. *Journal of Creation,* 11 (2): 137-154.

42. The Kayenta Formation is a geologic layer in the Glen Canyon Group that is spread across the Colorado Plateau province, including northern Arizona, northwest Colorado, Nevada, and Utah. This rock formation is particularly prominent in southeastern Utah, where it is seen in the main attractions of a number of national parks and monuments. These include Zion National Park, Capitol Reef National Park, the San Rafael Swell, and Canyonlands National Park. The Kayenta Formation frequently appears as a thinner dark broken layer below Navajo Sandstone and above Wingate Sandstone (all three formations are in the same group). Together, these three formations can result in immense vertical cliffs of 2000 feet or more. Kayenta layers are typically red to brown in color, forming broken ledges. From Wikipedia, the free encyclopedia.

43. The antediluvian period (preflood) was mild and food was plentiful, but postdiluvian climatic conditions (post-flood) were too severe for some animals that were acclimated to the milder climate. See chapters 4 and 10. Although adaptive specialization significantly improves the survival potential of the species, it makes many species prone to extinction if environmental conditions change. Remember that within a new species (subgroup), the original gene pool has been downsized, providing a survival advantage to a specific cllimatic condition (e.g., warm, subtropical), but such adaptation would be a distinct disadvantage if climatic conditions should ever change (i.e., post-flood ice age). See chapter 3, Natural Selection and Extinction.

44 Johnson, J.S., Tomkins, J. and Thomas, B. (October 2009). Dinosaur DNA research: Is the tale wagging the evidence? *Acts & Facts*, 38 (10), Dallas, TX: Institute for Creation Research, 4-6.

Chapter 8

Radioisotope Dating - Evidence for a Young Earth

"Know this first of all, that in the last days mockers will come with their mocking, following after their own lusts, and saying, 'Where is the promise of His coming? For ever since the fathers fell asleep, all continues just as it was from the beginning of creation.' For when they maintain this, it escapes their notice ["*deliberately forget,*" NIV] *that by the word of God the heavens existed long ago and the earth was formed out of water and by water, through which the world at that time was destroyed, being flooded with water. But the present heavens and earth by His word are being reserved for fire, kept for the day of judgment and destruction of ungodly men."*—2 Peter 3:3–7, NAS

Over the past fifty years we have heard that scientists can prove the earth is millions of years old with radioisotope dating—that the "absolute age" of rocks and organic materials (bones and flesh) can be calculated with precision. In reality, radioisotope dating methods suffer from various *unprovable assumptions*.

Is radioisotope dating reliable—can the technique and results be trusted? For decades now, creation scientists have demonstrated through research that the answer is clearly *No*. Results of radioisotope dating have been shown to be inconsistent and often bizarre. Many scientists do not recognize or rely on radioisotope methods, but this is seldom discussed or understood by the media or the general public.

Although the subject of radioisotope dating may seem a bit daunting, the dating method is rather straightforward. Before reviewing radioisotope dating methodology, however, it is important to have a basic understanding of the terms 'element,' 'atomic number,' 'atomic mass' (or atomic weight), and 'isotopes.' Refer to the text box and Radioisotope Dating in Appendix A, Glossary.

> An **element** consists of one type of atom distinguished by its atomic number. An **atomic number** is the number of protons found in the nucleus of an atom. The **mass number**, sometimes referred to as the **atomic weight**, is the number of protons and neutrons in the nucleus of an atom.

Isotopes are several different forms of the same element, each having a different atomic mass or weight. Isotopes of the same element have the same number of protons (atomic number) but different number of neutrons. Therefore, isotopes have different mass or weight (neutrons plus protons).

For example, there are three naturally occurring isotopes of carbon—carbon-12, -13, and -14. Carbon-12 has an atomic number of 6 and is stable—the nucleus comprises six protons and six neutrons, or an atomic weight of 12. Carbon-13 has an extra neutron, or atomic weight of 13, and carbon-14 has two extra neutrons, or atomic weight of 14. Carbon-12 and -13 are stable isotopes whereas carbon-14 is unstable and experiences decay. About 99% of all natural carbon is carbon-12, about 1% is carbon-13, and trace amounts are found as carbon-14.

On the other hand, elements with atomic numbers greater than 82 such as uranium undergo radioactive decay into smaller, more stable elements. Uranium can decay into an isotope of another element at a certain rate. Nuclear decay involves the breakdown of the atom on a subatomic scale by emitting radiation (alpha and beta particles, and gamma rays)—thus, the original atom becomes an entirely different kind of atom or element.

For example, uranium-238 (with an atomic number of 92 and mass number of 238) decays into thorium-234 (with an atomic number of 90 and mass number of 234), which itself is unstable, and the process continues until the atom changes into lead-206 (with an atomic number of 82 and mass number of 206), which is stable. The initial isotope is called the "parent" and the final isotope is called the "daughter."

If we can measure the rate of decay from uranium to lead (which involves precise measurements), and if we can measure how much parent and daughter isotopes are present in a given rock sample, then supposedly we can calculate the age of the rock. No, not exactly—various *assumptions* are made including constant rate of decay, no loss or gain of isotopes during earth's history, and known amount of parent and daughter isotopes at the beginning. These assumptions are discussed in a later section, but first, we will discuss another form of radioisotope dating called radiocarbon dating or carbon-14 dating, which is the most widely recognized form of dating fossils.

This chapter reviews the 2006 book, *Thousands ... Not Billions,* by Dr. D. DeYoung, and a few other books and articles on radioisotope dating. DeYoung's book summarizes eight years of research by a group of seven distinguished scientists who in 1997 began a research project known as RATE (Radioisotopes and the Age of the Earth).[1]

Uranium-238 to Lead-206 Decay

Decay Isotopes	Radiation Particles	Half Life
Uranium-238 (parent)	Alpha	4.47 billion years
Thorium-234	Beta	24 days
Protactinium-234	Beta	6.7 hours
Uranium-234	Alpha	240,000 years
Thorium-230	Alpha	75,380 years
Radium-226	Alpha	1,602 years
Radon-222	Alpha	3.8 days
Polonium-218	Alpha	3.1 minutes
Lead-214	Beta	27 minutes
Bismuth-214	Beta	20 minutes
Polonium-214	Alpha	0.000164 sec.
Lead-210	Beta	22 years
Bismuth-210	Beta	5 days
Polonium-210	Alpha	138 days
Lead-206 (daughter)	None	stable

In 1997 a group of seven scientists met to review the assumptions and procedures used in estimating the ages of rock strata. This research project, lasting more than eight years, is identified as RATE (Radioisotopes and the Age of the Earth).

Alpha (α) particles are equivalent to the nuclei of helium atoms—that is, two protons and two neutrons. Beta (β) particles are single electrons which carry a negative charge. Gamma (Y) rays are a form of electromagnetic radiation.

Carbon-14 Dating

Although many people are under the misconception that carbon-14 dating can date any object into the millions of years, it is valid only for "recent" times—that is, within 5,000 years. Carbon-14 cannot be used to date sedimentary rocks or most fossils (mineralized plants and animals; organic structures replaced by silica), but it can be used to date fossils that contain carbon. Only organic materials can be dated using carbon-14—for example, flesh, charcoal, wood, non-mineralized bone, and carbonate deposits including marine and freshwater shells. Sedimentary rocks, such as limestone, sandstone, and shale, cannot be dated with radioisotope methods because these rocks comprise clastic fragments and minerals from preexisting rocks that were eroded and redeposited.

> Radiocarbon dating has proven accurate to about 4,000 years, when artifacts could be calibrated to known dates, and dates older than 5,000 years could not be calibrated since there is no known historical material beyond that time. See Appendix H.

> Although there are a few "proposed schemes" to potentially determine the age of certain minerals in sedimentary rocks, such minerals cannot be dated for reasons cited above—also, dating methods are unreliable and "are seldom considered helpful."[2]

How does carbon-14 dating work? Two forms of carbon, carbon-12 (stable form) and carbon-14 (unstable form), are found in carbon dioxide (CO_2). Although carbon-14 is an unstable form, advocates of the radiocarbon method assume that carbon-14 and carbon-12 are "thoroughly mixed" in the atmosphere and the production and decay rates of carbon-14 have remained stable throughout most of earth's history. (Likewise, it is assumed the ratio of these carbon isotopes, $^{14}C/^{12}C$, has remained stable.) Both forms of carbon are taken up by plants during plant photosynthesis and, in turn, plants are eaten by animals.

> Refer to the previous section for a general description of the three naturally occurring isotopes of carbon—carbon-12, -13, and -14.

When a tree or animal dies, it ceases to take in new carbon of any form. Carbon-12 remains stable within the dead animal or plant tissue, but the unstable carbon-14 continues to decay. So at the point of death, the ratio between carbon-14 and carbon-12 ($^{14}C/^{12}C$) begins to decrease. By knowing the decay rate of carbon-14, or half-life, the amount found in the fossil can be measured to determine its age—*the less carbon-14, the older the fossil.*

> **How is carbon-14 formed?** When cosmic radiation strikes nitrogen-14 in the atmosphere, N-14 is converted into C-14—in turn, C-14 decays back into N-14 with a half life of 5,730 years.

The primary weaknesses of this dating technique are assumptions based on uniformity theory and evolutionary doctrine:[3]

- Carbon-14 and carbon-12 are thoroughly mixed in the atmosphere and they remain unchanged in their relationship—that is, $^{14}C/^{12}C$ has been in equilibrium throughout earth's history. This assumption cannot be tested—however, there is substantial evidence the concentrations and proportions were *much lower* in the past.
- The production and decay rates of carbon-14 have always been stable or constant. Again, this assumption cannot be tested—however, there is solid evidence the concentration of carbon-14 has been *increasing over time*.

So what would affect the concentrations and proportions of carbon-14 to carbon-12, and the production and decay rates of carbon-14? The answer is found in the decay of the earth's magnetic field and a worldwide flood.

Effects of the Second Law of Thermodynamics and a Worldwide Flood

Carbon-14 is produced when cosmic radiation in the atomoshere strikes nitrogen-14—in turn, carbon-14 decays back into nitrogen-14 with a half-life of 5,730 years. The production rate of carbon-14 is affected by the amount of cosmic rays penetrating the earth's atmosphere which, in turn, are affected by the strength of the earth's magnetic field which deflects cosmic rays. Precise measurements during this past century have shown a steady decay in the earth's magnetic field resulting in an increase in carbon-14 production.

Scientists have calculated the total energy stored in the earth's magnetic field (in trillions of kilowatt-hours) is decreasing with a half-life of 1,465 years (± 165 years). Also, such phenomenon is consistent with the Second Law of Thermodynamics and a worldwide flood. See chapter 9, Rapid Magnetic Field Reversals and Decay, for additional information and citations.

The rate of carbon-14 production today is about 18% higher [18 to 25% higher] than the rate of decay.[4] Although the assumption of equilibrium provides a benchmark for carbon-14 at the time of death, "It is now admitted by all investigators that *equilibrium does not exist—that the C-14 concentration is constantly increasing*."[5] [Italics added] This implies less carbon-14 in the past—so animals or plants that died hundreds or thousands of years ago would *date much older than their true age.*

With this in mind, let's examine what would have happened during a worldwide flood with massive tectonic upheaval. The flood would have *buried and sealed enormous amounts of carbon* in the form of plants that became coal and oil (see chapters 4 and 6), thus, exhausting the total carbon (carbon-12 and carbon-14) available in the biosphere—that portion of the earth where life occurs.

> Plants absorb carbon dioxide and use it to grow—but when plants decay or burn, carbon dioxide is released back into the atmosphere. Decaying plants also produce methane, a greenhouse gas more potent than carbon dioxide.

The *normal production of carbon dioxide by decaying plants would have been greatly reduced* for a time because of burial under millions of tons of water-deposited sediment. Carbon dioxide in the atmosphere (in the form of carbon-12 and carbon-14) would have been significantly reduced and the *lower* carbon-14 level in plants and animals during and after the flood would yield ages much older than their true ages. See Appendix H with graphical depiction of a young earth scenario.

> Measurement of carbon-14 in historically dated objects (for example, seeds placed in the graves of dated tombs) enables the level of carbon-14 in the atmosphere at that time to be estimated and the dating technique to be validated. *Even with historical calibration, archaeologists do not regard carbon-14 dates as reliable because of frequent differences.*
>
> Scientists actually do not put their trust in this technique unless the results are confirmed by historical records. "No matter how 'useful' it is, though, the radiocarbon method is still not capable of yielding accurate and reliable results. There are gross discrepancies, the chronology is *uneven* and *relative*, and the accepted dates are actually *selected* dates."[6]

Carbon-14 in Ancient Fossils and Rocks

As we have seen, the half-life of carbon-14 is 5,730 years and anything over 100,000 years is undetectable by current measurement techniques (AMS, or accelerator mass spectrometer). With this half-life, carbon-14 should not exist within supposedly ancient fossils and other materials. Yet, RATE scientists have found detectable carbon-14 in ancient fossils, petrified wood, graphite, and calcite from around the world—and in coal formations supposedly 300 million years old and in diamonds presumed to be many millions, if not billions of years old.[7]

Such analysis of diamonds by the RATE project is the first ever reported.[8] The presence of carbon-14 in "ancient" fossils, rocks, coal, and diamond samples is clearly at odds with millions of years as advocated by secular geologists and evolutionists.

An octahedral shape of this rough diamond crystal is typical of the mineral.

> **Half-life** is the time required for the mass of an atom to decay to half of its initial value. Given a half-life of 5,730 years (measured by precise instruments), we know that only a quarter of a carbon-14 atom will be left at the end of 11,460 years, an eighth of the carbon-14 at the end of 17,190 years, and so forth, with no detectable amount of carbon left at the end of 100,000 years. If a sample contains carbon-14, it is *not* millions of years old.

According to RATE scientists, traces of carbon-14 found in coal and diamonds are dated to about 44,000 and 57,000 years. Remember from the previous section that 1) carbon-14 concentration in the atmosphere is constantly increasing (which implies less C-14 in the past), and 2) during a worldwide flood, carbon dioxide in the atmosphere would have been significantly depleted; that is, the flood would have buried the world's vegetative biomass that would have significantly exhausted total carbon in the biosphere.

> **Diamonds** are a crystalline form of pure carbon and the hardest occurring mineral, and they are believed to originate in the upper mantle of the earth where pressure and temperature are extreme. They are presumed by secular geologists to be many millions, if not billions, of years old (based on radioisotope dating of inclusions occasionally found inside the diamonds). Yet, diamonds have detectable carbon-14.

Therefore, fossils that formed during and after the flood with a *lower* carbon-14 level would yield ages much older than their true ages. Because of the lower carbon-14 at the time of formation, the actual age of coal and diamonds is estimated by RATE scientists to be "several thousand years."[9]

Suffice it to say there has been no viable explanation for the presence of carbon-14 in coal and diamonds and in deep geological strata, all supposedly hundreds of millions of years old. No matter how deep the strata, researchers have been unable to find carbon (stable forms: carbon-12 and -13) without carbon-14 (unstable form). RATE scientists conclude the presence of carbon-14 is strong evidence for a young earth. (See Appendices F and H for more information.)

Dating Igneous and Metamorphic Rocks

Igneous and metamorphic rocks, which were once extremely hot and have cooled into solid rock, are dated using the following methods: uranium-238 to lead-206, uranium-235 to lead-207, potassium-40 to argon-40, thorium-232 to lead-208, and rubidium-87 to strontium-87. These methods are normally viewed as capable of providing an "absolute age" since solidification of magma—that is, the age of rocks between cooling and present day. Dating igneous rocks yields the supposed age of the original rock (magma), and dating metamorphic rocks yields the supposed age since metamorphism (when intense heat and pressure were applied to "pre-existing" rocks).

> The initial isotope is called the "parent" and the final isotope is called the "daughter."

How does dating of igneous and metamorphic rocks work? Elements with atomic numbers greater than 82 such as uranium undergo radioactive decay into smaller, more stable elements. As described at the beginning of this chapter, nuclear decay involves the breakdown of the atom on a subatomic scale by emitting radiation—thus, the "parent" isotope eventually decays into "daughter" isotopes. The radioisotope method is based on a comparison of parent (e.g., U-238) and daughter isotopes (e.g., Pb-206) while making certain unverified assumptions.

In order for the method to work, advocates assume 1) the rate of decay of isotopes is constant, 2) no parent or daughter isotopes were gained or lost over the supposed millions of years of decay, and 3) no daughter isotopes were present at the time of cooling of the original magma (that is, hot magma *resets the age clock to zero*). Of course, these assumptions are ***purely conjecture*** based on the notion that geological and meteorological processes have remained constant for millions or billions of years.

Advocates of these methods assume that *melting or liquid magma sets the age clock to zero*—that is, there are no daughter isotopes in the rock at the time of initial cooling or solidification. The ratio of parent to daughter in the rock is analyzed (again, using precise measurements) and given the half-life of the isotope, the time elapsed between cooling and present day is then calculated. If there are daughter isotopes in the sampled rock, then supposedly the rock can be dated—the greater the amount of daughter isotopes, the older the rock. Let's look at these assumptions.

Common Radioisotope Dating Methods		
Parent (unstable)	**Daughter (stable)**	**Parent Half-Life (million years)**
Carbon-14	Nitrogen-14	5,730 years
Uranium-238	Lead-206	4,470 my
Uranium-235	Lead-207	704 my
Potassium-40	Argon-40	1,280 my
Thorium-232	Lead-208	14,010 my
Rubidium-87	Strontium-87	48,800 my

Number of Half-Lifes Elapsed	Fraction Remaining	Percentage
0	1/1	100
1	1/2	50
2	1/4	25
3	1/8	12.5
4	1/16	6.25
5	1/32	3.125
6	1/64	1.5625
7	1/128	0.78125
N	$1/2^n$	$100\ (1/2^n)$

- ***Assumption 1*** (Constant rate of decay) — Although scientists have attempted to alter the rate of decay of isotopes in laboratory experiments, the rate has virtually remained constant. But because it is impossible to duplicate environmental conditions in the lab, and because the half-life of uranium-238 to lead-206 is extremely slow (half-life of 4.5 billion years is an immense expanse of time), many scientists are unconvinced the half-life has remained constant over this vast time period—especially during Creation when the universe was expanding, stretching out, and accelerating at inconceivable rates and during a worldwide flood with massive tectonic upheaval. In support of this, RATE scientists and other researchers have found strong evidences of accelerated nuclear decay in the past—these evidences are discussed in a later section, Accelerated Nuclear Decay.

  Dr. V. R. Cupps, Ph.D., a nuclear physicist, explains that "Recent experimental evidences verify that the decay rates of radioisotopes can vary significantly from the currently accepted values—by as much as 10^9 times faster (that's a billion times faster) when exposed to certain environmental factors"[10]—factors such as distance between the earth-sun and solar activity. Dr. Cupps further comments, "One cannot help but wonder what this might say about nuclear decay processes inside stars or large exoplanets."

- ***Assumption 2*** (No loss or gain of parent or daughter isotopes, assumes a closed system) — This assumes that neither the parent nor daughter isotopes (including the reactive and mobile intermediate isotopes, radium-226 and radon-222) were gained or lost through migration or contamination (mixing of magma) over the supposed many millions of years of earth's history. This, in turn, assumes no tectonic activity (that is, no uplift, folding, and thrusting of the earth's crust in forming mountain ranges). With such mixing, rocks are "open systems" (not cogenetic) which invalidates the radiometric technique. During any tectonic event, the earth's crust would be in such turmoil that leaching or movement of isotopes would have been very common. Even under the constraints of uniformity theory (that is, geological and meteorological processes have remained constant during earth's history), this assumption is *fictitious*.

> The parent/daughter ratio (P/D) is used to determine or calculate the age of rock. If parent or daughter isotopes were added or removed as a result of migrating magma, or if daughter isotopes were present at the start of solidification of magma, the P/D ratio would have nothing to do with age—the ratio would be *meaningless*.

- ***Assumption 3*** *(Known amounts of daughter isotopes present at the start)* — This assumption considers the original quantity of daughter isotopes was zero at the time of cooling of the original magma. If some daughter material (for instance, lead-206) was "inherited" or present at the start—for example, abundant daughter isotopes from the time of creation when the universe was expanding and accelerating, and during a worldwide flood with cataclysmic restructuring of the earth's crust, the rock would have the appearance of age when it was just formed. (Also, this assumption relies on the second assumption; if the second assumption is incorrect, it invalidates the third assumption.)

  Interestingly, this assumption can be tested today—rocks from recent eruptions can be tested for the amount of daughter isotopes. If the dating process is accurate, then the age derived from testing should be too young to measure. Remember that advocates of this method propose that "melting resets the age clock to zero."

Unreliable Results

Many people believe that radioisotope techniques (such methods as carbon-14, uranium-lead, potassium-argon, and rubidium-strontium) are generally in agreement and the earth is millions of years old. However, the reality is these various techniques produce results that *don't agree with known dates or each other (discordant), and don't agree with stratigraphic or fossil studies (discrepant).* The following are just a few examples:

> *Marlstone Rock Bed near Banbury, England*, is considered to be a Jurassic limestone dating "189-million years old" according to ammonite and belemnite index fossils but radiocarbon dating of wood found in the limestone dates 23,000 years. Uniformitarians assume that limestone was slowly deposited over millions of years on a shallow, tranquil ocean floor—a place normally absent of fossil wood. Factoring in the aging effect of depleted carbon by the flood, the radiocarbon date is consistent with fossils being buried with limestone 4,000 years ago[11] (see Appendices E and F). Additionally, wood found in sandstone and volcanic rock supposedly many millions of years old, according to geologic time scale, provides radiocarbon ages of just thousands of years.[12]
>
> *Sunset Crater,* located just northeast of Flagstaff, Arizona, is known to be a recent volcano with Native American artifacts found in basaltic rocks. According to the National Park Service, this volcano erupted between AD 1040 and 1100. This was verified by tree ring measurements that date the eruption at AD 1065. However, potassium-argon (K-Ar) dating provides ages of 210,000 and 230,000 years.[13] If the K-Ar dating process were accurate, then the radioisotope age should be about 904 years (date of analysis in 1969 minus 1065).
>
> *Mt. Rangitoto of New Zealand* erupted less than 300 years ago according to radiocarbon dating of polystrate trees, but potassium-argon dating provides an age of 485,000 years.[14] If the potassium-argon (K-Ar) dating process were accurate, then the radioisotope age should be about 300 years.
>
> *Mt. Ngauruhoe of New Zealand* erupted in 1949, 1954, and 1975, but potassium-argon dating ranges from 270,000 years to 3.5 million years, rubidium-strontium dating at 133 million years, samarium-neodymium at 197 million years, and lead-lead at 3,908 million years.[15] If these dating processes were accurate, these radioisotope ages should indicate recent eruptions with ages of less than 50 years.
>
> *Mt. Hualalai of Hawaii* is known to have erupted in 1800, but various radioisotope dating methods report ages between 140 million years and 2.96 billion years.[16] If the radioisotope dating process were accurate, then the radioisotope age should be about 168 years (date of analysis minus 1800).
>
> *Mt. Kilauea of Hawaii*, another known lava flow less than 200 years old, yielded a potassium-argon age of about 21 million years plus or minus 8 million years.[17] If the dating process were accurate, then the radioisotope age should be about 200 years.
>
> *Mt. St. Helens, Washington,* erupted in 1980 but potassium-argon dating provides an age of about 2.8 million years.[18] If the dating process were accurate, then the radioisotope age should be about 15 years. (date of analysis in 1995 minus 1980).

Cardenas Basalts of the Uinkaret Plateau (Grand Canyon) are datable by radioisotope techniques—and these basalts have been well studied by Dr. Steven A. Austin.[19]

As indicated in the diagram, younger basaltic lava was extruded by volcanoes at the top of the plateau sometime *after* the canyon was eroded. Based on historical accounts, it appears that Native Americans actually witnessed these eruptions within the last few thousand years.[20]

Nevertheless, radioisotope techniques date the younger 2,000-year-old basalt between 1,270 million to 2,600 million years! More amazingly, these younger rocks date older than the underlying, much older rocks (Precambrian basalts). Obviously, *something is wrong with these radioisotope techniques.*

Volcanic Basalt Rocks of the Uinkaret Plateau

Younger Volcanic Rocks

6 K-Ar model ages	0.01 — 117 million years
5 Rb-Sr model ages	1270 — 1390 million years
1 Rb-Sr isochron age	1340 million years
1 Pb-Pb isochron age	2600 million years

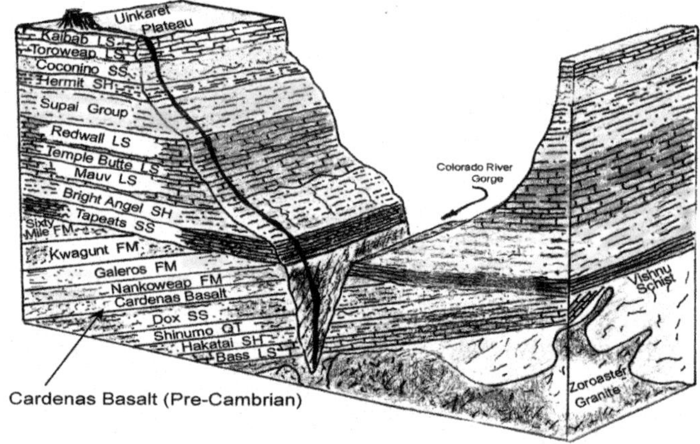

Older Volcanic Rocks

5 K-Ar model ages	791 — 853 million years
6 Rb-Sr model ages	980 — 1100 million years
1 K-Ar isochron age	715 million years
1 Rb-Sr isochron age	1070 million years

Data Source: Austin, S. A. (Ed.). *Grand Canyon: Monument to Catastrophe*, 126.

The question is: Why should anyone believe any radioisotope date if dates from recently solidified rocks indicate vast ages? Why do recent lavas indicate millions of years? Daughter isotopes are present at the start contrary to assumption 3, and isochron dating (see next section) is simply unable to differentiate between inherited daughter isotopes and non-inherited daughter isotopes.

In the beginning when the universe was expanding and accelerating at inconceivable rates, daughter isotopes (Pb-206 and Ar-40) would have been present. Large concentrations of daughter isotopes were mixed into the upper mantle during the first few days of Creation—and then later mixed into the earth's crust during the flood. This would erroneously represent millions or billions of years of age. The bottom line is this: Ancient and recent lava flows have the same chemistry derived from the earth's upper mantle. P/D ratios have nothing to do with age of the rock—they are just ratios.

So, if excess daughter isotopes are found in "recent lava," why do scientists remain so committed and place so much validity and reliability on these same methods for "unknown" ages? It is because of their commitment to evolutionary doctrine and an old earth—they are committed to one system of interpretation and contrary evidence is routinely rejected or ignored.

Dr. Andrew Snelling, ICR professor of geology, studied published uranium-lead ratios and ages from Koongarra uranium deposits in the Northern Territory of Australia. He wrote in his conclusion:[21]

> The above evidence conclusively demonstrates that the U/Pb system... has been so open with repeated large scale migrations of the elements that it is impossible to be sure of the precise status/history of any piece of pitchblende selected for dating. Even though geochronologists take every conceivable precaution when selecting pitchblende grains for dating, in light of the above evidence, no one could be sure that the U and Pb they are measuring is 'original' and unaffected by the gross element movements observed and measured.
>
> Those pitchblende grains dated have always contained Pb, both within their crystal lattices and as microscopic inclusions of galena, making it impossible to be sure that all the Pb was generated by radioactive decay from U. In addition, the pitchblende grains don't have uniform compositions so that 'dating' of sub-sections of any grain would tend to yield widely divergent U/Pb ratios and therefore varying 'ages' within that single grain. A logical extension of these data & conclusions is to suggest as others already have that *U/Pb ratios may have nothing to do with the age of a mineral.* So that in spite of the 'popular' dating results looking sensible, the evidence clearly indicates that these dates are ***meaningless.*** [Italics and bold added]

Isochron Dating

Isochron techniques were developed when scientists recognized difficulties with the assumptions of radiometric dating: 1) constant rate of decay, 2) no loss or gain of parent or daughter isotopes (closed system), and 3) known amounts of daughter isotopes at the start of cooling. This method cannot test for assumption 1—that is, rate of decay. Scientists have attempted to address assumptions 2 and 3 but, according to RATE scientists, "It appears that all three of these dating essentials commonly fail at some level."[22]

Regarding assumption 1, the RATE project has found substantial physical evidence that decay rates were much higher in the past (see next section)—and that accelerated nuclear decay would have been associated with creation and a worldwide flood. Obviously, accelerated decay would completely invalidate the isochron technique. A higher decay rate in the past—during creation and to a lesser extent, during a worldwide flood—would yield an excessive amount of daughter isotopes (decayed end products) in a very short period of time, thus millions or billions of years in just a few days, invalidating the radioisotope technique.

Assumptions 2 and 3 are plagued with multiple flaws. Isochron dating is simply unable to differentiate between inherited daughter isotopes (daughter isotopes present at the start) and isotopes that decayed from existing parents present at the start. Evidence indicates that P/D ratios have nothing to do with age of the rock—they are just ratios. An explanation of the isochron technique and the primary reasons for discordance is provided in Appendix I.

In summary, scientists are unable to look at a rock specimen and tell you if the amount of daughter isotopes was the result of accelerated nuclear decay, or if there was mixing of parent or daughter isotopes from crustal rocks, or if there was inheritance of daughter isotopes at the start of cooling or at the time of creation. The bottom line is, radioisotope dates are ***meaningless***.

Accelerated Nuclear Decay

The primary assumption of radioisotope dating is a constant rate of radioactive decay based on uniformity theory—that is, a doctrine which maintains that natural processes have remained constant during earth's history. It assumes radioactive decay has never been interrupted by worldwide cataclysmic events.

There is no disagreement with the amount of daughter isotopes found in rock formations, but rather the constancy of the rate of decay. There is ample evidence to suggest that nuclear decay was accelerated on two occasions: a much larger portion during creation week of Genesis 1 and a smaller portion during the year of the Flood (Genesis 6-8). Such enormous cosmological and tectonic events would cause accelerated nuclear decay[23] which, in turn, would result in a profusion of daughter isotopes. This would erroneously provide results in the millions or billions of years—and would thereby invalidate radioisotope dating techniques.

Constant rate of decay does not take into account original creation when the universe was expanding, stretching out, and accelerating at inconceivable rates from an extremely hot, dense phase and does not take into account a worldwide flood which included restructuring of the earth's crust.

Is there evidence in favor of accelerated decay? According to RATE scientists, the answer is ***Yes.*** Evidence includes zircon crystals containing a large component of helium atoms and an abundance of radiohalos and fission tracks. These are microscopic areas of damage within rocks that result from an accelerated decay process. Other evidences include rapid magnetic field reversals and decay (discussed in chapter 9) and new findings of variations in isotope ratios in the meteorite, Allende.

The First Law of Thermodynamics states that matter/energy cannot be created or destroyed—this law confirms that *creation* is no longer occurring, but it also implies that *creation existed at sometime in the past!* The Second Law states that matter/energy is decaying or running down.

Some people may object to "accelerated nuclear decay" prior to the "curse" in Genesis 3:14–19. Nuclear processes would have operated in the beginning to naturally facilitate the creation process—for example, the sun and other stars were heated by nuclear fusion reactions.

Before sin entered the world, the world was good (Genesis 1:31); God "upheld" and *continuously restored* everything in the beginning (Colossians 1:17 and Hebrews 1:3). It was not until the "curse" that creation was replaced with the First Law and the Second Law (unrestrained universal decay). See chapter 2, Biblical Explanation, and chapter 9, Authority of the Bible.

Although there is strong physical evidence for accelerated decay, there remains the question of excessive heat and radiation from radioactive decay during creation and the flood. Cooling of the earth would have certainly occurred as the universe was expanding and "stretching out"—excessive heat and radiation would have quickly dissipated into "the fabric of space itself" similar to cooling of a refrigerator by a compressed gas.[24] Expansion of space would have also occurred to a limited extent during the Flood.

The "*stretching out*" of the heavens is confirmed in the Bible in Isaiah 42:5 (NIV), Jeremiah 10:12 (NIV), and Zechariah 12:1-3 (NIV). See chapter 9, Authority of the Bible. During the Flood, Noah and his family would have been protected from excessive heat and radiation by water covering the planet (see chapter 4).

Helium in Zircon Crystals

Zircons are tiny crystals that often occur in the mineral biotite that are observed in crystalline granite as dark mica flakes. Within these zircon crystals, uranium atoms decay to lead-206 through a series of intermediate decay steps (see table at beginning of this chapter). During the decay process, some of the emitted alpha particles combine with electrons to form helium.[25]

RATE scientists have been able to calculate the amount of nuclear decay and determine the total amount of helium produced in the zircon. Although zircons have been dated into the billions of years by modern radioisotope techniques, astonishingly, zircons were found to hold "58 percent of the total helium generated by past nuclear decay, and most of the 42 percent that left the zircon appears to have gone no farther than into the surrounding biotite mineral."[26] According to RATE scientists, if zircon crystals are billions of years old, the internal helium atoms in the crystal structure should have slipped away long ago.

Evidence does not support the confinement of helium within zircons but rather, diffusion laws require free helium atoms to dissipate or migrate upward.[27]

> Granite is one of the most abundant rocks on earth; massive volumes of granite comprise large portions of the continents and the interior or central core of most mountain ranges. Mineral components of granite include quartz, feldspar, biotite, and hornblende. See Appendix D.

> Helium atoms are produced during radioactive decay of uranium-238 to lead-206. During the decay process, each emitted alpha particle consists of two protons and two neutrons which is equivalent to the helium atom (1 uranium atom produces 8 alpha particles, or 8 helium atoms, during the decay process).

Although rocks contain 1.5 billion years of nuclear decay products (that is, daughter isotopes), measured helium retention and diffusion experiments by RATE scientists and others gives an estimated age for zircons of *6,000 years (plus or minus 2,000)*[28]—which agrees with literal biblical history. In other words, inert helium gas has been leaking for only 6,000 years, not for 1.5 billion years!

RATE scientists conclude that "helium diffusion measurements show that such high concentrations of helium simply cannot be sustained for more than a few thousand years. The only way we can reconcile the observed amount of uranium decay with the observed levels [high concentration] of helium retention is with one or more periods of accelerated nuclear decay in the earth's recent past." RATE helium diffusion experiments "*give strong evidence* for accelerated decay of the uranium atoms inside zircon crystals, and a young age for the earth."[29] [Italics added]

The presence of large quantities of helium in zircons is a major challenge to old age doctrine and, as Dr. D. R. Humphreys states, "these data are powerful *evidence*" for a young earth.[30]

> Helium exists as a gas; it is the second lightest element and is relatively small. Helium atoms are chemically inert—that is, they do not combine with other atoms to form molecules. As gas particles, these atoms are in constant motion and therefore, helium is "difficult to confine and behaves as a 'slippery' material."[31]

Radiohalos and Fission Tracks

Radiohalos are rings of color, or radiation damage, formed by the decay of radioactive atoms in crystalline granite. As previously stated, uranium-238 (unstable parent isotope) decays through a series of intermediate steps to lead-206 (daughter isotope). In the decay process, alpha particles are emitted by various isotopes and penetrate the surrounding minerals in all directions—in a concentric fashion as halos.

> The center of radiohalos is the parent U-238 atom, commonly found in the mineral biotite. In the decay process (U to Pb), eight isotopes emit alpha particles—these isotopes are U-238, U-234, Th-230, Ra-226, Rn-222, Po-218, Po-214, and Po-210. (See decay table at the beginning of this chapter.) Each isotope can be identified by the distance traveled from the center of the radiohalo.

According to RATE scientists, an *abundance* of uranium and polonium radiohalos in granite rock formations found worldwide is evidence for accelerated nuclear decay.[32] Also, an *abundance* of very short-lived polonium halos *without* uranium-238 atoms found worldwide implies accelerated decay and instantaneous creation of granites.[33] [Precambrian granite is associated with Creation, and Paleozoic-Cenozoic granite is associated with the last phase of the Flood; see chapter 5, Young Earth Interpretation, and chapter 7, Standard Geologic Column and Time Scale.]

Polonium is a product of uranium and thorium, and polonium halos should exist in conjunction with these parent halos. It appears that parentless polonium was not derived from uranium or thorium, and the existence of these parentless halos is evidence the earth was formed in a very short period of time.

Fission tracks result when heavy, unstable atoms of uranium-238 spontaneously split into atoms of palladium (a rare element). A single fragment can produce a significant track of damage in the crystal structure. Although spontaneous fission of uranium-238 rarely occurs today, "fission tracks are common in minerals and natural glass"[34] found in crystalline granite. According to RATE scientists, fission tracks indicate large-scale acceleration of nuclear decay.

> Natural glass forms when magma cools and solidifies too quickly to crystallize. Examples are volcanic glass, quartz, obsidian, and pumice.

> In today's world, only one U-238 atom undergoes a fission reaction for every two million U-238 atoms which undergo alpha decay, a very slow process.[35]

An abundance of radiohalos and fission tracks indicates a recent creation. According to RATE scientists, "This follows because the host rocks [granite] have not experienced serious heating since the track and halo formation. Just hundreds of degrees are sufficient to erase the crystal defects, yet they remain. It is difficult to imagine the rock formations remaining cool over vast ages of time with accompanying episodes of volcanic and tectonic activity. In the young earth view, the radiohalos and tracks remain relatively recent and freshly made."[36]

RATE scientists conclude that the presence and abundance of radiohalos and fission tracks provide *strong evidence of accelerated nuclear decay* within months of each other and a recent creation, not hundreds of millions of years as required by standard geologic time scale.

> "As with radiohalos, the fission tracks disappear when a rock sample is heated above an annealing temperature"[37]—that is, the temperature in which crystal defects in rocks (i.e., radiohalos or fission tracks) disappear after the rock cools and solidifies.

Magnetic Field Reversals and Decay

Rapid magnetic field reversals and decay provide evidence for a worldwide flood and young earth, and are consistent with the Second Law of Thermodynamics.

Such evidence has been thoroughly studied by Dr. Russell Humphreys and has been substantially verified by others as discussed in the first section of chapter 9.

Meteorites – Resetting the Clock

Evolutionists typically believe that the earth and solar system are about 3 to 5 billion years old—more precisely, 4.5 billion years based on radioisotope techniques such as uranium to lead and lead to lead. These methods assume a constant rate of decay—the first or primary assumption in radioisotope dating.

In a recent study, research scientists at Arizona State University (using Pb-Pb dating technique in specialized situations such as meteorites) detected variations in isotope ratios (ratios $^{238}U/^{235}U$ are long considered invariant in meteorites and a cornerstone of the Pb-Pb dating technique) from different inclusions of the Allende meteorite. Any deviation from this ratio would, at the very least, cast doubt on the Pb-Pb age of the sample. In the abstract, it was stated, "This range [between isotope ratios] implies substantial uncertainties in the ages that were previously determined by lead-lead dating of CAIs..."[38] In conclusion, the researcher expresses "uncertainty of the age of the solar system."

> The **Allende meteorite** is the largest carbonaceous chondrite ever found on earth. It fell to the earth in Mexico on February 8, 1969. This meteorite is often described as "the best-studied meteorite in history." It is notable for possessing abundant, large calcium-aluminium-rich inclusions (CAIs), which are among the oldest objects formed in the Solar System.[39]

Research data suggest the rate of decay is not constant. As Brian Thomas, M.S., states in an ICR article, "Without a stable rate of decay, the [radioisotope] clock is broken, no matter how precisely the relative amounts of isotopes are measured."[40] Such findings imply accelerated nuclear decay and fit well with findings of creation research scientists.

Also, the meteorite has an excessive amount of daughter isotopes (Pb-lead) and a deficiency of parent isotopes (U-uranium and Th-thorium).[41] When the universe was expanding from an extremely hot, dense phase when matter and energy were concentrated, daughter isotopes such as lead would certainly have been present—created or derived from a variety of nuclear processes. To assume otherwise is wishful thinking.

> Daughter isotopes found in rock formations on earth, as well as meteorites, were adopted or derived by accelerated decay. In either case, the presence of abundant daughter isotopes at the start would erroneously represent millions or billions of years of age. See Assumption 3 in Appendix I.

Summary

Carbon-14 dating assumes that carbon-14 and carbon-12 are thoroughly mixed in the atmosphere–they remain unchanged in their relationship (C-14/C-12 is in equilibrium) and the production/decay rates of carbon-14 is stable—based on uniformity theory. It is now admitted by most, if not all, investigators that such equilibrium does not exist. There is substantial evidence carbon-14 has been increasing over time—the rate of production today is about 18% to 25% higher than the rate of decay. This means less carbon-14 in the past—so animals or plants that died hundreds or thousands of years ago (with less C-14) date much older than their true age.

According to old age advocates, carbon-14 should not exist within supposedly ancient fossils, coal, or diamonds (sampled worldwide)—all supposedly hundreds of millions or billions of years old. This is important because carbon-14 is undetectable after 100,000 years. Because of less carbon-14 at the time of formation, as described above, the actual age of coal and diamonds is estimated by RATE scientists to be several thousand years.

Radioisotope dating of igneous rock is based on three assumptions for the method to work but such assumptions "commonly fail at some level."[42] Most notably, this technique assumes the decay rate remained constant (assumption 1) during the supposed 4.5 billion years of earth history. This assumption does not take into account original creation and restructuring of the earth's lithosphere during a worldwide flood. Such events would have accelerated nuclear decay resulting in a profusion of daughter isotopes. This, in turn, would erroneously produce results in the millions or billions of years in just a few days.

Accelerated decay would completely invalidate the radioisotope dating technique–yet secular scientists cling to this assumption although there is powerful evidence for accelerated decay: helium in zircon crystals, radiohalos, fission tracks, rapid magnetic field reversals, and findings of unstable decay in the meteorite, Allende.

Also, the technique assumes a closed geologic system (assumption 2)—that is, there was no uplift, folding, and thrusting of the earth's crust. This assumption is considered by most geologists to be fatally flawed due to evidence of immense deformation of the earth's crust (worldwide) which would have resulted in the gain or loss of daughter isotopes in the rock.

Scientists have attempted to account for inherited daughter isotopes at the start of cooling (assumption 3) by using the isochron technique, but the technique is unable to differentiate between inherited daughter isotopes (isotopes already present at the start) and non-inherited daughter isotopes (isotopes decayed from parents within the rock from the start). Also, this assumption relies on the validity of assumption 2. Known amounts of daughter isotopes present at the start is invalidated by discordant dates and, most notably, by consistently finding vast ages from recently solidified magma. This indicates that something is fundamentally wrong with the dating technique.

In the beginning daughter isotopes (Pb-206 and Ar-40) would have been present. Large concentrations of daughter isotopes were mixed into the upper mantle during Creation–and then later mixed into the earth's crust during the flood. This would erroneously represent millions or billions of years of age. The bottom line is this: "Ancient" and "recent" lava flows have the same chemistry derived from the earth's upper mantle. Although instruments can measure the amount of parent and daughter isotopes with precision, P/D ratios have *nothing to do with age of the rock or mineral*–they are just ratios.

Radioisotope dating suffers from multiple unprovable assumptions—so why do scientists cling to this dating technique, and contend as fact what they cannot prove? This question is addressed in the epilogue of this book—but such evidences have been consistently rejected or ignored by secular geologists because they imply *Biblical creation and a young earth.*

notes: Chapter 8

1. DeYoung, D. (May 2006). *Thousands ... Not Billions*. Master Books, Green Forest, AR. The seven RATE scientists included two geologists (Steven Austin and Andrew Snelling), a geophysicist (John Baumgardner), three physicists (Eugene Chaffin, Don DeYoung, D. Russell Humphreys), and a meteorologist (Larry Vardiman, chairman of RATE). Steven Boyd, a biblical Hebrew scholar, also joined the RATE effort. All team members hold an earned doctorate. The RATE project was sponsored and promoted by leading creation science organizations that included the Institute for Creation Research and the Creation Research Society.

2. Morris, J.D. (1994). *The Young Earth*. Green Forest, AR: Master Books, 51; and Morris, J.D. (2007), 48.

3. Carbon-14 dating (November 2008, last modified). In Encyclopedia of Creation Science. Creation Wikipedia. Retrieved May 2009, from http://creationwiki.org/Carbon-14_dating.

4. Ibid. See Whitelaw, R.L. (1993). A review and critique of pertinent creationist writing. 1950-1990. *Creation Research Society Quarterly*, 29 (4): 170-183; Cook, M.A. (1986). Nonequilibrium radiocarbon dating substantiated. *Proceedings of the First International Conference on Creationism*, vol. 2, Pittsburg, PA: Creation Science Fellowship, 59-68; and Stansfield, W.D. (1977). *Science of Evolution*. New York: Macmillan Publishing Co., 83.

5. Morris, J.D. (1994), op.cit., 65; and Morris, J.D. (2007), 64. Also see Catchpoole, D., Sarfati, J., and Wieland, C. (2008). *The Creation Answers Book*. (D. Batten, Ed.). Atlanta, GA: Creation Book Publishers, 69-71; DeYoung (2006), op, cit., 59; and Wieland, C. (April 1979). Carbon-14 dating – explained in everyday terms. *Creation*, 2 (2): 14-18.

6. Lee, Robert E. (1981; reprinted September 1982). Radiocarbon, ages in error. *Anthropological Journal of Canada*, 19 (3): 125. Also cited in Morris, 1994, 67.

7. DeYoung (2006), op. cit., 46-62. Also see Baumgardner, J. (2003). Measurable ^{14}C in fossilized organic materials, presented August 2003 at the *Fifth International Conference on Creationism*. Copyright 2003 by Creation Science Fellowship, Inc., Pittsburg, PA. Retrieved June 2008, from http://globalflood.org/papers/2003ICCc14.html.

8. Ibid., 175.

9. Ibid., 59. Also see Humphreys, D.R. (September-October 2014). Carbon-14 is now the creationist's friend. *Creation Matters* (A Publication of the Creation Research Society), 19 (5): 1-3.

10. Bosch, F. et al. (1996). Observations of bound-state β – Decay of fully ionized 187Re-187Os cosmochronometry. *Physical Review Letters*, 77 (26): 5190-5193; Cardone, F. et al. (2009). Piezonuclear decay of thorium. *Physics Letters A*, 373 (3795): 1956-1958; and Jenkins, J.H., Mundy, D.W., and Fischbach, E. (2010). Analysis of environmental influences in nuclear half-life measurements exhibiting time-dependent decay rates. *Nuclear Instruments and Methods in Physics Research Section A: Accelerators, Spectrometers, Detectors and Associated Equipment*, 620 (2-3): 332-342; as cited in Cupps, V.R. (October 2014). Clocks in rocks? Radioactive dating Part 1. *Acts & Facts*, 43 (10), Dallas, TX: Institute for Creation Research, 8-11. Also see Wile, J.L. (September 1, 2013). More evidences for variable radioactive half-lives. *Proslogion*. Retrieved September 2014, from http://blog.drwile.com/?p=2477.

"An excellent literary argument" for a young earth is presented by Steven Boyd: Boyd, S. (2005). Statistical Determination of Genre in Biblical Hebrew: Evidence for a Historical Reading of Genesis 1:1-2:3. In *Radioisotopes and the Age of the Earth, Volume II*, edited by L. Vardiman, A. Snelling, and E. Chaffin. Dallas, TX: Institute for Creation Research; as cited in Cupps, V.R. (October 2014). Clocks in rocks? Radioactive dating Part 1. *Acts & Facts*, 43 (10), Dallas, TX: Institute for Creation Research, 8-11.

11. Snelling, A.A. (March 2000). Geological conflict: Young radiocarbon date for ancient wood challenges fossil dating. *Creation*, 22 (2): 44-47.

12. Snelling, A.A. (1999). Dating dilemma: Fossil wood in ancient sandstone, creation, radioactive 'dating' failure. *Creation*, 21 (3): 39-41 and Snelling, A.A. (1997). Radioactive 'dating' in conflict. *Creation*, 20 (1): 24-27. Also cited in Wieland, C. (2001). *Stones and Bones*. Green Forest, AR: Master Books, 37.

13. Morris (1994), op. cit., 54. Also see Dalrymple, G.B. (1969). 40 Ar/36 Ar analyses of historical lava flows. *Earth and Planetary Letters*, 6: 47-55. Refer to Sunset Crater Volcano at http://www.nps.gov/sucr/.

14. Ibid. Also see McDougall, I. et al. (1969). Excess radiogenic argon in young subaerial basalts from Auckland volcanic field, New Zealand. *Geochemica et Cosmochemica Acta*, 33: 1485-1520.

15. DeYoung (2006), op. cit., 132-133; and Catchpoole, D., Sarfati, J., and Wieland, C. (2008). *The Creation Answers Book*. (D. Batten, Ed.). Atlanta, GA: Creation Book Publishers, 76. Also see Snelling, A. (1998). The cause of anomalous potassium-argon 'ages' for andesite flows at Mt. Ngauruhoe, New Zealand, and the implications for potassium-argon 'dating.' *Proceedings of the Fourth International Conference on Creationism*, 503-525.

16. Morris (1994), op. cit., 55. Also see Funkhouser, J.G., and Naughton, J.J. (July 1968). Radiogenic helium and argon in ultramafic inclusions from Hawaii. *Journal of Geophysical Research*, 73 (14): 4601-4607.

17. Ibid. Also see Nobel, C.S., and Naughton, J.J. (October 1968). Deep-ocean basalts: Inert gas content and uncertainties in age dating. *Science,* 162 (11): 265-266.

18. Morris, J.D. (2003). *Is the Big Bang Biblical?* Green Forest, AR: Master Books, 135; and Morris, J.D. (2000). *The Geology Book*. Green Forest, AR: Master Books, 52.

19. Austin, S.A. (Ed.). (1994). *Grand Canyon: Monument to Catastrophe*. Institute for Creation Research, El Cajon, CA, 120-126; and Morris (1994), op. cit. 57-60.

20. Morris (1994), op. cit., 59.

21. Snelling, Andrew A. (June 1981). The age of Australian uranium. *Creation*, 4 (2): 44-57. Permission received on August 20 and 27, 2009. Also cited in Morris, *The Young Earth*, 53; and Snelling, A.A. (1994). U-TH-PB dating: An example of false isochrons. Creation Science Foundation, presented July 1994 at the *Third International Conference on Creationism*, Pittsburgh, PA. Copyright 1994 by Creation Science Fellowship, Inc., Pittsburgh, PA. Retrieved April 2008, from http://www.icr.org/research/index/researchp_as_uthpbdating/.

In an article, "Radiometric Backflip, Bird Footprints Overturn 'Dating Certainty'" (2015, *Creation* magazine, vol. 37, no. 1, 26-28), Jonathan O'Brien writes, "...long-age radioisotope dating does not give real dates...[and they] are demonstrably unreliable and ever-changing. They aren't real." www.Creation.com.

22. DeYoung (2006), op. cit., 139.

23. Ibid., 142-154.

24. Ibid., 152-153; Vardiman, Larry (December 2007). RATE in review: Unresolved problems. *Acts & Facts*, 36 (12), Dallas, TX: Institute for Creation Research, 6; and Humphreys, D.R. (2000). *Accelerated nuclear decay: A viable hypothesis?* In Vardiman, L., Snelling, A.A., and Chaffin, E.F. (Eds.), *Radioisotopes and the age of the earth: A young-earth creationist research initiative.* El Cajon, CA: Institute for Creation Research and St. Joseph, MO: Creation Research Society, 333–379.

25. Ibid., 68-70.

26. Ibid., 71, 176.

27. Gentry, R.V., Glish, G.L., and McBay, E.H. (October 1982). Differential helium retention in zircons: Implications for nuclear waste contaminant. *Geophysical Research Letters*, 9 (10): 1129-1130; as cited in Humphreys, D.R. (June 2005). Evidence for a young world. Institute for Creation Research, ICR *Impact,* Article 384 (Item #10).

28. Humphreys, D.R. (December 2003). New RATE data support a young world. Institute for Creation Research, ICR *Impact,* Article 366*;* Humphreys, D.R. et al. (June 2004). Helium diffusion age of 6,000 years supports accelerated nuclear decay. *Creation Research Society Quarterly*, 41 (1): 1-16; and DeYoung (2006), op. cit., 76, 176. Also see Humphreys, D.R. (June 2005). Evidence for a young world. ICR *Impact,* Article 384 (Item #10), op. cit.

29. DeYoung (2006), op. cit., 78.

30. Humphreys (2003), op. cit., 1. Also see Humphreys, D.R. (June 2012). Zircons: God's tiny nuclear laboratories. *Creation Matters* (A Publication of the Creation Research Society), 17 (2): 1-2.

31. DeYoung (2006), op. cit., 68.

32. Ibid., 88-92, 95, 176-177. Also see Snelling, A.A. (March 2006). Radiohalos: Startling evidence of catastrophic geologic processes on a young earth. *Creation*, 28 (2):46-50; and Chaffin, E.F. (2013). Fifty years of physics: Some observations regarding radiohalos and magnetic fields. *Creation Research Society Quarterly*, 50 (2): 89-92.

33. Ibid., 93-97; 176-177; Morris (1994), op. cit., 62-64; Humphreys, D.R. (June 2005). Evidence for a young world. ICR *Impact,* Article 384 (Item #9), op. cit.; and Snelling, A.A. (2010). Polonium radiohalos: Still "A Very Tiny Mystery," Institute for Creation Research. Retrieved from http://www.icr.org/article/polonium-radiohalos-still-a-very-tiny-mystery/.

34. Ibid., 100.

35. Ibid.

36. Ibid., 106. Also see R.V., Radioactive halos. *Annual Review of Nuclear Science*, 23: 347-362; Gentry, R.V. (April 1974). Radiohalos in a radiochronological and cosmological perspective. *Science*, 184: 62-66, and Snelling, A.A., and Armitage, M.H. (2003). Radiohalos – A tale of three granite plutons. *Proceedings of the Fifth International Conference on Creationism*, 2: 243-267.

37. Ibid., 86, 100.

38. Brennecka, G.A. et al. (January 2010). $^{238}U/^{235}U$ variations in meteorites: Extant ^{247}Cm and implications for Pb-Pb dating. *Science*, 327 (5964): 449-451.

39. Allende Meteorite. Retrieved from http://www/census.gov/ and Population growth. Retrieved from http://en.wikipedia.org/wiki/Allende_meteorite.

40. Thomas, B. (January 2010). It's official: Radioactive isotope dating is fallible. Article 5161, Institute for Creation Science. Retrieved from http://www.icr.org/article/5161/.

41. Morris (1994), op. cit., 61.

42. DeYoung, op. cit., 139.

Chapter 9

Other Evidence for a Young Earth and Universe

"Thus the heavens and the earth were completed in all their vast array. By the seventh day God had finished the work He had been doing; so on the seventh day He rested from all His work. And God blessed the seventh day and made it holy, because on it He rested from all the work of creating that He had done."—Genesis 2:1–3, NIV

A Young Earth

There is substantial evidence for a young earth, which has been verified by empirical science using precise instrumentation and mathematics. Such evidences, commonly referred to as geochronometers, include rapid magnetic field reversals and decay, helium in zircon crystals, lack of erosion of the continents and not enough sediment in ocean basins, lack of salt in the sea, carbon-14 in diamonds and coal, dinosaur fossils with soft tissue and blood cells, lack of human fossils, and many relict geomorphic indicators of a worldwide flood as described in the Book of Genesis. Also, there is much evidence for a young universe.

There are over 200 scientific geochronometers[1]—that is, time clocks that indicate that the earth, solar system, and universe are young.

Rapid Magnetic Field Reversals and Decay

Secular geologists believe that a supercontinental landmass existed 1 to 2 billion years ago and this landmass began to slowly drift apart about 200 million years ago. There is much evidence of the splitting: physiographic fit of the continents and alignment of major fault zones (rifts).

Plate tectonics is a theory that describes continental movement in which the earth's crust, or lithosphere, is segregated into crustal plates floating on a semi-fluid asthenosphere. Oceanic plates diverge or separate as molten magma from the mantle rises between the plates and cools to form new basaltic rock. The youngest rock is nearest the rift—a divergent plate boundary, with older rocks further away. Also, the principle of superposition would apply; that is, older rocks would lie below younger rocks.

From a biblical and young earth perspective, the separation or splitting apart of this ancient landmass took place about 4,400 years ago during the Flood event described in Genesis 6-8. As described in chapter 4, the first phase of this event was rapid seafloor spreading or rifting.

The **earth's magnetic field** is a magnetic dipole with one pole near the North Pole and the other near the South Pole. The earth's magnetic field is created by **electric currents** in the earth's core. Iron within rocks aligns with the magnetic current when heated and this orientation is preserved when cooled.

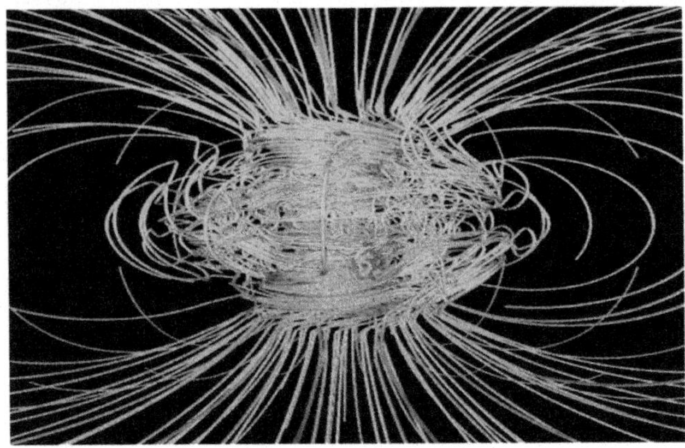

Computer simulation of the Earth's magnetic field in a normal period between reversals.

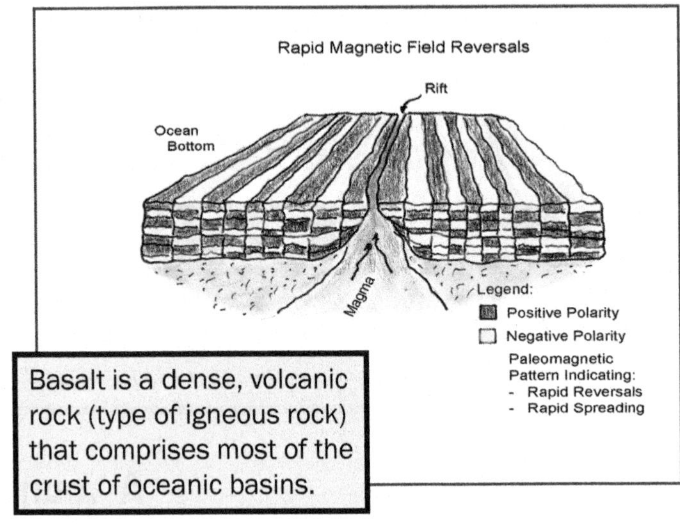

Basalt is a dense, volcanic rock (type of igneous rock) that comprises most of the crust of oceanic basins.

During this rapid spreading some rock minerals acquired magnetism from the earth's magnetic field, recording the direction of magnetic fields in geologic strata at the time of cooling. If seafloor spreading was uninterrupted over vast periods of time as maintained by uniformity doctrine, core samples would possess a very smooth, broad magnetic record of normal and reverse polarity. However, evidence indicates *chaotic and rapid field reversals*—not gradual sea floor spreading with slow reversals over millions of years.

As magma rose to the surface and cooled as dense basaltic rock, the spreading oceanic crust—both horizontal and vertical—was magnetized in reverse directions with no consistent patterns in a very short period of time.[2] According to Dr. J. D. Morris, this phenomenon can be "best understood as resulting from rapid reversals coupled with rapid [sea floor] spreading"[3]—evidence that is contrary to "slow-and-gradual" over vast periods of time. Magnetic field reversals have been investigated by independent researchers and found to be extremely rapid.[4] Such evidence supports a worldwide flood and is contrary to uniformity theory.

Another phenomenon indicating a young earth is deterioration and energy loss in the earth's magnetic currents. Based on measurements of the strength of the magnetic field, it has been shown that the earth's magnetic currents are *decaying*—which is consistent with the Second Law of Thermodynamics. Scientists have calculated the total energy stored in the earth's magnetic field (in trillions of kilowatt-hours) is decreasing with a half-life of 1,465 years (± 165 years).[5] Evolutionary doctrine has a difficult time explaining how the earth could have maintained this decay rate for hundreds of millions or billions of years. In fact, life would have been unsustainable 20,000 years ago given the present decay rate.[6]

A **geomagnetic reversal** is a change in the orientation of Earth's magnetic field—that is, the positions of magnetic north and magnetic south become interchanged.

Dr. Russell Humphreys, a physicist at the Sandia National Laboratories in New Mexico, developed the only model that accommodates all the known physical data. Contrary to old age doctrine, this model explains rapid reversals and energy loss during the Flood, and a steady decay of the magnetic field in the post-flood period.[7] This model replicates historic and present paleomagnetic field data and is "based on sound physics, and its predictions have been substantially verified."[8] The evidence is overwhelmingly on the side of a young earth and is consistent with the Second Law.

Paleomagnetism is the study of the remnant magnetization in rocks; the earth's magnetic field as it existed in the past. Paleomagnetic dating involves collecting rock samples to determine how the earth's magnetic poles were oriented when the rocks were formed.

Thousands of reversely polarized ancient rock specimens, including man-made artifacts such as bricks and pottery with known dates, have been studied throughout the world.

Continental Erosion and Ocean Sediments

Uniformitarians maintain the belief the earth is 3 to 5 billion years old—and that a super-landmass (Pangaea) began to drift apart 200 million years ago.[9] In the old earth model, continents were *slowly* crumpled and folded, and mountain chains and plateaus were lifted thousands of feet over tens of millions of years. Uniformitarians believe that existing physical processes (i.e., water, wind, rain, freezing and thawing, solar heating, chemical weathering, tidal action, etc.) account for today's geologic land features—essentially, hydrogeologic processes have remained relatively constant.

Denudation is the removal of rock and sediment by erosion and weathering—this process leads to a reduction of elevation and relief in landforms and landscapes.

Let's take a brief look at continental erosion on a global scale. In our present world, the sediment loading in rivers and streams has been measured at 27.5 billion tons per year and the volume of land above sea level has been measured at 383 million billion tons.[10] Assuming uniformitarian processes, simple calculation indicates the continents would be denuded to sea level in about 14 million years (383 million billion tons divided by 27.5 billion tons per year). But secular geologists believe the original supercontinent and oceans existed 1 to 2 billion years ago.

One might conclude that the continents were slowly uplifted while eroding to their present level over hundreds of millions of years, but this logic is contrary to geomorphic evidence (see chapter 5). Land features and rock formations thought to be on the earth hundreds of millions of years ago, as described by secular geologists, *are still on the surface of the earth today!*

An increase in population and activities of man in the last few thousand years would have had a minor overall effect on continental erosion.[11]

But more convincing evidence for a young earth is the small amount of sediment found in today's ocean basins. Research scientists know the sediment load entering the ocean is about 27.5 billion tons per year, and that ocean bottom sediments average approximately 300 million billion tons.[12] Again, a simple calculation indicates that it would have taken about 11 million years (300 million billion tons divided by 27.5 billion tons per year) to fill the *deep* ocean basins—or about 15 million years (417 million

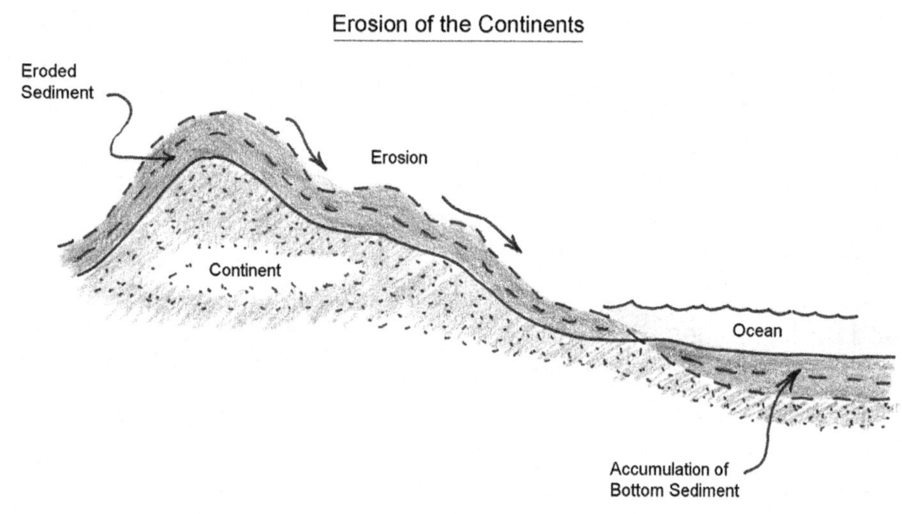

billion tons divided by 27.5 billion tons per year) to fill the *deep and shallow* ocean basins. If ocean basins are 1 to 2 billion years old as commonly believed by uniformitarians, they would be *full of sediment*. This is certainly not the case.

Measurements of continental erosion and ocean bottom buildup simply do not correspond with uniformity doctrine and an old earth, but conform to a catastrophic worldwide flood event as described in the Bible. At the time of the flood, deposition and erosion would have occurred on a massive scale, resulting in most of the landforms and sediment deposits we see today.

This does not suggest the earth is 15 million years old but rather, that *uniformity processes do not provide legitimate answers to questions concerning our physical world*. At the time of the flood (as discussed in chapters 4 and 10), deposition and erosion would have occurred on a massive scale and much of the ocean's sediment would have been added during this period.

One could argue that the ocean crust is being continually subducted beneath continental plates (see phase diagrams in chapter 4), but studies indicate that subduction rates account for only 15 percent of the rates of continental erosion and accumulation within ocean basins. This would result in a minor change of the calculated age.[13]

Salt Concentration in Seawater

Evolutionists have typically assumed that life evolved in a brackish sea about 3 to 5 billion years ago. Seawater comprises salts such as sodium, chloride, magnesium, sulfate, calcium, potassium, and other trace salts. Research scientists know that salt is entering the sea (by means of erosion plus a few other input sources) much faster than it is escaping (by means of sea spray plus a few other output sources). See input and output sources in text box. Assuming the oceans are millions, if not billions of years old, one might reasonably conclude that oceans would be extremely salty by now, far greater than the actual concentration of 35 parts per thousand.

This argument was formalized by Drs. Steven A. Austin and D. Russell Humphreys by identifying rates of addition and removal of salt during recent times and throughout the past.[14] Using modern input and output values and assuming that hydrogeologic processes remained relatively constant according to uniformity doctrine, the ocean's salt content would have accumulated in 32 million years. Using extreme minimum input values and extreme maximum output values, and assuming uniformity throughout the ages, the maximum age is 62 million years. Again, this does not suggest the earth is 62 million years old—however, the ocean basins could not possibly be 1 to 2 billion years old as claimed by evolutionists and old earth theorists.

Seawater in the world's oceans has a salinity of 3.5 percent, or 35 parts per thousand (average). This means that every 1 kg of seawater has approximately 35 grams of dissolved salts. Dissolved salts include the ions of chloride (Cl^-), sodium (Na^+), magnesium (Mg^{2+}), sulfate (SO_4^{2-}), calcium (Ca^{2+}), potassium (K^+), and other trace salts.

This study assumes there was no salt in the sea at the time of creation. If salt was already present, which is much more likely, then the earth would be much younger. These numbers assume that hydrogeologic processes have remained constant over time; that is, there have been no large, unaccounted for additions or deletions of salt. *But at the time of a worldwide flood, erosion would have occurred on a massive scale and much of the ocean's present salt concentration would have been added during this short period of time.* The data indicates that the *earth is far too young to be compatible with old earth scenario as maintained by secular geologists and evolutionists.*

Input sources: rivers, ocean-floor sediments, glacial ice and sediment, atmospheric dust, volcanic dust and aerosols, coastal erosion, groundwater seepage, seafloor hydrothermal vents, etc.

Output sources: sea spray, ion exchange, burial of pore water, deposition as halite, etc., which account for only 27% of sodium input.[15] The rest of the salt simply accumulates in the ocean.

Other Geochronometers

Although the following examples of a young earth were discussed in chapters 7 and 8, these geochronometers bear repeating in this chapter.

Humans and Dinosaurs

According to evolutionists, man became "fully" human (*Homo sapiens*) about 250,000 to 185,000 years ago.[16] Supposing evolution was true and humans lived on earth this long ago without a worldwide flood, many billions, if not trillions, of humans would have lived and died. One could reasonably assume that "human" family groups would have buried or attempted to preserve remains or artifacts. But fossil evidence indicates that only 0.0125 percent of the total number of fossils represents vertebrates (not necessarily humans; see chapter 7) and only one percent of these consists of a single bone![17] So, where are the human fossils? A worldwide flood would certainly provide a reasonable answer.

Contrary to old earth doctrine, there is considerable empirical evidence that dinosaurs roamed the earth within the past 6,000 years, not millions of years as maintained by evolutionists and old earth advocates. Soft tissue has been recovered from dinosaur fossils that supposedly walked the earth 68 to 248 million years ago during the Triassic, Jurassic, and Cretaceous periods.[18,19,20] If these tissues and bones were really many millions of years old, they would have completely disintegrated. Such findings have confounded those who believe that dinosaurs lived many millions of years ago. Refer to chapter 7.

Additionally, great fossil graveyards found worldwide with dinosaur fossils and other creatures violently mixed together are indicative of rapid burial and a worldwide flood. Also, dinosaur tracks in mudflats (see photos, chapter 7) are relatively fresh with little or no signs of erosion—certainly incompatible with many millions of years of age.

Carbon-14 in Coal and Diamonds

The half-life of carbon-14 is 5,730 years—and anything over 100,000 years is undetectable by modern measurement techniques. With this half-life, carbon-14 should not exist within supposedly ancient fossils and other materials. Yet, carbon-14 has been detected in "ancient" fossils and petrified wood from samples from around the world—and in coal formations supposedly 300 million years old, and in diamonds presumed to be many millions, if not billions of years old.[21] The presence of carbon-14 in ancient fossils, rocks, coal, and diamond samples is very strong evidence that the *earth is just a few thousand years old*, not millions or billions of years as uniformitarians might have you believe. See chapter 8, Carbon-14 in Ancient Fossils and Rocks.

Many other evidences and laws of science completely contradict evolution. They include the First and Second Laws of Thermodynamics, Law of Biogenesis, no transitional types in the fossil record, and the fact that no one has ever observed evolution or explained the cause and origin of matter and energy, and the “breath of life.”

A Young Universe

The universe is defined as everything that physically exists: the entirety of space and time with all forms of matter and energy. As far as we know, the universe comprises at least 100 billion galaxies which are gravitational bound systems containing billions of multiple star systems.

The Milky Way, estimated to be 100,000 light years across, is a "barred" (central bar shaped structure) spiral galaxy that is the home galaxy of our Solar System that includes the Earth and Sun. The distance from earth to the outer reaches of the universe in all directions (assuming the earth is the center of the universe) has been estimated at 10 to 20 billion light years, although some scientists believe it may be more, perhaps 156 billion light years.[22] These numbers are just approximations—no one really knows the extent of the universe.

One of the primary differences between creationists and evolutionists is the issue of time—that is, the age of the universe and the earth. Evolutionists and secular scientists claim that the universe is very old and came into existence via the Big Bang about 10 to 20 billion years ago (based on the distance of 10 to 20 billion light years). They also believe that the earth is approximately 3 to 5 billion years old.[23] Assuming the Big Bang theory is correct and the speed of light is a mathematical constant (immutable; cannot be modified), secular scientists believe the radius of the universe in light years is synonymous with its age in years.

The following are several "naturalistic" explanations as to why the universe may appear to be 10 to 20 or more billion light years across and still be only 6,000 years old!

> A **galaxy** is a massive, gravitationally bound system consisting of stars, interstellar gas and dust, and dark matter orbiting around a common center of mass. Our **Solar System**, located in the galaxy known as the **Milky Way**, consists of the Sun and celestial objects that include eight planets: Mercury, Venus, Earth, Mars, Jupiter, Saturn, Uranus, Neptune, and two dwarf planets, Pluto and Eris, their 166 known moons, five other dwarf planets, and billions of small bodies. See Appendix A, Glossary.

> The estimation of distance and age is based on theoretical cosmology models and astronomical observations. It should be emphasized not to confuse distance as measured in light years as equal to time. It is only a *distance based on known speed of light* of 186,282.397 miles per second or 670,616,629.4 miles per hour in a vacuum, or 5,865,696,000,000 miles per year (5.866 trillion miles per year). So, 20 billion light years X 5.866 trillion miles/year = 117 billion trillion miles.

Milky Way Galaxy

Milky Way Galaxy with two major stellar arms wrapping from the ends of a central bar of stars

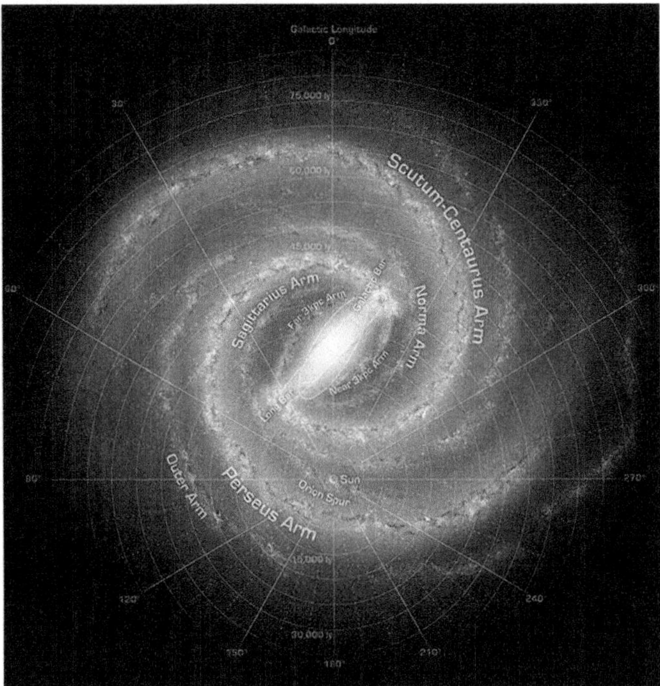

Spiral galaxy

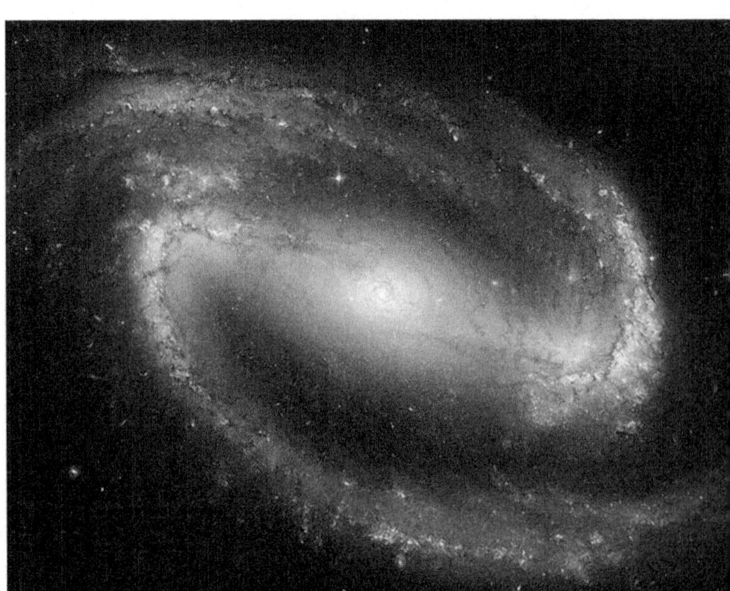

Barred Spiral galaxy

Superluminal Speed and Decay of Speed of Light

The most obvious solution for a young universe is demonstrating that light can be accelerated faster than the speed of light. In astronomy, superluminal motion is the apparently faster-than-light motion seen in some radio galaxies, quasars, and microquasars.[24] All of these sources of faster-than-light are thought to be associated with black holes (see section, Other Interesting Phenomena).

Superluminal motion is being studied by the NEC Research Institute in Princeton, New Jersey, and other research institutes and laboratories throughout the world. Astrophysicists have supposedly demonstrated that light (group of light waves with no mass) can be accelerated and even slowed down.[25] Although there are claims of superluminal motion, published research has not been accepted by the scientific community.

Another explanation is tied to the Second Law of Thermodynamics which stipulates that matter/energy degrade over time—energy in the universe is decaying or running down. Light would be no exception—that is, the speed of light we can measure today would certainly have diminished over time. The decay of the speed of light (c-decay) refers to a proposition by some astrophysicists that the speed of light was millions of times faster immediately following the creation of the universe approximately 6,000 years ago. By carefully selecting the decay rate, one can construct a universe that is billions of light years across, yet only a few thousand years old.

A **quasar** is an extremely powerful and distant active galatic nucleus (compact region at the center of a galaxy which has a much higher than normal luminosity). A **microquasar** is a smaller cousin of a quasar. See Appendix A, Glossary.

A **black hole** is a theoretical region of space in which the gravitational field is so powerful that nothing, not even electromagnetic radiation or visible light, can escape its pull. See Appendix A, Glossary.

The concept of *c*-decay was first proposed by Barry Setterfield in a 1981 article for the Australian creationist magazine, *Creation*.[26] He selected a number of historical measurements of the speed of light, and demonstrated that it has declined in a smooth curve over a span of 325 years. This implies the speed of light could have been nearly infinite at the time of creation, thus allowing light to traverse a 20 billion light-year distance in just a few thousand years.

There has been much debate among astrophysicists and physical scientists in creationist circles concerning the evidence and whether the statistical data actually support *c*-decay. Currently, "general support by the creationist community... is not warranted by the data upon which the hypothesis rests."[27] Creation Ministries International (CMI), The Institute for Creation Research (ICR), and Answers in Genesis (AiG) believe that this proposal has not been satisfactorily addressed in the research. Most creation research organizations currently prefer the hypothesis of gravitational time dilation and distant starlight.

Gravitational Time Dilation and Distant Starlight

Let's take a look at the elementary equation of speed, distance, and time.

Speed = distance/time (s = d/t)
Time = distance/speed (t = d/s)

Assuming the distances across the universe and Milky Way are correct and the speed of light *(c)* is constant, then the only variable left untouched is *time.*

General relativity is the geometric theory of gravitation in physics—it essentially states that *time is not a constant*. Two things are believed to distort time in relativity theory—one is speed (as speed increases, time decreases) and the other is gravity (less gravitational pull, time increases or passes faster).

General relativity is the geometric theory of gravitation in modern physics, published by Albert Einstein in 1916. It unifies special relativity and Newton's law of universal gravitation and describes gravity as a property of the geometry of space and time, or spacetime. Such "predictions have been confirmed in all observations and experiments to date."[28]

Albert Einstein's theory has been proclaiming to the world over this past century that gravity distorts time. For example, it has been demonstrated that precision clocks at the top of tall buildings run faster (where gravity is less, time increases) as compared with precision clocks at the bottom (where gravity is greater, time decreases).

Mathematics and physics associated with general relativity are compatible with the concept of gravitational time dilation.

Einstein first described the term gravitational time dilation in 1907. It is a phenomenon that explains the effect of time in regions of space with different gravitational potential; that is, the greater the gravitational pull, the slower time passes. This phenomenon was investigated by Dr. Russell Humphreys and described in his book, *Starlight and Time*.[29] His cosmological model describes the earth as several thousands of years old, but the outer edge of an expanding and rotating three-dimensional universe as billions of years old with varying ages in between.

At the outer fringes of the universe where there is less gravitational pull—the faster time passes. Thus, billions of years would be available for light to reach the earth while less than an ordinary day is passing on earth. In other words, light from the extremities of the universe has the potential of reaching earth in a relatively short period of time (from earth's perspective). This means that the earth is relatively young, not billions of years old.

High-precision test of general relativity by the Cassini space probe (artist's impression): radio signals sent between the earth and the probe (green wave) are delayed by the warping of space and time (blue lines) due to the Sun's mass.

Other Interesting Phenomena

Einstein said that light is bent by the force of gravity as it travels by "heavy objects," meaning stars and galaxies. Today we know that this is correct. NASA has photos taken by the Hubble astronomy satellite that clearly show that light is bent by gravity as it travels through space. When the concentration of matter is very dense, gravitational distortion can be so immense that even light cannot escape—and such concentrations of matter are typically termed a "black hole."

General relativity equations show that the enormous gravitational pull will bend light back on itself and time will literally stand still. If space is curved as predicted by Einstein (now we are in the realm of quantum physics, spacetime, parallel universes, and wormholes), then light could travel 10 to 20 billion light years in just a few thousand years—or one might say that the other side of the universe is just a stone's throw away.

In his book, *Journey Into Eternity*, Dr. Grant Jeffrey points out that the relative distance between the nucleus of the atom, comprising protons and neutrons, and the smaller electrons in their orbits is astonishing. "Particle physics reveals that the distance from the nucleus to the electron circling it is equivalent to the vast distance between our sun and our earth of approximately 93 million miles. In other words, the atoms that make up everything apparently 'solid' in our universe are in fact composed of 99.999999999% empty space."[30]

Atoms also comprise "elementary particles" called quarks and leptons that are infinitely small with no dimension or mass. So then why do most solids that comprise virtually empty space feel "rock solid"? It is because the energy that ties the atom particles together is immensely powerful. In Colossians 1:16-17 (NAS), "*...all things were created by him and for him. He [Christ] is before all things, and in him all things hold together.*"

This is pointed out because matter, energy, time, distance, and speed are not exactly what they appear. As Dr. Jeffrey explains, "When we take into consideration that our universe, our earth, and our bodies are composed primarily of empty space and tiny particles of energy spinning at the speed of light, it is easier to understand how the nonmaterial human spirit could survive the physical dissolution of the human body."

Also, let's not forget the Anthropic Principle (discussed in chapter 2) that demonstrates there is substantial evidence the universe is fine-tuned for our existence. The bottom line is, if any of the constants of physics and chemistry were just slightly different, life would not be possible.

A **wormhole** is a theoretical connection between two parallel universes. One end of this type of wormhole is a black hole, pulling in matter, and the other is a white hole, emitting matter. A **black hole** acts as a vacuum, sucking up any matter that crosses the spacetime boundary (event horizon) whereas a **white hole** acts as a source that ejects matter from its spacetime boundary into another dimension outside the parent universe. See Glossary.

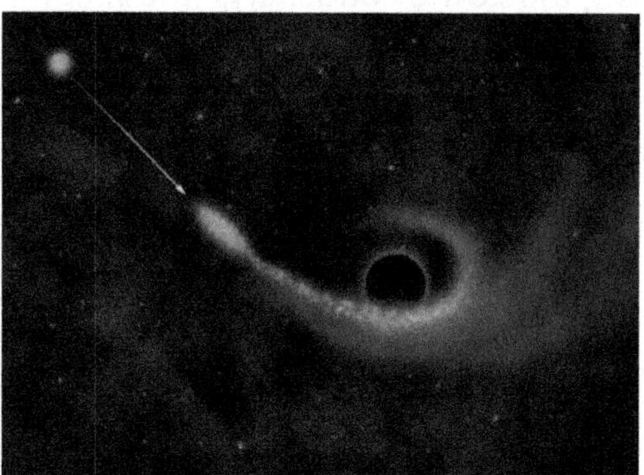

Artist's conception of a super massive black hole tearing apart a star

General relativity is an essential tool in modern astrophysics. It provides the basis for the current understanding of black holes—regions of space where gravitational attraction is so strong that not even light can escape.

The Big Bang and Its Problems

The Big Bang theory is the idea that the universe expanded from a primordial hot, dense soup at some time in the past and continues to expand to this day at the speed of light. Although evolutionists recognize the perfect order that exists in the universe, they claim that such perfection came from a miniscule starting point of chaos and somehow evolved to higher levels of complexity by random chance.

The single greatest problem for the Big Bang theory is the *Cause* of the universe and the *Origin* of matter and energy within space—how our cosmic continuum of space, matter, and time suddenly came into existence. The fact is, astrophysicists and other scientists don't have the faintest idea. They have never been able to explain the origin of the original infinite mass and energy and *Why* there is a universe in the first place.

How did the universe progress from nothing without a causal agent (that is, a Divine Creator), thus denying the First and Second Laws of Thermodynamics, and the Law of Causality (cause and effect; every effect has a cause), and how did life and human consciousness come into existence? Regarding the Law of Causality, "the following reasoning stands up to scrutiny:

- Everything *which has a beginning* has a cause
- The universe has a beginning
- Therefore the universe has a cause."[31]

As authors explain in *The Creation Answers Book*, "It is important to stress the words 'which has a beginning.' The universe requires a cause because it had a beginning... God, unlike the universe, had no beginning, so does not need a cause. In addition, Einstein's general relativity [theory], which has much experimental support, shows that time is linked to matter and space. So, time itself would have begun along with matter and space at the beginning of the universe. Since God, by definition, is the creator of the whole universe, He is the creator of time. Therefore, He is not limited by the time dimension He created, so He has no beginning in time. Therefore, He does not have, or need to have, a cause."[32]

> God is outside and inside the universe, and created matter and energy; time is linked to matter and energy; so therefore He is not limited by time.

God is the Creator—the power within and outside the universe. The Bible teaches there is only one true God who is completely loving, just, and holy. *"I am the Alpha and the Omega, the First and the Last, the Beginning and the End."* (Revelation 22:13, NIV). As God told Moses, "*I AM WHO I AM*" (Exodus 3:14, NIV).

> Genesis 21:33 (NIV), *"the name of the Lord, the Eternal God."* Habakkuk 1:12, 3:6 (NIV), *"His ways are eternal."* 1 Chronicles 16:36, 29:10; Nehemiah 9:5; Psalm 103:17, 106:48 (NIV), *"everlasing to everlasting."* Hebrews 1:8 (NIV), *"O God will last for ever and ever."* Revelation 1:8, 21:6, 22:13 (NIV), *"I am the Alpha and the Omega, the Beginning and the End."* Revelation 22:5 (NIV), *"And they will reign for ever and ever."*

The First Law implies that original creation is no longer occurring and the Second Law implies that the universe was created with plenty of usable energy and is now running down like a windup clock (see chapter 2). As stated in *The Creation Answers Book*, "...the universe cannot have existed forever, otherwise it would *already* have exhausted all usable energy and reached what is known as 'heat death.' For example, all radioactive atoms would have decayed, every part of the universe would be the same temperature, and no further work would be possible. So the best solution is that the universe must have been created with a lot of usable energy, and is now running down."[33]

Evolutionists have to "explain how nothing became everything" and therefore "they must invoke their own supernatural 'first act'."[34] They make-believe the universe created itself out of nothing—a belief that is in clear violation of the First and Second Laws of Thermodynamics. What we see are shrinking stars and decay; not dust particles evolving into higher levels of complexity by some "unknown process." The bottom line is that evolutionists will never find satisfactory answers because they are looking for a "naturalistic" solution that does not exist. The tools of science cannot find God—see Moral Law in chapter 3. They refuse to consider a supernatural source that would provide satisfactory answers to Cause, Origin, and Why there is a universe.

Evidence of a Young Universe

The following are just a few geochronometers that conflict with the idea that the universe is 10 to 20 billion years old. These time restraints are far less than the 3 to 5 billion years required by evolutionary theorists.

- The observed rotation speeds of stars, interstellar gas and dust, and dark matter of the Milky Way (as well as other galaxies) are too fast for an old universe.[35] This "winding-up dilemma" remains a puzzle to evolutionists although they believe that our universe is about 10 to 20 billion years old.
- Observation of gas and dust remnants from supernovas (violently exploding stars) is consistent with 7,000 years worth of supernovas.[36]
- Old earth theorists believe that comets should be the same age as the Solar System, about 5 billion years, yet many comets have an age of less than 10,000 years because of their "melting rate."[37] There is no satisfactory explanation to explain the large number of comets.
- Other evidence includes the shrinking sun (0.1% per century, the sun would have been much too large and hot billions of years ago) and the receding moon (6 inches per year; too close to orbit earth millions of years ago); and the existence of uranium-236 (half life of 23.44 million years) and thorium-230 (75,380 years half life) on the moon.
- Blue stars are strong confirmation of a young universe consistent with 6,000 years.[38] They are more massive, brighter and burn their hydrogen fuel more quickly, and cannot last even a million years. Secular astronomers assume blue stars formed recently—school textbooks even have photos of "stellar nurseries" but the fact is, "star formation has never been observed." If the universe is 15 billion years old, there shouldn't be any blue stars, but they are found throughout the universe.

As described in chapter 2, the Law of Increasing Entropy (implicit in the Second Law) states that usable energy in the universe available for work is decaying or running down. Ultimately, when all the energy in the universe has been degraded, the entire universe will be cold and without order (heat death).

Because the universe is winding down, logically it would mean the universe was created with plenty of usable energy at the beginning. The question one might logically ask: "Who wound up the clock?" Applied to the whole universe, the First and Second Laws are a fundamental contradiction to the "chaos to cosmos, all by itself" doctrine of evolution.

Keeping Everything in Perspective

It is important to keep in perspective the immense complexity of space, matter, time, distance, and speed, and more exotic concepts such as spacetime, gravitational time dilation, frames of reference, geometry of space, string theory and other dimensions, black holes, white holes, and wormholes. What does all this mean? It simply means that most of this is hypothetical and far beyond the understanding of even the very brightest astrophysicist.

An astrophysicist at the University of Rochester once said: "General relativity consists of ten interwoven equations. Along with these ten equations come hundreds of others that must also be solved. Each of these mathematical expressions can be hideously complicated, with term after term appearing in forms that provide no simple means of solution. A single equation can fill many pages...the equations of GR [General Relativity] are intractably woven together. They snake through each other in a deeply nonlinear way, forming a kind of mathematical Gordian knot."[39]

As Dr. Henry Morris wrote, "Perhaps some of the inhabitants of Bubbleland can deal with such complexity, but I simply have to wonder and doubt ...[yet scientists, as a whole, often] off-handedly dismiss the strong and simple evidences for creation and the saving gospel of Christ does not generate trust, not to mention their naïve commitment to the pseudo-evidence for both cosmic evolution and organic evolution, including human evolution."[40]

Authority of the Bible

We should always keep in mind that all theories of men are subject to revision or abandonment in the light of future discoveries. The authority of the Bible should never be compromised by mankind's "scientific" theories and hypotheses. Consider the words of Charles Haddon Spurgeon in 1877:[41]

> What is science? The method by which man tries to conceal his ignorance. It should not be so, but so it is. You are not to be dogmatical in theology, my brethren, it is wicked; but for scientific men it is the correct thing. You are never to assert anything very strongly; but scientists may boldly assert what they cannot prove, and may demand a faith far more credulous than any we possess. Forsooth, you and I are to take our Bibles and shape and mould our belief according to the ever-shifting teachings of so-called scientific men. What folly is this! Why, the march of science, falsely so called, through the world may be traced by exploded fallacies and abandoned theories. Former explorers once adored are now ridiculed; the continual wreckings of false hypotheses is a matter of universal notoriety. You may tell where the learned have encamped by the *debris* left behind of suppositions and theories as plentiful as broken bottles.

God stated in Genesis 1:1-2 (NAS), "*In the beginning God created the heavens and the earth. And the earth was formless and void, and darkness was over the surface of the deep;*

and the Spirit of God was moving over the surface of the waters." This appears to be an instantaneous act of creation including all the light at the same time.

Also, the expansion or "*stretching out*" of the universe is supported by Biblical scripture. In Isaiah 42:5 (NIV), it states "*This is what God the Lord says, he who created the heavens and* ***stretched them out.***" Jeremiah 10:12 (NIV) states, "*But God made the earth by his power; he founded the world by his wisdom and* ***stretched out the heavens*** *by his understanding.*" [Bold added] And Zechariah also emphasized the "*stretching out*" of the heavens in the end-of-days prophecy, prophecy we are witnessing today.

> This is an effortless stretching out of one's hand like the stretching out of a "paper daisy chain." The matter-energy did not come from within the universe but *flowed into* the universe.

> *This is the word of the Lord concerning Israel. The Lord, who* ***stretches out the heavens****, who lays the foundation of the earth, and who forms the spirit of man within him declares: I am going to make Jerusalem a cup that sends all the surrounding peoples reeling. Judah will be besieged as well as Jerusalem. On that day, when all the nations of the earth are gathered against her, I will make Jerusalem an immovable rock for all the nations. All who try to move it will injure themselves* (Zechariah 12:1-3, NIV). [Bold added]

Other related Biblical scriptures include Job 9:8; Psalm 104:2; Isaiah 40:22, 40:28, 44:24, 45:12, 45:18, 48:13, 51:13; and Jeremiah 51:15.

As confirmed by the Anthropic Principle, the universe is not a random or chance event. According to Isaiah 45:18 (NIV), "*For this is what the Lord says—he who created the heavens, he is God; he who fashioned and made the earth, he founded it; he did not create it to be empty, but* ***formed it to be inhabited***—*he says, I am the Lord, and there is no other.*" [Bold added]

Summary

Evidences of a global flood and young earth are substantial and have been verified using precise instrumentation and mathematics. They include rapid magnetic field reversals and decay, lack of continental erosion and ocean sediments, and lack of salt in the sea. Other evidences found throughout the world (as discussed in previous chapters) include:

- Massive mountain ranges with no visual signs of denudation (erosion) to offset supposed slow and gradual uplift over many millions of years (chapter 5)
- Massive batholiths and monoliths with great peneplains, mountain overthrust (low angle reverse thrust), uplift and tilting of formations, overfit valleys and raised flood terraces, massive alluvial plains, and dry lake basins and salt lakes consistent with a worldwide flood (chapter 5)
- Tight folds of igneous and sedimentary rock strata when layers were soft and pliable, and presence of polystrate fossils and clastic dikes (chapter 6)
- Sharp, undisturbed bedding planes commonly found between sediment formations and layers supposedly millions of years old (chapter 6)
- No transitional fossils or living forms and extreme rarity of human fossils (chapter 7)
- Dinosaur fossils with soft tissue and blood cells, and fossil graveyards with well-preserved fossils violently mixed together in all countries of the world (chapter 7)
- Carbon-14 found in coal and diamonds supposedly millions, if not billions of years old (chapter 8)
- Evidences of accelerated nuclear decay consistent with creation and a worldwide flood (chapter 8)
- Helium in zircon crystals (chapter 8)

Astronomical solutions in support of a young universe include superluminal speed, the decay of speed of light consistent with the Second Law, but most notably, gravitational time dilation as described by Albert Einstein in 1907. Evidences for a young universe include rapid rotation of stars, large number of comets, our shrinking sun, and young blue stars found throughout the universe.

The single greatest problem for the Big Bang theory is the *cause* of the universe and the *origin* of matter and energy within space—how our cosmic continuum of space, matter, and time suddenly came into existence. Secular scientists have never been able to explain the origin of the original infinite mass and energy and *why* there is a universe in the first place. Evolutionists cannot explain away the First and Second Laws of Thermodynamics which completely contradict "chaos to cosmos, all by itself" doctrine.

The theories of men are subject to revision or abandonment in the light of future discoveries, but the truth and certainty of the Bible is everlasting and should never be compromised by mankind's "scientific" theories and hypotheses.

Evolution is contrary to the First and Second Laws of Thermodynamics whereas creation is consistent with such laws (see chapter 2).

There are no known biological processes for evolution to higher levels of organization and complexity—gene mutations are overwhelmingly destructive and none are uphill (that is, unequivocally beneficial) in the sense of adding new genetic information to the gene pool (see chapter 3).

Enormous limestone formations and huge coal and oil deposits are indicative of a worldwide flood—*not billions of years*. Such processes can be satisfactorily explained by known geophysical and geochemical processes (see chapters 5 and 6).

There are no transitional types—there is not one single example of evolution! Evolutionists look for "the" missing link. But there should be billions of examples of transitional forms with transitional structures if evolution were true, but there are none (see chapter 7).

---

notes: *Chapter 9*

1. McMurtry, Grady S. (2008). The nine great 'proofs' for evolution: and why they are all false! (proof #8). Creation World Ministries. Retrieved July 2008, from www.creationworldview.org/articles_view.asp?id=53.

2. Morris, J.D. (1994). *The Young Earth*. Green Forest, AR: Master Books, 78-80. Also see Hall, J.M., and Robinson, P.T. (May 11, 1979). Deep crustal drilling in the North Atlantic Ocean. *Science*, 204: 573-586; and Catchpoole, D., Sarfati, J., and Wieland, C. (2008). *The Creation Answers Book*. (D. Batten, Ed.). Atlanta, GA: Creation Book Publishers, 164.

3. Ibid.; Humphreys, D.R. (June 2005). Evidence for a young world. Institute for Creation Research, ICR *Impact,* Article 384 (Item #6); and Catchpoole, Sarfati, and Wieland (2008), op. cit., chapter 11. Also see Geomagnetic reversal (May 2, 2010, last modified). In Wikipedia, the free encyclopedia, p. 3, External events. Retrieved May 2010, from http://en.wikipedia.org/wiki/Geomagnetic_reversal.

4. Coe, R.S., Prevot, M., and Camps, P. (April 1995). New evidence for extraordinarily rapid change of the geomagnetic field during a reversal. *Nature*, 374: 687-692; Humphreys, D.R. (1986). Reversals of the earth's magnetic field during the Genesis flood. *Proceedings of the First International Conference on Creationism*, 2: 113-126; as cited in Humphreys (June 2005), op. cit., Item #6; and Coe, R.S., and Prevot, M. (1989). Evidence suggesting extremely rapid field variation during a geomagnetic reversal. *Earth and Planetary Science Letters*, 92: 292-298; as cited in Morris (1994), op. cit., 80–81. Also see Catchpoole, Sarfati, and Wieland (2008), op. cit., chapter 11.

5. Humphreys (June 2005), op. cit., Item #6; and Morris (1994), op. cit., 80-83. Also see Humphreys, D.R. (June 2002). The earth's magnetic field is still losing energy. *Creation Research Society Quarterly*, 39 (1): 3-13.

6. Ibid. Also see Humphreys, D.R. (1991). Physical mechanism for reversals of the earth's magnetic field during the flood. *Proceedings of the First International Conference on Creationism*, 2: 129-142; as cited in Humphreys (2005), op. cit., Item #6.

7. Ibid.

8. Ibid. Also see Coe, R.S., Prevot, M., and Camps, P. (April 1995), op. cit., 687-692.

9. Thompson, G.R., and Turk, J. (1991). *Modern Physical Geology*. Philadelphia, PA: Saunders College Publishing, 359-360.

10. Nevins, S.E. (1973). Evolution: The oceans say no! Institute for Creation Research, ICR *Impact*, Article 8; and Morris (1994), op. cit., 89. Also see Roth, Ariel A. (1986). Some questions about geochronology. *Origins*, 13 (2): 64-85; and Milliman, J.D. and Syvitski, J.P.M. (1992). Geomorphic/tectonic control of sediment discharge to the ocean: The importance of small mountainous rivers. *The Journal of Geology*, 100: 525-544; as cited in Humphreys, D.R. (June 2005) (Item #4), op. cit.

11. Morris (1994), 89. Also see Judson, S. (1968). Erosion of the land—or what's happening to our continents? *American Scientist*, 56: 356-374.

12. The average uncompacted ocean sediment thickness of deep ocean basins is 0.40 mile, or 2,100 feet (640 meters). Because ocean sediment commonly contains 50% water by volume, compacted sediment thickness would be approximately 0.20 mile. With total area of oceans about 139.4 million square miles, the compacted ocean sediment would be about 28 million cubic miles. With an average sediment density of 10.7 billions tons/cubic mile, the ocean sediments are approximately 300 million billion tons. See Morris (1994), 90; and Nevins (1973).

According to another researcher, the average uncompacted ocean sediment thickness of deep ocean basins is about 0.37 mile (600 meters); thus, compacted sediment thickness would be approximately 0.19 mile. With total area of oceans about 139.4 million square miles, the compacted ocean sediment would be about 26.5 million cubic miles. With an average sediment density of 10.7 billions tons/cubic mile, the ocean sediments are approximately 284 million billion tons. See Vardiman, Larry (1996). *Sea-Floor Sediment and the Age of the Earth*. El Cajon, CA: Institute for Creation Research, 25-26.

The average uncompacted ocean sediment thickness of ocean basins (deep and shallow basins) is 0.56 mile, or 2,950 feet (899 meters); thus, compacted ocean sediment worldwide would be 0.28 mile. With total area of oceans about 139.4 million square miles, the compacted ocean sediment would be about 39 million cubic miles. With an average sediment density of 10.7 billions tons/cubic mile, the ocean sediments are approximately 417 million billion tons. See Morris (1994), 90; and Nevins (1973).

13. Morris (1994), op. cit., 90. Also see Nevins (1973) and Roth (1986), op. cit; and Hay, W.W. et al. (December 1988). Mass/age distribution and composition of sediments on the ocean floor and the global rate of sediment subduction. *Journal of Geophysical Research*, 93 (B12): 14,933-14,940.

14. Austin, S.A., and Humphreys, D.R. (1991). The sea's missing salt: A dilemma for evolutionists. *Proceedings of the Second International Conference on Creationism*, 2: 17-33. Also see Morris (1994), op. cit., 85-87; and Sarfati, J.D. (1999). Salty seas: Evidence for a young earth. *Creation*, 21 (1): 16-17.

15. Ibid.

16. The word *homo* is Latin for "human." There is no real distinction between *H. erectus* and *H. sapiens*—in modern taxonomy, *Homo sapiens* is the only extant (actually existing) species of its genus, *Homo*.

17. Morris, J.D. (2003). *Is the Big Bang Biblical?* Green Forest, AR: Master Books, 108-109; and Morris (1994), 70.

18. Catchpoole, Sarfati, and Wieland (2008), op. cit., 255. Also see Wieland, C. (September 1997). Sensational dinosaur blood report! *Creation*, 19 (4): 42-43; Wieland, C. (1999). Dinosaur bones: Just how old are they really. *Creation*, 21 (1): 54-55; Schweitzer, M., and Staedter, T. (June 1997). The real Jurassic Park. *Earth*, 6 (3): 55-57; and Schweitzer, M., Wittmeyer, J.L., Horner, J.R., and Toporski, J.K. (March 2005). Soft tissue vessels and cellular preservation in Tyrannosaurus rex. *Science*, 307: 1952-1955; Thomas, B. (October 2008). Dinosaur soft tissue: Biofilm or blood vessels? *Acts & Facts*, 37 (10), Dallas, TX: Institute for Creation Research, 14; and Johnson, J.S., Tomkins, J. and Thomas, B. (October 2009). Dinosaur DNA research: Is the tale wagging the evidence? *Acts & Facts*, 38 (10), Dallas, TX: Institute for Creation Research, 4-6.

19. Scientists recover T. rex soft tissue. The Associated Press (2008). Retrieved April 2008, from http://www.msnbc.msn.com/id/7285683/. Also see Fields, H. (May 2006). Dinosaur shocker, *Smithsonian Magazine*.

20. Catchpoole, Sarfati, and Wieland (2008), op. cit., 255. Also see Batten, D. (1997). Buddy Davis – The creation music man (who makes dinosaurs). *Creation*, 19 (3): 49-51; and Davies, K. (1987). Duckbill dinosaurs (Hadrosauridae, Ornithishia) from the north slope of Alaska. *Journal of Paleontology*, 61 (1): 198-200. Humphreys, D.R. (Item #8), op. cit.

21. DeYoung, D. (May 2006). *Thousands ... Not Billions*. Green Forest, AR: Master Books, 46-62. Also see Baumgardner, J. (2003). Measurable ^{14}C in fossilized organic materials, presented August 2003 at the *Fifth International Conference on Creationism*. Copyright 2003 by Creation Science Fellowship, Inc., Pittsburg, PA. Retrieved June 2008, from http://globalflood.org/papers/2003ICCc14.html.

22. Elert, Glenn (2004). Diameter of the known universe. Retrieved from http://hypertextbook.com/facts/2002/CarmenBissessar.shtml.

23. McMurtry (2008), op. cit. (proof #8). Also see Thompson and Turk, (1991), op. cit., 9.

24. Superluminal Motion (January 3, 2008). In Wikipedia, the free encyclopedia. Retrieved December 2009, from http://en.wikipedia.org/wiki/Superluminal_motion.

25. McMurtry (2008), The nine great proofs..., op. cit. (proof #8); and McMurtry, G.S. (2008). The decay in the speed of light and the truth about red shift. Retrieved July 2008, from www.creationworldview.org/articles_view.asp?id=7.

26. Setterfield, Barry (March 1981). The velocity of light and the age of the universe. *Creation*, 4 (1): 38-48. Retrieved July 2008, from http://creation.com/contents-all-creation-magazines; and Catchpoole, Sarfati, and Wieland (2008), op. cit., 90-91.

27. Aardsma, Gerald A. (1988). Has the speed of light decayed? Institute for Creation Research. Retrieved July 2008, from http://www.icr.org/index.php?module=articles&action=view&ID=283.

28. General Relativity (January 8, 2009, last modified). In Wikipedia, the free encyclopedia. Retrieved December 2009, from http://en.wikipedia.org/wiki/General_relativity.

29. Humphreys, D.R. (1994). *Starlight and Time*. Green Forest, AR: Master Books, 11-13, 101-106. Also see Catchpoole, Sarfati, and Wieland (2008), op. cit., chapter 5; and Faulkner, D. (2004). *Universe by Design*. Green Forest, AR: Master Books. Dr. Faulkner provides an in-depth explanation of cosmology and creation.

30. Jeffrey, G.R. (2000). *Journey Into Eternity*. Toronto, Ontario: Frontier Research Publications Inc., 95-96. Also see Ledger, Lynchburg, Virginia. George Caylor's article titled "God's Building Blocks." Retrieved August 2009, from http://www.OnTheRightSide.com.

31. Catchpoole, Sarfati, and Wieland, op. cit., 18.

32. Ibid.

33. Ibid., 18-19.

34. Bates, G. Who made God? Can there be an uncreated creator? CD Produced by Creation Ministries International, Creation.com.

35. Scheffler, H. and Elasser, H. (1987). *Physics of the Galaxy and Interstellar Matter*, Springer-Verlag, Berlin, 352-353, 401-413, as cited in Humphreys, D.R. (June 2005). Evidence for a young world. Institute for Creation Research, ICR *Impact,* Article 384 (Item #1).

36. Davies, K. (1994). Distribution of supernova remnants in the galaxy. *Proceedings of the First International Conference on Creationism*, 2: 113-126; as cited in Humphreys (2005), Item #2, op. cit. (Davies article can be ordered from http://www.icc03.org/proceedings.htm.)

37. Steidl, P.F. (1983). Planets, comets, and asteroids. *Design and Origins in Astronomy*, 73-106, as cited in Humphreys (2005), Item #3, op. cit. Also see Whipple, F.L. (September 1976). Background of modern comet theory. *Nature*, 263: 15-19.

38. Lisle, Jason (September 2012). Blue stars confirm recent creation. *Acts & Facts*, 41 (9), Dallas, TX: Institute for Creation Research, 16.

39. Frank, Adam (October 2000). Teaching Einstein to dance: The dynamic world of general relativity. *Sky and Telescope*, 100: 53. Also cited in Morris, Henry M. (June 2001). The cosmic bubbleland. Institute for Creation Research, ICR *Impact,* Article 767. Retrieved July 2008, from www.icr.org/article/767. Permission received on October 2, 2009.

40. Morris, Henry M. (June 2001). The cosmic bubbleland, op. cit. Copyright © 2001 Institute for Creation Research, used by permission. Morris, H. 2001. The cosmic bubbleland. *Acts & Facts*, 30 (6).

41. Spurgeon, C.H. (May 1877). *The Sword and the Trowel (*Inaugural Address - 13th Annual Conference of the Pastors College), Pilgrim Publications, 197.

Chapter 10

The Biblical Account of Noah's Flood

"So they went into the ark to Noah, two by two of all flesh in which was the breath of life."—Genesis 7:15, NAS

The Book of Genesis is the first book of Scripture. Historically, Jews and Christians believe that Moses was the author of Genesis and the next four books of the Bible. It describes the beginning of creation—the heavens and earth, light and darkness, sun, moon and stars, seas and skies, and animals and vegetation—and the creation of human beings in the image of God.

It describes the destruction of the earth by flood, and the antediluvian and postdiluvian periods. It describes migrations of groups of people (descendants of Noah and his family) following the flood, society and civilizations (i.e., Mesopotamian life and culture), and marriage and family. Additionally, it includes such topics as construction and map-making, sale and purchase of land, legal and social customs, and care of cattle and sheep.

It is a book of relationships between God and man, and God and nature, and states that there is only one sovereign God. It describes sin, blessings and the Edenic curse, and conditions for redemption (salvation or rescue), and gives humanity its first understanding of God's condition for redemption from forces of evil. It contains the oldest definition of faith (Genesis 15:6).

And by divine choice, God has exercised unlimited freedom to intervene and change human customs and plans, and made promises to His chosen people (the Jews of Israel) by pledging His faithfulness and love, and holding them to their promises of faithfulness and love to Him.

Creation of Light by Gustave Doré (1832-1883). The engraving depicts a literal representation of Genesis 1:1 ("Let there be light").

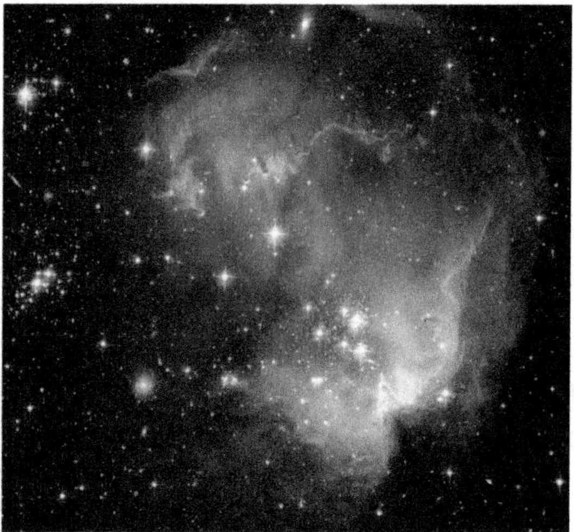

Stars in nearby galaxy. Genesis Creation Reveals that God is Eternal.

Creation – Genesis 1

Day 1 – Earth, Space, Time, Water, and Light

"In the beginning God created the heavens and the earth. And the earth was formless and void, and darkness was over the surface of the deep; and the Spirit of God was moving over the surface of the waters." (Genesis 1:1-2, NAS) This initial act of creation included the earth's core, crust, and water.

"Then God said, 'Let there be light'; and there was light. And God saw that the light was good; and God separated the light from the darkness. And God called the light day, and the darkness He called night. And there was evening and there was morning, one day." (Genesis 1:3-5, NAS)

In verse 3, God made light—but it didn't say God made the sun. The sun is not needed for day and night but, rather, light and a rotating earth. Thus, if we have light from one direction and a revolving earth, there can be day and night. On the first day, light was created to provide day and night until God made the sun on day 4.

So where did the light come from? Why do people argue there could be no light without the sun? When people try to rationalize based on their finite minds and logic, they do not understand their own limitations and God's infinite power and glory. Revelation 21:23 (NAS) says, *"And the city [Jerusalem] has no need of the sun or of the moon to shine upon it, for the glory of God has illumined it, and its lamp is the Lamb."* This tells us that following the Second Coming of Jesus Christ, the sun will not be needed, as the glory of God will light the heavenly city. In Deuteronomy 4:19, God warned the Israelites not to worship the sun like the pagan cultures but to worship God who created the sun.

Day 2 – Atmosphere

"Then God said, 'Let there be an expanse in the midst of the waters, and let it separate the waters from the waters.' And God made the expanse, and separated the waters which were below the expanse from the waters which were above the expanse; and it was so. And God called the expanse heaven. And there was evening and there was morning, a second day." (Genesis 1:6-8, NAS)

On the second day, the atmosphere was formed. The atmosphere as we know it today (that is, the troposphere, stratosphere, mesosphere, thermosphere, ionosphere, and exosphere) would not have existed during initial creation. However, Genesis 2:4-6 implies periodic rainfall, a normal hydrologic cycle, and moderate seasonal variations during the pre-flood period (See next section, The Antediluvian Period, and chapter 4.)

Day 3 – Dry Land, Seas, and Plants

"Then God said, 'Let the waters below the heavens be gathered into one place, and let the dry land appear'; and it was so. And God called the dry land earth, and the gathering of the waters He called seas; and God saw that it was good." (Genesis 1: 9-10, NAS)

On the third day, land was made to appear above the surface of the waters. This was a great orogeny as landmasses (rocks and the primitive earth, or the primeval supercontinent, Pangaea) were uplifted above the waters. This was the first of two worldwide tectonic upheavals—the other was the great orogeny associated with Noah's flood.

"Then God said, 'Let the earth sprout vegetation, plants yielding seed, and fruit trees bearing fruit after their kind, with seed in them, on the earth'; and it was so. And the earth brought forth vegetation, plants yielding seed after their kind, and trees bearing fruit, with seed in them, after their kind; and God saw that it was good. And there was evening and there was morning, a third day."
(Genesis 1:11-13, NAS)

On the same day, God made vegetation.

Day 4 – Sun, Moon, and Stars

"Then God said, 'Let there be lights in the expanse of the heavens to separate the day from the night, and let them be for signs, and for seasons, and for days and years; and let them be for lights in the expanse of the heavens to give light on the earth'; and it was so. And God made the two great lights; the greater light to govern the day, and the lesser light to govern the night; He made the stars also. And God placed them in the expanse of the heavens to give light on the earth, and to govern the day and the night, and to separate the light from the darkness; and God saw that it was good. And there was evening and there was morning, a fourth day." (Genesis 1:14-19, NAS)

On the fourth day, God established the sun, moon, and stars with respect to the earth.

Day 5 – Sea Creatures and Flying Creatures

"Then God said, 'Let the waters teem with swarms of living creatures, and let birds fly above the earth in the open expanse of the heavens.' And God created the great sea monsters, and every living creature that moves, with which the waters swarmed after their kind, and every winged bird after its kind; and God saw that it was good. And God blessed them, saying, 'Be fruitful and multiply, and fill the waters in the seas, and let birds multiply on the earth.' And there was evening and there was morning, a fifth day." (Genesis 1:20-23, NAS)

On the fifth day, God created all sea creatures and birds after their kind.

Day 6 – Land Animals and Man

"Then God said, 'Let the earth bring forth living creatures after their kind: cattle and creeping things and beasts of the earth after their kind'; and it was so. And God made the beasts of the earth after their kind, and the cattle after their kind, and every thing that creeps on the ground after its kind; and God saw that it was good." (Genesis 1:24-25, NAS)

On the sixth day, God created all land animals after their kind.

"Then God said, 'Let us make man in our image, according to our likeness; and let them rule over the fish of the sea and over the birds of the sky and over the cattle and over all the earth, and over every creeping thing that creeps on the earth.' And God created man in His own image, in the image of God He created him; male and female He created them. And God blessed them; and God said to them, 'Be fruitful and multiply, and fill the earth, and subdue it; and rule over the fish of the sea and over the birds of the sky, and over every living thing that moves on the earth.' Then God said, 'Behold, I have given you every plant yielding seed that is on the surface of all the earth, and every tree which has fruit yielding seed; it shall be food for you; and to every beast of the earth and to every bird of the sky and to every thing that moves on the earth which has life, I have given every green plant for food'; and it was so. And God saw all that He had made, and behold, it was very good. And there was evening and there was morning, the sixth day." (Genesis 1:26-31, NAS)

On the same day, God created man (male and female) after His own image and gave them dominion over all living things.

> The Hebrew word *yom* is translated "day" and when used with a number, such as "six days," it always means a literal day. In Genesis 1:5, 8, 13, 19, 23, 31, the phrase *"there was evening and there was morning"* is used in each of the six days of creation, and in the remainder of Genesis, *yom* is used numerous times, and each time it translates as an ordinary 24-hour solar day.[1]

God, author of Scripture, could not have been more concise—He created earth, the universe and all that exists in six solar days! It was an instantaneous act of creation done in a purposeful sequence of events.

Some may ask, "If God is omnipotent and created the entire universe and all that exists instantaneously, why did He need six days?" The answer is found in Exodus 20:8-11 (NAS).

> *"Remember the sabbath day, to keep it holy. Six days you shall labor and do all your work, but the seventh day is a sabbath of the Lord your God; in it you shall not do any work, you or your son or your daughter, your male servant or your female servant or your cattle or your sojourner who stays with you. For in six days the Lord made the heavens and the earth, the sea and all that is in them, and rested on the seventh day; therefore the Lord blessed the sabbath day and made it holy."* (Exodus 20:8-11, NAS)

The answer is that God provided a plan or design for our work week. We are to work six days and rest one day, just as He did. The seventh day is a day of rest, which honors His perfect work of creation.

Day 7 – Day of Rest

"Thus the heavens and the earth were completed, and all their hosts. And by the seventh day God completed His work which He had done; and He rested on the seventh day from all His work which He had done. Then God blessed the seventh day and sanctified it, because in it He rested from all His work which God had created and made." (Genesis 2:1-3, NAS)

From a scientific perspective, during the six days of creation, matter and energy were being created. This newly created matter and energy were being organized into complex, energized systems that are a contradistinction to the First and Second Laws of Thermodynamics of today (that is, matter and energy cannot be created or destroyed; and unrestrained universal decay as a dominating trend in nature). See chapter 2. In other words, the universe and biological systems were "wound up" or energized by creation at the beginning and everything was upheld.

Before sin entered the world, the world was good (Genesis 1:31); God *"upheld"* and *continuously restored* everything in the beginning (Colossians 1:17 and Hebrews 1:3)—but when sin entered the world (because of Adam—Genesis 3:6), God *cursed the world* (Genesis 3:14-19), so the perfect creation began to *degenerate—that is, suffer death and decay* (Romans 5:12, 6:23, and 8:22). It was not until the "curse" that creation was replaced with the First and Second Laws. As indicated in Genesis 3:3 and Romans 5:12, there was no death in the animal kingdom until Adam sinned.

According to Genesis 5:5, Adam lived 930 years. Although the life span of Eve is unknown, they would have had many children that intermarried without producing deformed offspring. The degeneration and dying process started slowly because of minimal "mutational load" in the beginning (see chapter 3). God clearly said the process of creation no longer operates—a fact which is thoroughly verified by the two universal laws of thermodynamics. Today we are in a world that *"groans and suffers"* (Romans 8:22, NAS) because of the Edenic Curse.

The Antediluvian Period
(4004 BC – 2385 BC)

The original ancient continent was created by God as a single landmass about 6,000 years ago: *"Let the waters below the heavens be gathered into one place, and let the dry land appear."* (Genesis 1:9, NAS) Pangaea, as defined in geology textbooks, is a primeval supercontinent composed of all the major landmasses (as described in chapter 4). The antediluvian continent had mountains (Genesis 7:20), rivers (Genesis 2:10), and seas (Genesis 1:10), although topography was much less prominent—the oceans weren't so deep and the mountains weren't so high (that is, low-lying mountains).[2]

Genesis 1:7 (NAS) states, *"And God made the expanse and separated the waters which were below the expanse from the waters which were above the expanse."* This describes a division of waters—oceans on the surface of the earth and clouds of condensed water vapor in the atmosphere—separated by an expanse of atmosphere in which birds were to fly (Genesis 1:20) and through which the sun, moon, and stars were to provide light for the earth (Genesis 1:17).

> Air contains an invisible gas called water vapor that is generated by evaporation and removed by condensation as warm air rises and cools forming clouds. Clouds are condensed water vapor—a visible mass of water droplets or ice crystals. Because cool air cannot hold as much water vapor as warm air, water vapor condenses into drops heavy enough to fall from the atmosphere as rain.

Genesis 2:4-6 (NIV) states *"...When the Lord God made the earth and the heavens—and no shrub of the field had yet appeared on the earth and no plant of the field had yet sprung up [before the creation of man], for the Lord God had not sent rain on the earth...but streams came up from the earth and watered the whole surface of the ground."* This implies that periodic rainfall and a normal hydrologic cycle (evaporation and condensation) would have prevailed during the 1600 year pre-flood period. There would have been moderate seasonal variations throughout Pangaea (Genesis 1:14, 8:22).

Substantial evidence has been found indicating the climate along the coastline of northern Siberia and Alaska and islands of the Arctic Ocean was at one time very mild and food was plentiful.[3] This polar region once flourished with tropical plants such as "mangroves, palm trees, Burmese lacquer trees, and groups of trees that now produce nutmeg and Macassar oil."[4] Such tropical vegetation was indigenous to this time period. Consistent with the antediluvian period, a warm and moist climate is further verified by "widespread discoveries of surface coal and petrified wood."[5]

The Antediluvian People

It is important to understand the state of affairs during the 1,600-year antediluvian period. Because of the fall of Adam and Eve, human beings had acquired the knowledge of good and evil (Genesis 3:7, 22); more specifically, the knowledge of evil. This was the Age of Conscience, the second of the seven ages or dispensations[6]—God's method of dealing with humanity throughout history.

During this time, people had no institution of government and law, and could only distinguish between good and evil by their own conscience and personal Divine revelations. Mankind was given the individual and moral responsibility to pursue goodness and virtue—and turn away from evil.

God provided man with a means of atonement; that is, animals were allowed to be sacrificed and their blood shed so mankind could be reconciled to their Creator. However, during this period, humanity abandoned its relationship with God—and ultimately, they were consumed with wickedness and violence continually. *"Now the earth was corrupt in the sight of God, and the earth was filled with violence."* (Genesis 6:11, NAS) The human race became morally depraved, which is thoroughly described in Genesis 6.

> *"Then the Lord saw that the wickedness of man was great on the earth, and that every intent of the thoughts of his heart was only evil continually. And the Lord was sorry that He had made man on the earth, and He was grieved in His heart. And the Lord said, 'I will blot out man whom I have created from the face of the land, from man to animals to creeping things and to birds of the sky; for I am sorry that I have made them.'"* (Genesis 6:5-7, NAS)

> *"Now the earth was corrupt in the sight of God, and the earth was filled with violence. And God looked on the earth, and behold, it was corrupt; for all flesh had corrupted their way upon the earth."* (Genesis 6:11-12, NAS)

W. Graham Scroggie has graphically depicted the biblical picture of antediluvian humanity.[7]

> The appalling condition of things is summed up in a few terrible words, words which bellow and burn: *wickedness, evil imagination, corruption*, and *violence*; and these sins were *great, widespread*, 'in the earth,' *continuous*, 'only evil continually,' *open* and *daring*, 'before God,' *replete*, 'filled,' and *universal*, 'all flesh.'... This is an astounding event! After over 1,600 years of human history the race was so utterly corrupt morally that it was not fit to live; and of all mankind only four men and four women were spared, because they did not go with the great sin drift.

This period had reached a point of continuous utter depravity. The degradation and wickedness of the antediluvians have been affirmed by an astonishing collection of Scriptural testimony (Genesis 6:1-6, 11-13; Luke 17:26-27; 1 Peter 3:20; 2 Peter 2:5; and Jude 14-15).

Noah and His Family

As emphasized in Scripture, all mankind was destroyed in the Flood *except* Noah and his family. Because Noah was righteous and did not yield to the depraved sin-drift of that age, God established a covenant with Noah—that he and his family would be the only ones to escape God's judgment. The relevant passages are as follows:

> *"But Noah found favor in the eyes of the Lord. These are the records of the generations of Noah. Noah was a righteous man, blameless in his time; Noah walked with God."* (Genesis 6:8-9, NAS)

> *"Then God said to Noah, 'The end of all flesh has come before Me; for the earth is filled with violence because of them; and behold, I am about to destroy them with the earth.'"* (Genesis 6:13, NAS)

> *"And behold, I, even I am bringing the flood of water upon the earth, to destroy all flesh in which is the breath of life, from under heaven; everything that is on the earth shall perish. But I will establish My covenant with you; and you shall enter the ark—you and your sons and your wife, and your sons' wives with you."* (Genesis 6:17-18, NAS)

As stated in 1 Peter 3:20 (NAS), *"...when the patience of God kept waiting in the days of Noah, during the construction of the ark, in which a few, that is, eight persons, were brought safely through the water."* Noah and his family (wife, three sons and three wives) were the only people saved from the onslaught of judgment waters. One righteous family was left to repopulate the world while all ungodly humanity perished in the flood.

Comparison to the Second Coming

The history of civilization as described in the early chapters of Genesis is confirmed in several New Testament passages. In Matthew 24:36-44 and Luke 17:26-27, the Lord Jesus Christ compared the days of Noah and the Flood with the future Second Coming and final judgment of humanity—a time when some will be left and the rest will be taken.[8] In Hebrews 11:7 (NAS), we are told that Noah was *"...warned by God about things not yet seen"*—referring to the supernatural origin of the Flood.

The parallels between the days of Noah and the Second Coming are described in 2 Peter.

> *"Know this first of all, that in* [1] *the last days mockers will come with their mocking, following after their own lusts, and saying,* [2] *'Where is the promise of His coming?* [3] *For ever since the fathers fell asleep, all continues just as it was from the beginning of creation.' For when they* [ungodly men] *maintain this,* [4] *it escapes their notice* [*"deliberately forget,"* NIV] *that by the word of God the heavens existed long ago and the earth was formed out of water and by water,* [5] *through which the world at that time was destroyed, being flooded with water.* [6] *But the present heavens and earth by His word are being reserved for fire, kept for the day of judgment and destruction of ungodly men."* (2 Peter 3:3-7, NAS)

*Bracketed numbers - refer to text box

Noah's task was to construct the ark and to warn the world of God's coming judgment (Hebrews 11:7, 1 Peter 3:20, 2 Peter 2:5). During the construction of the ark, multitudes *laughed and scoffed* at Noah's warnings, just as many people today scoff as tele-evangelists such as Hal Lindsey (www.hallindsey.org), Jack Van Impe (www.jvim.org), John Hagee (www.jhm.org), and others (e.g., Mac Brunson, www.fbcjax.com, and John MacArthur, www.gty.org) who preach from the Bible on the fulfillment of end-time prophecy with the Second Coming of Christ, and warn of Armageddon.

In times of judgment, God provides a means of escape: Noah's Ark during the great flood and belief in the Lord Jesus Christ during the final judgment in the last days (1 Thessalonians

> Apostle Peter's prophetic words predict the rise of evolutionary doctrine in the last days. It foretells that ungodly men will [1] ridicule the Bible, [2] deny the Second Coming of Jesus the Messiah, [3] believe in uniformitarianism, [4] deny the Genesis account of creation, [5] deny the worldwide flood, and [6] receive judgment in the last days. (See end note 8 for an explanation of the Second Coming.)

4:16-17; 1 Corinthians 15:51-54). The Scripture describes the Flood as a unique, never-to-be repeated event in earth's history (Genesis 8:21–9:17), but comparable in magnitude and significance with the final judgment of the world as described in Matthew 24:9-26, Mark 13:1-37, Luke 21:5-36, 2 Peter 3, and the Book of Revelation. (See epilogue for more information.)

It is perplexing when people state that it is impossible for the Red Sea to part, or the earth to flood, or the world to be created in six literal solar days. People are unable to concede that their minds are finite, and their attitudes are arrogant and corrupt, and never give a thought to the possibility and reality of an infinite God who created the heavens and earth and all that exists (Jeremiah 10:12-16, 51:15-19, and Isaiah 40:28, NIV)—omnipotent (Jeremiah 32:27, 48:15, 51:57, NIV) and omnipresent (Jeremiah 23:24, NIV)—a Creator who will ultimately judge ungodly people.

Scientists and many theologians "deliberately forget" that God is not only living and personal, but He is absolutely capable of achieving exactly the kinds of miracles, blessings, and judgments testified to in Scripture.

Four Horsemen of the Apocalypse (1887) by Viktor Vasnetsov
Four horsemen are entities that bring war, famine, pestilence, and death.

Christian Eschatology and Book of Revelation. Eschatology is the study of the end of the age and the destiny of man as it is revealed in the Bible, which is the source for all Christian eschatology studies.

Noah's Ark

The Design of Noah's Ark

In Genesis 6:14-16, God commanded Noah to build an ark of gopher wood—an ark 450 feet long (300 cubits) by 75 feet wide (50 cubits) and 45 feet high (30 cubits).[9] Essentially, it was a flat-bottomed barge (the word ark signifies "box") that was "designed for floating, not for sailing."

With regard to its proportions and design, "a model was made by Peter Jansen of Holland, and Danish barges called *Fleuten* were modelled after the ark. These models proved that the ark had a greater capacity than curved or shaped vessels. They were very sea-worthy and almost impossible to capsize.... The stability of such a barge is great and it increases as it sinks deeper into the water. The lower the centre of gravity the more difficult it is to capsize."[10]

In a study of the stability of the ark, Dr. Henry Morris concluded "the Ark as designed was highly stable, admirably suited for its purpose of riding out the storms of the year of the great Flood."[11]

The available floor space of this multi-decked barge was well over 95,000 square feet with a volume of 1,396,000 cubic feet, which is equivalent to about 520 modern railroad boxcars or eight freight trains with 65 boxcars each.[12] A barge of this enormous size, with its many thousands of compartments (Genesis 6:14), was more than sufficient to house *"two of every kind"* of air-breathing animal in the world at that time.[13]

Remember that the original "kind" of dog, horse, or bird can represent hundreds, if not thousands, of different species seen today by natural selection, as discussed in chapter 3. The remaining space on the ark would have been for storage of food and supplies, and accommodations for Noah's family.

How long did it take Noah to build the ark? In Genesis 5:32, Noah was 500 years old when he had three sons. When the floodwaters came upon the earth, Noah was 600 years old (Genesis 7:6). It took Noah less than 100 years, likely 80 to 100 years, to complete the ark.

Painting by Edward Hicks, 1780–1849

Gathering and Care of the Animals

Obviously, it would have been impossible for Noah and his family to assemble 30,000–40,000 "kinds" of animals into the ark. In Genesis 6:20 (NIV), God commanded that *"two of every kind of bird, of every kind of animal and of every kind of creature that moves along the ground will come to you to be kept alive. You are to take every kind of food that is to be eaten and store it away as food for you and for them."* And Genesis 7:15 (NAS) says that *"So they went into the ark to Noah, two by two of all flesh in which was the breath of life."*

Amazingly, many theologians and scientists continually resort to endless debate about how Noah gathered so many animals throughout Pangaea, and should just recognize the fact that God has absolute power to bring two of the basic "kinds" of air-breathing creatures to the ark. It is back to the same problem of recognizing who God is—an eternal, omnipotent, and omnipresent Creator. If God created the heavens and earth and all that exist, then gathering two of every animal was well within His capability!

> People of that time must have been completely bewildered and alarmed by the ominous procession of thousands of animals heading toward the ark.

How did God control the animals in the ark so Noah and his family would not be burdened with the impossible task of tending to all their needs? It is believed that God imposed on the animals a year-long hibernation or aestivation period.

Understandably, with violent seas and exceedingly strong winds for a full year, there were minimal food consumption and bodily functions—most animals slept because of the continuous motion. (This is common for sailors who spend long periods of time at sea.) Also, breeding was naturally curtailed for the same reasons.

In such a confined space, and with noises of the sea and wind, the animals took refuge in a year-long period of sleep. It was not until after Noah brought the creatures out of the ark that God commanded them to *"breed abundantly on the earth, and be fruitful and multiply on the earth."* (Genesis 8:17, NAS)

Dinosaurs on Noah's Ark

Most people today think that the existence of dinosaurs and their extinction is a great mystery—but dinosaurs are only a mystery if one accepts evolutionary and old earth doctrine. There is *no mystery surrounding dinosaurs if one accepts the biblical account.*

According to the Bible, all animal kinds, including dinosaurs, were created approximately 6,000 years ago on day six of creation. Because dinosaurs and Adam and Eve were created on day six, dinosaurs and mankind would have lived at the same time. Because there was no death until Adam sinned (Genesis 3:3; Romans 5:12, 14, 6:23; and 1 Corinthians 15:21-22), fossilization of dinosaurs—and all other animals—would have occurred sometime after the Edenic Curse of Genesis 3.

Before Noah's Flood, all animals including man and dinosaurs were vegetarian. Genesis 1:30 (NAS) states, *"'...and to every beast of the earth and to every bird of the sky and to every thing that moves on the earth which has life, I have given every green plant for food'; and it was so."* Many might ask, how can this be—so many animals have large teeth? Although an animal has large teeth, it doesn't necessarily describe its behavior or mean it was a meat eater—it just means it had large teeth.[14]

Just a few examples of vegetarian animals with large teeth are pandas, fruit bats, and bears. Bears have teeth similar to tigers but many, if not most, are vegetarian.

As described in Genesis 6, representatives of all kinds of land animals, including dinosaur kinds, boarded the ark. Some people argue that dinosaurs were too big to fit in the ark and there were just too many varieties. Contrary to popular belief, most dinosaur kinds were small—the size of a bear, small horse, or sheep, and many were much smaller.[15] Also, there were only 55 kinds of dinosaurs with 668 varieties (species) within these kinds.[16]

Because animals were to repopulate the earth, it would have been necessary for God to send adolescents (young adults) to the ark—all animals, including the very large dinosaur "kinds," were not fully grown. All those that were left behind drowned in the flood and many of their remains became fossils.

After the flood during the Ice Age (see chapter 4), it was cold and harsh—the land was devastated and denuded of vegetation, and meat could be easily preserved. Thus God allowed man to eat meat and, consequently, animals developed a fear and dread of man. In Genesis 9:2-3, NIV, we read: *"The fear and dread of you will fall upon all the beasts of the earth and all the birds of the air, upon every creature that moves along the ground, and upon all the fish of the sea; they are given into your hands. Everything that lives and moves will be food for you. Just as I gave you the green plants, I now give you everything."*

The original "kind" of animal can represent hundreds, if not thousands, of different species seen today by natural selection, as discussed in chapter 3.

Humans and many animal kinds became predatory—a learned trait and a form of adaptation to the post-flood environment. Natural selection (see chapter 3) allowed certain kinds of dinosaurs and other animals coming off the ark, especially those with large teeth, to gain a predatory advantage in the new world. But with superior intelligence, man was by far the most proficient hunter.

Studies of bones of large dinosaurs (e.g., *Apatosaurus,* synonym *Brontosaurus*) show an adolescent (s-shaped pattern) growth spurt at about 5 years of age when animals were less than one tonne. Growth leveled off at 12 to 13 years old at 25 tonnes. The adolescent growth spurt of the *T. rex* (*Tyrannosaurus*) occurred at 14 to 18 years of age when the animal was just a half tonne. Growth leveled off at 20 years at 5.5 tonnes.[17] In comparison, today's African elephants weigh about 5 to 6 tonnes, and Asian elephants about 3 to 5 tonnes.

The Great Flood
(2385 BC)

Geologists are correct when they assert that a worldwide flood could not cover the Rocky Mountains, the Himalayas, and the Andes, and such a flood could not happen based upon current geologic processes and the present balance of oceanic and continental masses. Based on uniformity theory, secular geologists assume that such mountain ranges have been in place for hundreds of millions of years and a worldwide flood is impossible.

These same geologists (as well as some progressive theologians, who are profoundly influenced by the theory of evolution and uniformity), assert the Book of Genesis is mistaken if it subscribes to a mountain-covering flood and a young earth—but it is these geologists and progressive theologians who are mistaken. Although the supercontinent had mountains, its topography was much less pronounced as compared to how it exists today—the oceans weren't so deep and the mountains weren't so high.

Geomorphic evidence (geologic land features; see chapters 4, 5 and 6) strongly supports a worldwide flood, but such evidence has been *largely ignored or misinterpreted* in favor of evolution and an old earth. Supernatural creation and a young earth have not been disproved by science—they have simply been rejected in favor of false assumptions. The next two sections are a biblical review of the first three phases of Noah's flood.

Refer to chapter 4 and phase diagram of a worldwide flood, and view the Pangaea Flood video at www.CreationScienceToday.com.

Phases One and Two
Continental Separation, Uplift of Ocean Basins, and Flooding

In Genesis 7:11-12 (NIV), *"In the six hundredth year of Noah's life, on the seventeenth day of the second month—on that day all the springs of the great deep burst forth, and the floodgates of the heavens were opened. And rain fell on the earth forty days and forty nights."*

The Bible gives us indications about the sources of floodwaters and, eventually, drainage from the continents. Sources of water are given as *"all the springs of the great deep burst forth, and the floodgates of the heavens were opened."* This suggests chronology of events.

"All the springs of the great deep burst forth" initiated the splitting of the supercontinent (Pangaea) and tectonic *uplift of ocean basins*, and the beginning of catastrophic earthquakes and tsunamis that engulfed the continent, Pangaea. *"Floodgates of the heavens,"* or the release of water from above (intense rainfall), was brought about by a combination of two factors:

- Catastrophic (runaway) plate tectonics, associated with the breaking up or rifting of the preflood ocean floor, caused linear geysers of superheated steam and intense, continuous global rainfall, and
- Volcanic eruptions contained copious amounts of water vapor (70 to 90 percent of the plume component) that resulted in more intense rainfall.[18]

The uplift of ocean basins, accompanied by suboceanic earthquakes and tsunamis, is the primary mechanism that caused catastrophic flooding of the continents as described in chapter 4. This tectonic event continued for about six weeks (40 days and nights) until the flood *"...covered the mountains to a depth of more than twenty feet."* (Genesis 7:20, NIV) The earth remained inundated for another 110 days, or a total of 150 days (Genesis 7:24, 8:3), until *"Everything on dry land that had the breath of life in its nostrils died."* (Genesis 7:22, NIV)

Phase Three
Subsidence of Ocean Basins, Continental Uplift, and Drainage

Two passages of Scripture in widely separated Old Testament books (Genesis and Psalms) deal with the formation of present ocean basins and drainage.

First, Genesis 8:2-5 describes the formation of our present ocean basins and recession of water from the newly formed continents.

> *"Now the springs of the deep and the floodgates of the heavens had been closed, and the rain had stopped falling from the sky. The water receded steadily from the earth. At the end of the hundred and fifty days the water had gone down, and on the seventeenth day of the seventh month the ark came to rest on the mountains of Ararat. The waters continued to recede until the tenth month, and on the first day of the tenth month the tops of the mountains became visible."* (Genesis 8:2-5, NIV)

Because the *"springs of the great deep burst forth"* (Genesis 7:11, NIV) involved the uplift of ocean floors (second phase of the flood), the verse *"now the springs of the deep and floodgates of the heavens had been closed"* (Genesis 8:2) refers to the reversal of this action, the subsidence or lowering of the ocean basins (third phase of the flood). During the latter part of the year-long flood event, much deeper ocean basins were formed to serve as reservoirs as floodwaters drained from newly formed continents.

Second, Psalm 104:6-8 describes how the natural result of this subsidence of oceanic basins was uplift of great mountain ranges.

> *"Thou didst cover it with the deep as with a garment; the waters were standing above the mountains. At Thy rebuke they fled; at the sound of Thy thunder they hurried away. The mountains rose; the valleys sank down to the place which Thou didst establish for them."* (Psalm 104:6-8, NAS)

Mountains were lifted to balance the receding floodwaters that flowed into deeper oceanic basins. God supernaturally lowered the oceanic basins (using geologic processes) and floodwaters *"fled"* and *"hurried away"* into oceanic reservoirs—those places which *"Thou didst establish for them."* Uplift of the new continental landmasses in association with oceanic subsidence meant that floodwaters would have rapidly drained from the newly emerging continents and such torrential drainage would have resulted in relict erosional landforms we see today.

Soft sediment layers (formed by the vast composite of currents during the flood) were uplifted (folded and tilted) thousands of feet above sea level and soon thereafter these layers were intruded by great mountain ranges (massive igneous rock). Mountain ranges, batholiths, pancake (turbidite) sediment deposits, and erosional land features found throughout the world (evidentiary features described in chapters 4, 5 and 6) testify to these great tectonic events.

In Genesis 8:13-14 (NIV), *"By the first day of the first month of Noah's six hundred and first year, the water had dried up from the earth. Noah then removed the covering from the ark and saw that the surface of the ground was dry. By the twenty-seventh day of the second month the earth was completely dry."* Then God instructed Noah and his family to come out of the ark and to off-load every living creature that was on the ark.

Noah and his family faced a completely devastated earth—a mud-slick land surface, no vegetation (most of the vegetation had been buried and sealed under massive sediment deposits), great winds, cold temperatures, and a forbidding landscape. This was the beginning of the Postdiluvian Period and the Great Ice Age.

Chronology of Noah's Flood

As described in Genesis 7:12, it rained for 40 days and nights, and the world remained inundated for 150 days (Genesis 7:24, 8:3) before waters began to recede. According to Scripture the flood began when Noah was 600 years old (17th day, second month) and ended when Noah was 601 years old (first day of the first month)—thus, the flood lasted 314 days. The earth was completely dry by the 27th day of the second month (56 days later), or a total of 370 days. Mountain uplift and oceanic subsidence would have occurred during the final 164 days (314 days minus 150 days) of the flood.

Rainbow Covenant

A covenant pertaining to this worldwide flood, as compared with normal geologic processes throughout earth's history, is highlighted in Genesis 8:21-22 (NAS): *"I will never again destroy every living thing, as I have done. While the earth remains, seedtime and harvest, and cold and heat, and summer and winter, and day and night shall not cease."*

The rainbow covenant is emphasized in Genesis 9:12-17 (NAS) and Psalm 104:9 (NAS). (Also refer to Job 38:8-11 and Jeremiah 5:22.)

"And God said, 'This is the sign of the covenant which I am making between Me and you and every living creature that is with you, for all successive generations; I set My bow in the cloud, and it shall be for a sign of a covenant between Me and the earth. And it shall come about, when I bring a cloud over the earth, that the bow shall be seen in the cloud, and I will remember My covenant, which is between Me and you and every living

creature of all flesh; and never again shall the water become a flood to destroy all flesh. When the bow is in the cloud, then I will look upon it, to remember the everlasting covenant between God and every living creature of all flesh that is on the earth.' And God said to Noah, 'This is the sign of the covenant which I have established between Me and all flesh that is on the earth.'" (Genesis 9:12-17, NAS)

"Thou didst set a boundary that they may not pass over; that they may not return to cover the earth." (Psalm 104:9, NAS)

A distinction should be made between Psalm 104:6-9 which refers to Noah's Flood and Psalm 33:7 and Proverbs 8:22-31 which refer to the original creation, or the creation of Pangaea.[19] The original oceans of creation were not "*standing above the mountains*" as described in Psalm 104:6 (NAS) and they were not bound by *"a boundary that they may not pass over; that they may not return to cover the earth"* as described in Psalm 104:9 (NAS). This covenant assured mankind there would never again be a worldwide flood (Isaiah 54:9).

Rainbow Covenant, assurance that there will never be another worldwide flood (Genesis 9:12-17 and Psalm 104:9)

The Postdiluvian Period

The Great Ice Age

After the earth was flooded, "*God caused a wind to pass over the earth*" (Genesis 8:1, NAS), which implies significant meteorological changes—widespread temperature gradients for the first time between polar and equatorial regions. These changes initiated the onset of the Great Ice Age. In polar regions where tropical plants and animals once lived in abundance, copious amounts of snow began to fall and huge masses of snow and ice began to accumulate. "Once an ice sheet got started, it would probably grow rapidly and extensively."[20]

The Ice Age would last until oceanic volcanism subsided and oceans cooled, and trees and shrubs were reestablished on the newly formed continents. This likely would take place within 500 hundred years following the flood. During this period "large amounts of water were removed from the oceans [by the hydrologic cycle] and stored in the polar regions in the form of great ice caps... "[21] Geomorphic evidence indicates that "ocean levels were at least 400 feet lower than at present, possibly much more, as shown by such features as the continental shelves, sea-mounts, submerged canyons and terraces, etc."[22]

The onset of the Ice Age and lowering of sea level made it possible for animals coming off Noah's ark at Mount Ararat to migrate over land bridges to remote areas of the world (Genesis 8:15-9:7). For example, men and animals could freely cross the Bering Strait, which separates Asia and North America. Also, many animals would have migrated to remote and otherwise isolated areas of the world—for instance, kangaroos of Australia and lemurs of Madagascar. With the return of ocean waters to their present levels following the Ice Age, some animals would have become isolated and protected from competition and predation whereas other animals would become extinct.

For Noah and his family, life would have been harsh and difficult, but cold temperatures were not as severe as many might presume—moderately warm, wet winters (just below freezing) and moderately cold summers (just at or slightly above freezing). There is evidence that human societies lived near the edge of the ice sheet in Western Europe throughout the Ice Age (2384-1884 BC)—and it is likely that these people were direct descendants of the Mesopotamian civilization. Toward the end of the Ice Age, man began to experience the modern hydrologic cycle—and, for the first time, the beginnings of seasonal fluctuations. See chapter 4, The Great Ice Age.

Ararat is the tallest peak in Turkey. This snow-capped, dormant volcanic mountain is located near the northeast corner of Turkey, 10 miles (16 km) west of the Iranian border and 20 miles (32 km) south of the Armenian border.

The following bears repeating (see section, Dinosaurs on Noah's Ark): After the flood during the Ice Age, it was cold and harsh—the land was devastated and denuded of vegetation, and meat could be easily preserved. Therefore, God allowed man to eat meat and, consequently, animals developed a fear and dread of man as described in Genesis 9:2-3, NIV.

Humans and many animal kinds became predatory—a learned trait and a form of adaptation to the post-flood environment. Natural selection (see chapter 3) allowed certain kinds of dinosaurs and other animals coming off the ark, especially those with large teeth, to gain a predatory advantage in the new world. But with superior intelligence, man was by far the most proficient hunter.

Mesopotamia – The Early Civilization

Immediately following the Flood, God commanded the descendants of Noah to *"increase in number and fill the earth"* (Genesis 9:1, NIV), but within just a few hundred years people chose to locate in the area of Mesopotamia in direct disobedience to God. Mesopotamia (Greek meaning "between two rivers") is an area geographically located between the Tigris and Euphrates rivers, corresponding to modern-day Iraq and the western part of Iran (south of Mt. Ararat). The area called Sumer in southern Mesopotamia (see map of Mesopotamia) is commonly known as the "cradle of civilization."

This was a time described by God as the Age of Government which followed the Age of Conscience. (See end note 6 regarding dispensations.) A human organization known as civil government was divinely established to control the wickedness of men.

Under the rule of King Nimrod,[23] the people united in building a great city called Babel (Babylon) and erected an enormous tower. The Tower of Babel (2200 BC) was not built for the worship and praise of God but rather, it was dedicated to a false man-made religion, with a motive of celebrating man's

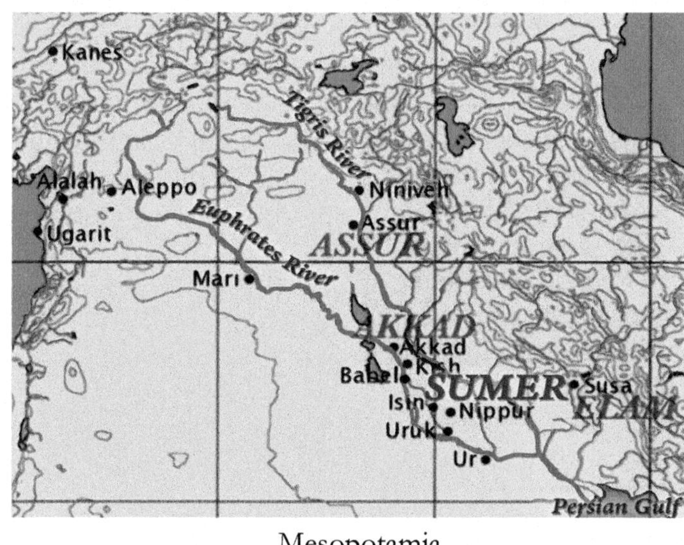

Mesopotamia

Modern Iraq and Surrounding Countries

crowning achievement. This is made clear in the words of Genesis 11:4 (NIV), *"Then they said, 'Come, let us build ourselves a city, with a tower that reaches to the heavens, so that we may make a name for ourselves and not be scattered over the face of the whole earth.'"*

Note how often plaques are placed on public buildings or monuments with the names of public officials (for example, mayor, head of public works, architect, and engineer) who were in power or had a hand in the building. *"Let us make for ourselves a name"* (Genesis 11:4, NAS) is the fundamental nature of man—the nature of an arrogant and immoral race. It exposes the basic philosophy of humanism: "Glory to man in the highest, for man is the master of things."[24] This is the central thought and motivation of humanism—glory to mankind.

The population that descended from Noah's family had one language and by living together in one location (Mesopotamia), they were disobeying God's command to *"fill the earth"* (Genesis 9:1). Because of this, God confused their languages and caused the people to break up into small groups and scatter throughout the earth (Genesis 11:5-9).

Why would God want to split the population into many subgroups? Because of the fallen nature of man, God knew that mankind as a centralized group would eventually rebel against the true God and begin to worship a human king and create pagan idols and religions. This has been the trend throughout history (as described in the Old and New Testaments), and it is a fast-emerging trend in today's world—a one-world government and religion excluding God the Creator.

Racial Groups

In Mesopotamia there was one language and one cultural group, and there was no barrier to marriage within the population. One population group living together in one location would keep skin color away from the extreme. Very dark or very light skin would occasionally appear, but people would be free to marry someone with lighter or darker skin color, ensuring the average color stayed roughly the same. Thus, unique traits or differences in appearance would never become dominant in the population. (See chapter 3, Race and Human Morphology.)

When separate languages were imposed, there were instantaneous barriers. People would not be inclined to marry someone they did not understand, so subgroups moved away to remote areas of the world. This, of course, is exactly what God intended with the confusion of languages.

Each subgroup would not have carried the same broad range of genes as the original population. One group would have more dark genes than light genes and more (or less) of certain genes for other physical traits such as nose shape, eye shape, hair color, height, etc. Because people within groups would intermarry, broad differences would no longer be averaged out as in the original "kind" population—so, there would be more extreme individual characteristics.

> Anthropologists once depicted Neanderthal man as a brutish, stooped, knuckle-dragging ape-man but, in fact, now admit that the so-called Neanderthals were fully human, highly skilled, and had an advanced social structure. The ape-man portrayal is another example of the large overdose of evolutionary prejudice.

As people moved to distant and isolated areas of the world, they faced different environments. Some groups would have migrated south to warmer, more tropical climates. Here light-skinned people would be affected by skin cancer and dark-skinned people would more easily survive. Conversely, if dark-skinned people moved to a cold region with little sunlight, they would not be able to produce enough vitamin D; thus, they would develop rickets and have fewer children. The so-called Neanderthals were likely dark-skinned people who migrated from Mesopotamia and lived along the edge of the ice caps of Western Europe.

To summarize, the dispersion in Mesopotamia separated a large original breeding group into smaller groups that scattered throughout the world. Smaller groups would have started out with different gene pools, and breeding within the group would have insured certain fixed features. As discussed in chapter 3, natural selection pressure in response to their specific environment would "thin out" the existing gene pool so that certain characteristics of each group would be dominant. Each of these smaller groups developed into races and cultures that we are familiar with today.

See Dr. C. Wieland's book, *One Human Family: The Bible, Science, Race & Culture*. Atlanta, GA: Creation Book Publishers, 2011, for a captivating and informative review of this subject (www.creation.com).

The Great Ice Age in the Bible

Although scripture reveals little about climatic conditions following Noah's Flood, there are certain references to an ice age in the Book of Job. As described in Job 1:1 (NIV), *"In the land of Uz there lived a man whose name was Job. This man was blameless and upright; he feared God and shunned evil."* Uz was a large mountainous region to the south-southeast of the Jordan River and the Dead Sea, which included Edom in the south. Most conservative Bible students agree that Job probably lived between the time of the Tower of Babel (2200 BC) and Abram (2166–1991 BC).

Job 37:9-10 (NIV) states, *"The tempest comes out from its chamber, the cold from the driving winds. The breath of God produces ice, and the broad waters become frozen."* In Job 38:22-23 (NIV) it states, *"Have you entered the storehouses of the snow or seen the storehouses of the hail, which I reserve for times of trouble, for days of war and battle?"* And in Job 38:29-30 (NIV) it states, *"From whose womb comes the ice? Who gives birth to the frost from the heavens when the waters become hard as stone, when the surface of the deep is frozen?"*

This Scripture is most likely a reference to the climatic conditions of the Great Ice Age—such effects are not experienced in the Middle East today.

Dinosaurs in the Bible

We've all heard the word "dragon"—dragons have long been considered creatures of legend, and such legends are abundant in most ancient cultures. Legends tell of Chinese-bred dragons that fit the descriptions of dinosaurs and many historical accounts of "dragons" are descriptions of what we call dinosaurs today. These include *Triceratops, Stegosaurus, Ankylosaurus,* theropod dinosaurs (genera include *Tyrannosaurus*, *Dilophosaurus*, and *Allosaurus*), and sauropod dinosaurs (genera include *Apatosaurus, Diplodocus,* and *Brachiosaurus).* See depictions below and photos in chapter 7, Dating Dinosaur Fossils.

If dinosaurs lived alongside man, memories of such creatures should be found in historical artifacts and records. There is historical and empirical evidence (i.e., pictographs and petroglyphs) that dragons were actually dinosaurs that lived with many other animal groups throughout the world and with man. Native Americans in Utah, for example, depict a sauropod-like creature on a rock face; a monastery in Cambodia (AD 1200) portrays carvings of real and fabled creatures on its walls, one a clear depiction of a *Stegosaur*; figurines from gravesites found in Mexico (dating BC) depict dinosaur-like creatures, one a *Triceratops*; and burial stones found in Peru (AD 500–1500) depict a dinosaur resembling a *T. rex.*[25] Dr. J. D. Morris concludes, "surely the evidence is clear—dinosaurs and humans lived at the same time, just as it says back in Genesis!"

The Bible talks about dragons as real animals as stated in Psalm 91:13 (KJV): *"Thou shalt tread upon the lion and adder* [cobra]: *the young lion and the dragon shalt thou trample under feet."* The word "dragon" (Hebrew: *tannim*) actually appears in the Old Testament at least 21 times. If one replaces "dragon" with "dinosaur," it mostly fits well in the biblical text.

According to the Bible, dinosaurs such as the *Apatosaurus* lived in the Jordan River valley as late as 2000 BC (during the Ice Age), as indicated in Job.

> *"Behold now, Behemoth, which I made as well as you; he eats grass like an ox. Behold now, his strength in his loins, and his power in the muscles of his belly. He bends his tail like a cedar; the sinews of his thighs are knit together. His bones are tubes of bronze; his limbs are like bars of iron. He is the first of the ways of God; let his maker bring near his sword."* (Job 40:15-19, NAS)

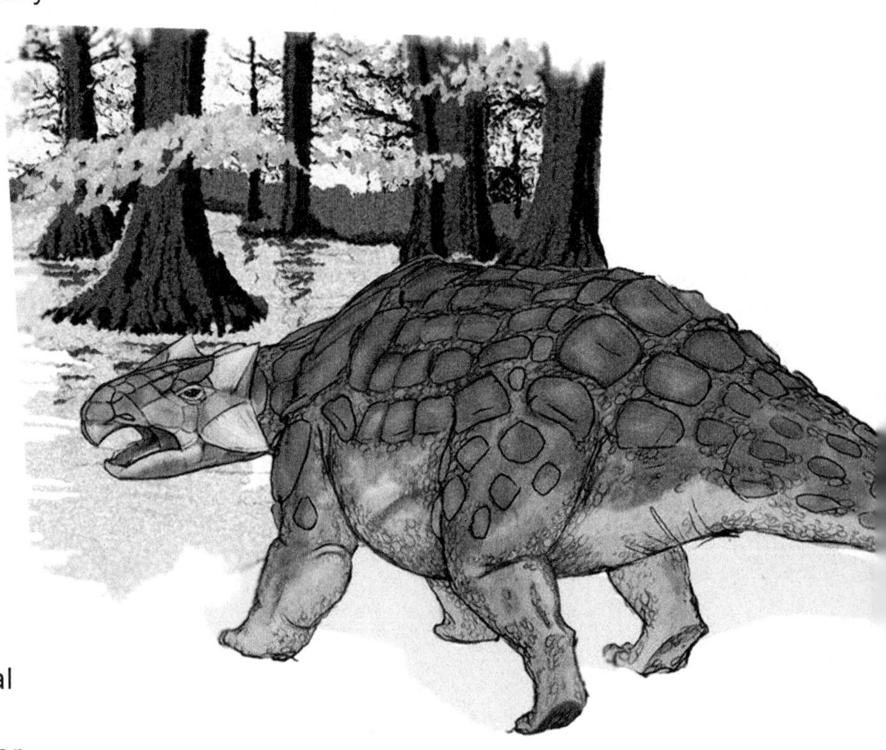

Ankylosaurus

The behemoth described in Job 40 is likely one of the big dinosaurs, such as *Apatosaurus or Brachiosaurus*. The *"tail like a cedar"* clearly indicates a large sauropod dinosaur. Also, Job 41:1, Psalm 74:13-14, and Isaiah 27:1 describe a leviathan, a large marine animal even more terrifying than the behemoth. The word *leviathan* has become synonymous with a large sea monster or creature.

Sauropods had long necks, long tails, small heads, and thick legs—and some species were notable for their enormous size.

Illustration of an *Apatosaurus* by Charles R. Knight, 1902

Illustration of a *Stegosaurus* by Othniel Charles Marsh, 1896

"Destruction of Leviathan." 1865 engraving by Gustave Doré

notes: Chapter 10

1. Stambaugh, James (October 1988). The meaning of "day" in Genesis. Institute for Creation Research, ICR *Impact,* Article 184; and Catchpoole, D., Sarfati, J., and Wieland, C. (2008). *The Creation Answers Book.* (D. Batten, Ed.). Atlanta, GA: Creation Book Publishers, chapter 2 (31, 36-37, 40). Also see Brown, Driver, and Briggs (1951). *A Hebrew and English Lexicon of the Old Testament.* Oxford: Clarendon Press, 398; Morris, J.D. (1994). *The Young Earth.* Green Forest, AR: Master Books, 28; Stambaugh, James (April 1991). The days of Creation: A semantic approach. *Journal of Creation,* 5 (1): 70-78.

2. Dillow, Joseph C. (1982). *The Waters Above.* Chicago, IL: Moody Press, 141. Also see Whitcomb, J.C. (1988). *The World That Perished.* Grand Rapids, MI: Baker Book House, 41-46.

3. Whitcomb, J.C. (1988). *The World That Perished.* Grand Rapids, MI: Baker Book House, 77, 80; and Dillow (1982), op. cit., 141, 335-353.

4. Wolfe, Jack A. (1978). *U.S. Geological Survey Report*; as cited in Whitcomb (1988), op. cit., 80; and Dillow, op. cit., 348.

5. *National Geographic* magazine (February 1963). Antarctica: New look at a continent (Researched by George W. Beatty), 123 (2): 288, 296; cf. Nov. 1971, 653. Also cited in Whitcomb, 1988, 83.

6. A dispensation in the Bible is a period of time in which God dealt/deals with mankind. The Scripture divides time (from the creation of Adam in Genesis to the "new heaven and a new earth" of Revelation 21:1) into seven unequal periods, usually called dispensations (Ephesians 3:2), although these periods are also called ages (Ephesians 2:7). These periods are 1) Innocence, which extends from the creation of Adam in Genesis 2:7 to the expulsion from Eden, 2) Conscience, which extends from expulsion from Eden to the flood, 3) Government, which extends from the flood to Abram, 4) Promise, which extends from Abraham to Moses, 5) Law, which extends from Moses and the Ten Commandments to the time of Jesus Christ, 6) Grace, which extends from the crucifixion of Christ to the Second Coming, and 7) Millennium, which extends from the Second Coming for another 1,000 years.

These periods are designated in Scripture by a change in God's method of dealing with mankind with respect to sin and man's responsibility. Each of the dispensations may be regarded as a new test, and each ends in judgment, marking mankind's complete failure in every dispensation. Five of these dispensations, or periods of time, have been fulfilled—we are living at the close of the sixth, and facing the seventh and last dispensation, the millennium.

7. Scroggie, W. Graham (1953). *The Unfolding Drama of Redemption.* London: Pickering & Inglis Ltd., I: 74, 77. Also cited in Whitcomb, J.C., and Morris, H.M. (1961). *The Genesis Flood.* Phillipsburg, NJ: The Presbyterian and Reformed Publishing Company, 18.

8. In this context, the Second Coming refers to the return of Jesus Christ in the end times: 1) once for His church (without setting foot on earth) just before the seven-year tribulation and 2) once with the church at the end of the tribulation to put an end to evil, pain, and suffering as a result of ever-increasing sin and moral decay of mankind, reclaim the earth as His own, and judge ungodly men.

The return of Jesus Christ for His church is described in 1 Corinthians 15:51-52 and 1 Thessalonians 4:13-18, and in Matthew 24:37-39 and Luke 17:26-27. The church, made up of all who have trusted in the Lord Jesus Christ as their Savior and accepted the gift of salvation, will not be present during the tribulation. This event will take the world by surprise and will occur just prior to the tribulation period. In times of judgment, God provides a means of escape: Noah's Ark during the great flood and belief in the Lord Jesus Christ just prior to the final judgment in the last days. Jesus Christ will return with the church at the end of the tribulation as the King of Kings and Lord of Lords (Revelation 19:11-16).

The end times are described by Jesus in the Olivet Discourse found in Matthew 24:1–25:46, Mark 13:1-37, Luke 21:5-36, and in the Book of Revelation. The seven-year tribulation is a time of unimaginable carnage: worldwide famine, pestilence, persecution, and nuclear warfare. Throughout Scripture, the tribulation is referred to as the Day of the Lord (Isaiah 2:12, 13:6-9; Joel 1:15, 2:1-31, 3:14; 1 Thessalonians 5:2); trouble or tribulation (Deuteronomy 4:30, Zephaniah 1:1); the great tribulation, which refers to the more intense second half of the seven-year period (Daniel 9:27, Matthew 24:21); time or day of trouble (Daniel 12:1, Zephaniah 1:15); and time of Jacob's trouble (Jeremiah 30:7).

Eschatology (the study of the end of the age and the destiny of man as revealed in the Bible) is well studied and understood by many distinguished theologians. Please refer to books concerning eschatology and the Book of Revelation by Hal Lindsey (*The New World Coming,* 1984), Jack Van Impe (*Revelation Revealed,* 1982), Josh McDowell (*Evidence That Demands a Verdict,* 1973), Tim LaHaye (*Charting the End Times,* 2001), and many others on these subjects.

9. Cubit is about the distance between thumb and another finger to the elbow on an average person, or 1½ feet. This is used in the Roman system of measures and in different Greek systems.

10. Ramm, Bernard. *The Christian View of Science and Scripture* (Grand Rapids: Wm. B. Eerdmans Pub. Co. (1954), 157. Also cited in Whitcomb (1988), op. cit., 24.

11. Morris, Henry M. (September 1971). The ark of Noah. *Creation Research Society Quarterly,* 8 (2): 144. Also cited in Whitcomb (1988), 24; see Ramm, op. cit., 156-158.

12. Whitcomb, J.C., and Morris, H.M. (1961). *The Genesis Flood.* Phillipsburg, NJ: The Presbyterian and Reformed Publishing Company, 67. Also see Whitcomb (1988), 25.

13. Ibid., 65-70.

14. Ham, K., Sarfati, J., and Wieland, C. (2000). *The Revised & Expanded Answers Book.* (D. Batten, Ed.). Green Forest, AR: Master Books, 247.

15. Dinosaur (May 2010). In Wikipedia, the free encyclopedia. Retrieved May 2010, from http://en.wikipedia.org/wiki/Dinosaur. Also see Catchpoole, Sarfati, and Wieland, op. cit., 185.

16. Catchpoole, Sarfati, and Wieland, op. cit., 248; Sarfati, J. (2004). *Refuting Compromise.* AR: Master Books, chapters 7-8; as cited in Sarfati, J. (2005). How did dinosaurs grow so big? *Creation*, 28 (1): 44-47.

17. Stokstad, E. (November 2004). Dinosaurs under the knife. *Science*, 306 (5698): 962-965; Erickson, G. et al. (July 2001). Dinosaurian growth patterns and rapid avian growth rates. *Nature*, 412 (6845): 429-433; and Erickson, G. et al. (August 2004). Gigantism and comparative life-history parameters of tyrannosaurid dinosaurs. *Nature*, 430 (7001): 772-775; as cited in Sarfati, J. (2005). How did dinosaurs grow so big? *Creation*, 28 (1): 44-47.

18. Spinetti, C., and Buongiorno, M.F. (September 2004). Volcanic water vapour abundance retrieved using hyperspectral data. *IEEE International Geoscience & Remote Sensing Symposium*, 2: 1487.

19. Whitcomb (1988), op. cit., 40.

20. Whitcomb and Morris (1961), op. cit., 294. Also see Brooks, C.E.P. (1949). *Climate Through the Ages* (2nd Ed.). McGraw-Hill, 31-45.

21. Ibid., 292.

22. Ibid., 294. Also see Russell, Richard J. (December 1957). Instability of sea level. *American Scientist*, 45: 414-430.

23. Nimrod's name is from the verb "let us revolt."

24. Stedman, Ray C. (1978). The tower of Babel. Retrieved April 2008, from http://www.ldolphin.org/babel.html (adapted from Stedman, Ray C. [1978]. *The Beginnings.* Waco Books). Also see http://www.raystedman.org/genesis/.

25. Morris, J.D. (June 2008). The dinosaur next door. *Acts & Facts*, 37 (6), Dallas, TX: Institute for Creation Research, 13. Also see Catchpoole, Sarfati, and Wieland, op. cit., 244-247; and Hoesch, W.A. (January 2008). The Hualapi and the flood. *Acts & Facts*, 37 (1), Dallas, TX: Institute for Creation Research, 16.

epilogue

Prior to the nineteenth century, most geologists and paleontologists believed that the earth's land features and fossil rocks were the result of the biblical global flood. With the publication of Darwin's *The Origin of Species* in 1859, the theory of evolution gained wide acceptance in university circles. Geologists began to assume evolution and old earth as an established fact—and subsequently, began to interpret geological land features and their fossils, and even the origin of life, based on the assumption of Darwinian evolution. As evolutionary doctrine spread, so did the idea of uniformitarianism—that is, natural physical processes have remained constant during earth's history.

For more than a century, such beliefs have been taught in schools and universities, and students (who are teachers today) rotely accept such doctrine without independent study and research of their own. They carry this erroneous belief throughout their lives without question. In today's classrooms, students continue to be taught evolution and the old earth model as "proven fact"—which is "one of the biggest, if not the biggest, stumbling block to people being receptive to the gospel of Jesus Christ."[1]

Evolutionary doctrine is completely contrary to the First and Second Laws of Thermodynamics, Law of Biogenesis, and Law of Causality. The laws of thermodynamics are telling us that *creation ended sometime in the past*—and it (creation) was replaced with the First Law (matter and energy cannot be created or destroyed); and the Second Law (*unrestrained* universal decay as a dominating trend in nature). Also, it is telling us that life was created from life, not from nonliving chemicals (dust or rocks); and the earth and universe had a Cause. Evolutionists have no explanation for cause.

No one has ever observed or demonstrated spontaneous generation of life from nonliving chemicals and for that matter, no one has ever observed any organism give rise to a different type of organism—it remains purely conjecture. Evolution has no known biological processes or mechanisms to form higher levels of organization and complexity. Gene mutations are overwhelmingly degenerative and none are uphill in the sense of adding new genetic information to the gene pool. No one has ever observed "uphill drift" of genetic information.

What we observe today are distinct kinds of plants and animals (mankind, horse kind, dog kind, cat kind, bear kind, etc.) with no transitional forms found anywhere in the world. If there were a slow evolutionary transition from amoeba to higher order animal to man over billions of years, one would expect to find billions upon billions of transitional fossils and living forms. So where are all the in-between fossils and living forms? This one fact alone should convince most rational people that evolution is a false doctrine—a great deception. Variations in animal populations are not evolutionary changes but reshuffling of genes within an "existing gene pool" that was present in the original kind of animal (or plant) in response to the environment.

Consider the unimaginable improbabilities of getting the evolutionary process started. The probability of getting an average-size protein of left-handed amino acids by random, natural processes is zero. And the probability of getting a living cell, synonymous to the most sophisticated supercomputer yet microscopic in size, is likewise zero. How did we get all the organs of the body—a brain, heart, lungs, stomach, colon, teeth, tongue, eyes, ears,

> Secular scientists contend as fact what they cannot prove—and they underestimate the God of creation and judgments. In the words of Jesus Christ, *"You are mistaken, not understanding the Scriptures or the power of God."* (Matthew 22:29, NAS)

skin, muscles, bones, a nervous and vascular system—all working together, interdependently. And the ingredient that "makes it alive" and "consciousness" can only be explained by a Divine Creator. Yet people are willing to accept unknown evolutionary processes—chaos to cosmos all by itself—without question.

Geologic landforms and sedimentary features found throughout the world confirm catastrophic upheaval and a global flood—not uniformitarianism. For example, massive mountain ranges with no visual signs of denudation (erosion) to offset supposed slow and gradual uplift over many millions of years; huge monoliths and gigantic igneous formations (batholiths); vast areas of igneous and metamorphic rock often exposed at the surface (continental shields); ancient super volcanoes such as the one that lies beneath Yellowstone National Park and found elsewhere throughout the world; mountain overthrust (low angle reverse thrusting); and rock strata most often uplifted, tilted, inverted, folded, or missing—all are the result of worldwide tectonic forces vastly more intense than those observed today.

Also found worldwide are sharp contacts between sediment layers and formations supposedly many millions of years old with no evidence of erosion, soil layers, or animal activity (bioturbation), and polystrate fossils and clastic dikes, all indicative of rapid burial and a worldwide flood. And dry falls and canyons, ancient dry lake basins and enclosed salt lakes, and vast alluvial plains, overfit valleys and raised river terraces are all consistent with a worldwide flood and drainage.

Secular scientists assume radioactive decay has never been interrupted by worldwide cataclysmic events and thus, radioisotope dating is assumed to be valid. To the contrary, there is very strong scientific evidence for accelerated nuclear decay and a young earth which includes helium in zircon crystals, radiohalos, fission tracks, and rapid magnetic field reversals and decay. Also, there are hundreds of other geochronometers for a young earth that include carbon-14 in coal and diamonds and in deep geological strata, all supposedly hundreds of millions of years old (researchers are unable to find carbon without carbon-14), and lack of continental erosion, ocean sediments, and salt in the sea.

Although many people reject evolution for many of these reasons, particularly with the knowledge that there are no transitional forms anywhere in the fossil record, the possibility of creation in six literal days and a young earth seems to be a difficult barrier for many to overcome. Again, this is because people today have been so well indoctrinated with the idea of evolution and an ancient earth by secular teachers or professors in public schools and universities, and are never exposed to the substantial evidence for a young earth and the many "fatally flawed" assumptions and "gaps and failures" in evolutionary doctrine.

Creation has not been disproved by science but has simply been "rejected or ignored" in favor of *false assumptions and poor science*. Consider the words of Charles Haddon Spurgeon in 1877,[2] part of which bears repeating.

> What is science? ... scientists may boldly assert what they cannot prove, and may demand a faith far more credulous than any we possess....the march of science, falsely so called, through the world may be traced by exploded fallacies and abandoned theories. Former explorers once adored are now ridiculed; the continual wreckings of false hypotheses is a matter of universal notoriety. You may tell where the learned have encamped by the *debris* left behind of suppositions and theories as plentiful as broken bottles.

The fact of the matter is, the vast majority of scientific evidence (i.e., geology, biology, physics, astronomy, anthropology, archaeology, etc.) supports creation and a young earth.

> The Bible speaks of people being corrupted in their thinking by turning their backs on God (Romans 1:18) and so spiritually blind that they cannot see the obvious (Acts 28:25-27). Also see 2 Corinthians 4:4.

why do people believe so strongly in evolution?

So why do people (scientists and laymen) believe so strongly in evolution? Essentially, there are four reasons:

1. In today's world, people are generally a product of a secular, naturalistic educational system. Evolution is taught in school (science, history, philosophy, and even religion courses) as "proven fact"—and then later in life, students perpetuate this false doctrine without question as teachers, journalists, and parents. Teaching evolution is upheld by the man-made concept of "separation of church and state" and, consequently, opposing arguments are rarely, if ever, presented to students (see chapter 1). For example, most students have never been taught or even considered the merits and scientific evidence of global catastrophism, which support creation and a young earth. Evolution is widely accepted because many leading scientists and educators, as well as some progressive-minded clergymen, voraciously assert evolutionary doctrine and exclaim that only the ignorant would believe otherwise.

2. In 2 Peter 3:5 (NIV and KJV), it states, *"they deliberately forget"* and *"they willfully are ignorant."* In today's world, the term "supernatural" or the concept of God or creation is considered outside the realm of real science. Life is here on earth, so secular scientists feel they must explain life "naturalistically"—consequently, they believe that evolutionary doctrine and ignoring data contrary to evolution is legitimate. Thus, any evidence that supports a young earth or creation is automatically rejected or ignored. Although there is overwhelming scientific evidence supporting a young earth and Divine Creation, such evidence has been ignored by secular scientists because they would have to admit that evolution is ***wrong***.

"The question is, just why do they need to counter the creationist message?" (This question is addressed in the Summary of chapter 2.) "The fact is that evolutionists believe in evolution because they want to. It is their desire at all costs to explain the origin of everything without a Creator..."[3] Reasons are further described in item #4 below.

3. Many scientists hold to the evolutionary doctrine for fear of being ridiculed. As one good naturalist put it, "Any hint of teleology [i.e., intelligent design or purpose] must be avoided."[4] A geneticist once said that the scientific community requires the scientist to maintain two insanities at all times. "One, it would be insane to believe in evolution when you can see the truth for yourself. Two, it would be insane to say you don't believe in evolution. All government work, research grants, papers, big college lectures—everything would stop. I'd be out of a job, or relegated to the outer fringes where I couldn't earn a decent living."[5]

4. But there is an even more ominous reason for belief in evolutionary doctrine. Humanity, ever since the rebellion of the first man, Adam, has had an inborn tendency to oppose the Creator's rule or sovereignty over their lives as foretold in Genesis 3:6 and Romans 5:12. See Prologue. Rebellious and idolatrous history of humanity is graphically portrayed in the Bible (pagan antediluvian and postdiluvian

Despite the impossible improbabilities of evolution, scientists continue to adhere to Darwinian evolution from primeval chaos to a state of higher organization and complexity.

times); the Dark Ages; World War I and World War II; and by today's continual worldwide ethnic conflicts.

The sin nature of man is also portrayed by the repugnant and exceedingly evil behavior of some clergymen over the ages—physical cruelty, collusion with political dictators, and abhorrent sexual abuses—behavior that has alienated many from the Christian faith. The problem has never been with the God of the Holy Bible but with the corruption and moral depravity of humankind. The Bible speaks of it in Romans, chapter 1.

Romans, chapter 1:

"The wrath of God is being revealed from heaven against all the godlessness and wickedness of men who suppress the truth by their wickedness, since what may be known about God is plain to them, because God has made it plain to them. For since the creation of the world God's invisible qualities—His eternal power and divine nature—have been clearly seen, being understood from what has been made, so that men are ***without excuse.*** *For although they knew God, they neither glorified Him as God nor gave thanks to Him, but their thinking became futile and their foolish hearts were darkened."* (Romans 1:18-21, NIV) [Bold added]

"Although they claimed to be wise, they became fools and exchanged the glory of the immortal God for images made to look like mortal man and birds and animals and reptiles. Therefore God gave them over in the sinful desires of their hearts to sexual impurity for the degrading of their bodies with one another. They exchanged the truth of God for a lie, and worshiped and served created things rather than the Creator—who is forever praised. Amen. Because of this, God gave them over to shameful lusts. Even their women exchanged natural relations for unnatural ones. In the same way the men also abandoned natural relations with women and were inflamed with lust for one another. Men committed indecent acts with other men, and received in themselves the due penalty for their perversion." (Romans 1:22–27, NIV)

"Furthermore, since they did not think it worthwhile to retain the knowledge of God, He gave them over to a depraved mind, to do what ought not to be done. They have become filled with every kind of wickedness, evil, greed and depravity. They are full of envy, murder, strife, deceit and malice. They are gossips, slanderers, God-haters, insolent, arrogant and boastful; they invent ways of doing evil; they disobey their parents; they are senseless, faithless, heartless, ruthless. Although they know God's righteous decree that those who do such things deserve death, they not only continue to do these very things but also approve of those who practice them." (Romans 1:28-32, NIV)

In spite of all known physical and biological laws and principles, and geologic evidence of a worldwide flood, *scientists continue to devise various schemes and false assumptions to explain and uphold evolution.*

The concept of creation and a Divine Creator is *completely avoided* in all scientific literature (except creation research journals, magazines, and creation-science books). Because of atheistic views of university academia today, the belief that life spontaneously generated

One of the primary reasons for abandoning faith in God is the widespread acceptance of evolutionary thinking—that everything made itself by natural processes and God is unnecessary. To avoid having to recognize God, *"professing themselves to be wise, they became fools."* (Romans 1:22, NAS)

from nonlife has been universally accepted as the explanation for all life forms including man. *The most impossible improbabilities are considered more probable than the preponderance of observational and empirical evidence for creation and a Divine Creator.*

Scientific evidence for creation is overwhelming and, based on science alone, it is completely illogical to believe anything else. Only a dedicated commitment to atheism, or a strong need for approval of atheistic university academia, could lead one to believe in evolutionary doctrine. As explained in chapter 10, the fundamental inclination of godless men and basic philosophy of humanism is that man created himself, has control over his own destiny, and can rely on his own superior "intellect"—that man is God. "Glory to man in the highest," for man created and is the master of all things.

As with the Tower of Babel, the ultimate motive is expressed in these words: *"Let us make for ourselves a name"* (Genesis 11:4, NAS), or *"so that we may make a name for ourselves."* (NIV) The apostle Peter prophesized in 2 Peter 3:5-6 this very thing when he said that those who would *"in the last days"* deny the flood [in support of evolution] not because of scientific evidence but rather because *"they deliberately ignore."* The impact of this statement may be better understood by this range of Bible translations:

> *"they willfully are ignorant of"* (KJV, King James Version)
>
> *"they willfully forget"* (ASV, American Standard Version)
>
> *"they deliberately forget"* (NIV, New International Version)
>
> *"they willfully shut their eyes to the fact that"* (Twentieth Century New Testament)
>
> *"they lose sight of the fact that"* (NEB, New English Bible)
>
> *"it escapes their notice"* (NASB, New American Standard Bible; footnote: "*they are willfully ignorant of this fact*")
>
> *"for this they willfully forget"* (NKJV, New King James Version)
>
> *"they are choosing to forget"* (JB, Jerusalem Bible)
>
> *"deliberately shutting their eyes to a fact that they know very well"* (J. B. Phillips, The New Testament in Modern English, 1958)

The Bible says that those who deny creation are *"without excuse."* (Romans 1:20, NIV) Paul warned that *"For the time will come when men will not put up with sound doctrine. Instead, to suit their own desires, they will gather around them* ***a great number of teachers to say what their itching ears want to hear.*** *They will turn their ears away from the truth and turn aside to myths."* (2 Timothy 4:3-4, NIV) [Bold added]

John C. Whitcomb, in his book *The Early Earth*, writes, "In recent years remarkable testimony has come from the pens of highly respected scientists who recognize that the evolution concept, in its broader aspect, rests upon a vanishing foundation. G. A. Kerkut of the Department of Physiology and Biochemistry at the University of Southampton, for example, notes that evolutionists often write as though they 'have had their views by some sort of revelation.' In spite of 'many gaps and failures' in their system [evolution], it is 'taken on trust' by a 'blind acceptance' and a 'closing of the eyes' to many important facts, thus revealing an 'arrogant' rather than truly scientific spirit. Attempts to bridge the gap between invertebrates and vertebrates for example, have resulted in 'science fiction' rather than discovery, and the possibility that life first began spontaneously is a 'matter of faith on the part of the biologist.'"[6]

> The magnificence of our world and universe clearly demonstrates a Creator.
>
> "*...God has made it plain to them. For since the creation of the world God's invisible qualities—His eternal power and divine nature—have made it clearly seen, being understood from what has been made, so that men are* ***without excuse.***" (Romans 1:19-20, NIV) [Bold added]

days of noah and the second coming

There are two earth-shattering events which cannot be explained on the basis of uniformitarianism. The first was the creation of the world and the second was the worldwide flood. It was the second of these two events, the Flood, which was the basis for Peter's comparison with the Second Coming of Jesus Christ and the final destruction of the world.[7] (See end note 7 for an explanation of the Second Coming.)

What can humanity expect in the near future—at the Second Coming of Christ? Like it was in the days of Noah, today's world is filled with political corruption, lying, slander, public displays of depravity, violent crimes, abortion, theft, adultery, drug-taking, drunkenness, gambling, and greed of all kinds, not to mention wars, rumors of wars, and terrorism. Does this sound familiar? (Refer to Romans 1:18-22.) Just like in the days of Noah (antediluvian period—Age of Conscience), *"The Lord saw how great man's wickedness on the earth had become, and that every inclination of the thoughts of his heart was only evil all the time."* (Genesis 6:5, NIV)

The teaching echoed by Christ (Matthew 24:37-39, Luke 17:26-27) and the apostle Peter (2 Peter 3:6-7) in the New Testament speaks of the destruction of mankind by flood and then by fire at the time of Christ's return.

> *"As it was in the days of Noah, so it will be at the coming of the Son of Man. For in the days before the flood, people were eating and drinking, marrying and giving in marriage, up to the day Noah entered the ark; and they knew nothing about what would happen until the flood came and took them all away. That is how it will be at the coming of the Son of Man."* (Matthew 24:37-39, NIV)

> *"Just as it was in the days of Noah, so also will it be in the days of the Son of Man. People were eating, drinking, marrying and being given in marriage up to the day Noah entered the ark. Then the flood came and destroyed them all."* (Luke 17:26-27, NIV)

The parallels between the days of Noah and the Second Coming are described by the apostle Peter as well:

> *"Know this first of all, that in the last days* ***mockers will come with their mocking****, following after their own lusts, and saying, 'Where is the promise of His coming? For ever since the fathers fell asleep, all continues just as it was from the beginning of creation.' For when they maintain this, it escapes their notice* [*"deliberately forget,"* NIV] *that by the word of God the heavens existed long ago and the earth was formed out of water and by water, through which the world at that time was destroyed, being flooded with water. But the present heavens and earth by His word are being reserved for fire, kept for the day of judgment and destruction of ungodly men."* (2 Peter 3:3-7, NAS) [Bold added]

In this passage of Scripture, the apostle Peter spoke of a day in the future when men would *scoff* at the possibility of Christ's Second Coming as a cataclysmic, global intercession by God into world affairs. And the reason for this skeptical attitude is because of pagan rebellion and unbelief in God, and blind obedience to evolutionary doctrine. (See chapter 10, Comparison to the Second Coming for more information.) Essentially, it is the belief that God has never judged the world through a worldwide flood. Because we are told by atheistic academia and evolutionists that this is so, most people show little or no concern for God's future judgment as foretold in the Bible.

> Many bible scholars believe that we are at that prophetic time in world history with the miraculous rebirth of the nation of Israel on May 14, 1948, and with Jerusalem in the hands of the Jewish people following the Six Day War (June 5-10, 1967). The State of Israel and Jerusalem under Israeli control are necessary prerequisites for the fulfillment of end-time prophecy.

why it matters

One of the primary reasons for abandoning faith in God is the widespread acceptance of evolutionary thinking—that everything made itself by natural processes and God is unnecessary. Why does it matter? It matters because the assumption of evolution is based on a LIE—*the greatest deception in modern history.*

It also matters because removing God and prayer from school and family life, including Christmas and Easter celebration, has led to a collapse of our ethical and moral foundation and is the primary reason for today's social ills. There seems to be a willful (many would term ferocious) determination by some to turn America into a godless, amoral, secular-humanist society.

> **There is only one unpardonable sin** as expressed in John 3:18 (NIV): "*Whoever believes in him is not condemned, but whoever does not believe stands condemned already because he has not believed in the name of God's one and only Son.*"

Most importantly, it matters because our *eternal salvation* depends on our belief and acceptance of the Lord Jesus Christ as our Savior. It is the most important decision in your life—it is a matter of where you will spend eternity. We are familiar with the Scripture of John 3:16 (NIV), *"For God so loved the world that He gave His one and only Son, that whoever believes in Him shall not perish but have eternal life."*

> Students continue to be taught evolution and old earth model as "proven fact," and they carry this false belief throughout their lives without question. And sadly, such false beliefs undermine traditional beliefs and cause many to turn away from the Creator God.

Dr. Hal Lindsey, in his weekly "The Hal Lindsey Report," said, "I'll share a few thoughts on one of the greatest crises we face in the West today. Jesus described it as *having eyes, but not seeing; having ears, but not hearing.* [This is Scripture taken from Luke 8:10-12.] Here's the danger; our scientific, medical, educational, governmental, and moral agendas today are determined *by people who are willfully blind—not stupid or naïve—but deliberately blind to the truth*. If we sit idly by and let them set the course for the future, we'll get what we, as a nation, deserve. God help us."[8] [Italics added]

Jesus Christ is our Heavenly Father from ancient times (from beyond time), *"His goings forth are from long ago, from the days of eternity."* (Micah 5:2, NAS) He is the Creator of the heavens and earth and all that exists (Jeremiah 10:12-16, 51:15-19)—and He created a sinless world where everything was perfectly upheld (John 1:1-3, Colossians 1:16-17, Hebrews 1:2-3). This perfection was lost by the rebellion of Adam.

Today, we are in a world that *"groans and suffers"* (Romans 8:22, NAS) because of the Edenic Curse. Unless we let go of our pride and arrogance, and realize that Jesus Christ is our Heavenly Father, we will never understand the significance of His death on the Cross and the work of redemption and we will be lost for eternity.

In the Age of Grace, Romans 10:13 (KJV) victoriously announces that *"whosoever shall call upon the name of the Lord shall be saved"*—it requires just one simple act before God in prayer. By allowing God in your life, and by acknowledging and repenting of sins, and believing in our Creator, the Lord Jesus Christ, anyone, no matter what sins an individual has committed, can obtain the free gift of eternal salvation. Believing the Lord Jesus Christ died and shed his blood for our sins on the Cross is a matter of where you will spend eternity—eternally and everlasting.

Moses said to Israel, *"I have set before you life and death ... choose life."* (Deuteronomy 30:19, NIV)

notes: epilogue

1. Ham, K. (2003). *Why Won't They Listen?* Green Forest, AR: Master Books, chapter 8. Retrieved November 2009, from http://www.answersingenesis.org/Home/Area/wwtl/chapter8.asp.

Alcorn, R. (2009). *If God Is Good – Faith in the Midst of Suffering and Evil.* Colorado Springs, CO: Multnomah Books, 11–12.

Although evolutionary doctrine remains an obstacle for many in receiving the gospel of Jesus Christ, people point to the problem of evil and suffering as their primary reason for not believing in God. A Barna poll asked, "If you could ask God only one question and you knew he would give you an answer, what would you ask?" The most common response was, "Why is there pain and suffering in the world?" (p. 11, Alcorn) Remarkably, this problem is bluntly and passionately articulated throughout the Bible. See Psalms 6:3, 10:1, 13:1; Habakkuk 1:2–3. The Bible never evades the problem of evil. "Six hundred times specific terms for *evil* appear, with thousands of other [particular] references to sin and wickedness...[including] Adam and Eve's fall, Cain's murder, Noah's flood, the tower of Babel, the patriarchs' sins, Job's tragedies, Egypt's oppression of Israel, David's psalms of lament, Israel's rebellion and exile, the suffering of the prophets, and the long, lonely wait for Messiah—it goes on relentlessly..." (p. 20–23, Alcorn) See Prologue, note 2.

Nowhere in the Bible does it state that Christians are free from pain and suffering (notably, between the fall in Genesis 3 and the future judgment in Revelation 20)—but to the contrary, *"For it has been granted to you on behalf of Christ not only to believe on him, but also to suffer for him..."* (Philippians 1:29, NIV) We all suffer the consequences of sin, *"For the wages of sin is death, but the gift of God is eternal life in Christ Jesus our Lord."* (Romans 6:23, NIV) But in Revelation 21:4 (NIV), it states that *"He will wipe every tear from their eyes. There will be no more death or mourning or crying or pain, for the old order of things has passed away."*

2. Spurgeon, C.H. (May 1877). *The Sword and the Trowel (*Inaugural Address – 13th Annual Conference of the Pastors College), Pilgrim Publications, 197.

3. Morris, H.M. The scientific case against evolution. Institute for Creation Research. Retrieved October 2014, from http://www.icr.org/home/resources/resources_tracts_scientificcaseagainstevolution/. (Dr. Henry M. Morris (1918-2006) was founder of the Institute for Creation Research.)

In the article by Dr. Morris, it is interesting to note the candid statements by various university professors concerning the 'belief in modern evolution.'

4. de Duve, Christian (September–October 1995). The beginnings of life on earth. *American Scientist*, 428. Also cited in Morris, J.D. (2003). *Is the Big Bang Biblical?* Green Forest, AR: Master Books, 79.

5. Ledger, Lynchburg, Virginia. George Caylor's article titled "The Biologist," appeared on February 17, 2000. Permission received from George Caylor on August 10, 2009. Also cited in Morris, J.D. (2003). *Is The Big Bang Biblical?* Green Forest, AR: Master Books, 70.

6. Whitcomb, J.C. (1972). *The Early Earth*. Grand Rapids, MI: Baker Book House, 21; and Kerkut, G.A. (1960). *Implications of Evolution.* New York, NY: Pergamon Press, 150, 153-155.

7. In this context, the Second Coming refers to the return of Jesus Christ in the end times: 1) once for His church (without setting foot on earth) just before the seven-year tribulation and 2) once with the church at the end of the tribulation to put an end to evil, pain, and suffering as a result of ever-increasing sin and moral decay of mankind, reclaim the earth as His own, and judge ungodly men.

The return of Jesus Christ for His church is described in 1 Corinthians 15:51-52 and 1 Thessalonians 4:13-18, and in Matthew 24:37-39 and Luke 17:26-27. The church, made up of all who have trusted in the Lord Jesus Christ as their Savior and accepted the gift of salvation, will not be present during the tribulation. This event will take the world by surprise and will occur just prior to the tribulation period. In times of judgment, God provides a means of escape: Noah's Ark during the great flood and belief in the Lord Jesus Christ just prior to the final judgment in the last days. Jesus Christ will return with the church at the end of the tribulation as the King of Kings and Lord of Lords (Revelation 19:11-16).

The end times are described by Jesus in the Olivet Discourse found in Matthew 24:1–25:46, Mark 13:1-37, Luke 21:5-36, and in the Book of Revelation. The seven-year tribulation is a time of unimaginable carnage: worldwide famine, pestilence, persecution, and nuclear warfare. Refer to end note 8 of chapter 10 for additional information.

8. The Hal Lindsey Report, Palm Desert, CA, by Hal Lindsey dated April 24, 2008. (hallindsey.com).

Appendix A
Glossary

Abiogenesis– a hypothetical term coined by evolutionists to describe how life arose spontaneously from nonlife. No one has ever observed or demonstrated spontaneous generation of life from nonlife.

Adaptation (commonly known as "survival of the fittest," natural selection, speciation, or variation)– the genetic process within populations which selects gene traits from an *existing gene pool* best suited for a specific environment. For example, DDT resistance in insects, change in dominant color, and long hair versus short hair (sometimes referred to by scientists as microevolution) is NOT evolutionary change but merely shuffling or recombination of the already *existing gene pool* that was originally present in a certain "kind" of animal or plant population. No one has ever observed uphill drift or the *addition of genetic information* to the gene pool. Genetic drift has *always* been "downhill."

Alcove– a small, wide canyon with an overhanging wall.

Alluvial plains (or alluvial desert)– are relatively flat landforms that, in most cases, have no stream or river drainage. Such plains are relict landforms created by the deposition of suspended sediment and receding floodwaters of a worldwide flood.

Alluvium– a general term describing all river-deposited sediment. It includes sediment deposits of riverbeds, river flood plains, river deltas, and alluvial fans. It is typically made up of a variety of materials, including fine particles of silt and clay, sand, gravel, cobbles, and sometimes, boulders.

Ammonites– an extinct group of marine animals belonging to the Cephalopod family, subclass Ammonoidea. The closest living relative is the subclass Coleoidea (octopus, squid, and cuttlefish). Ammonites were free-swimming, deep sea animals. They are considered index fossils and often linked to specific geological time periods by secular geologists.

Amino acids– complex genetic machinery; basic building blocks of protein molecules made up of carbon, hydrogen, oxygen, nitrogen, and sometimes sulfur. See proteins; enzymes.

Animal phyla– the largest generally accepted groupings of animals and other living things. The best known are the Molluscs, Porifera, Coelenterata, Playhelminthes, Nematodes, Annelids, Arthropods, Echinoderms, and Chordates (Chordata), the phylum to which humans belong. Although there are approximately 35 phyla, these nine include over 96% of animal species. The classification of living things is life, domain, kingdom, phylum, class, order, family, genus, and species.

Animals– a major group of multicellular organisms of the kingdom Animalia or Metazoa. Most animals are motile—a term which refers to the ability to move, consuming energy in the process. Animals are heterotrophs; they are dependent on other organisms (for example, plants) for sustenance.

Annealing temperature– the temperature at which crystal defects in rocks disappear after the rock has cooled and solidified. Defects such as radiohalos or fission tracks form after igneous rock has solidified but would disappear if the rock were reheated above the annealing temperature.

Antediluvian (syn. Prediluvian; Latin for *"before the deluge"*)– a term describing a period of time that preceded the Great Flood of Noah as described in the Book of Genesis in the Bible.

Anthropic Principle, The – the universe was designed in a very precise manner to support human life.

Anthropology– the study of the origin, the behavior, and the physical, social, and cultural development of humans.

Anticline– a bend in a rock formation that resembles an arch—that is, the bend or fold is concaved downward.

Aragonite– a carbonate mineral, one of the two common, naturally occurring polymorphs of calcium carbonate, $CaCO_3$. The other polymorph is the mineral calcite. (Polymorphism is the ability of a solid material to exist in more than one form or crystal structure.)

Asthenosphere– that portion of the upper mantle of the earth beneath the lithosphere where magma is formed. It extends from a depth of about 100 to about 350 kilometers below the surface of the earth. See diagram in Appendix D.

Atmosphere (Earth's atmosphere)– a layer of gases surrounding the earth and retained by the earth's gravity. It contains approximately 78.08% nitrogen, 20.95% oxygen, 0.93% argon, 0.038% carbon dioxide, trace amounts of other gases, and a variable amount (average around 1%) of water vapor. This mixture of gases is commonly known as **air.** The atmosphere protects life on earth by absorbing ultraviolet solar radiation and helps moderate temperature extremes (temperature decreases with height). The atmosphere comprises the following layers (top to bottom): exosphere, thermosphere, mesophere, stratosphere, and troposphere.

Bacteria (singular bacterium)– any of numerous groups of microscopic one-celled organisms of the phylum Schizomycota of the kingdom Monera.

Bacteriophage (informally, phage)– a virus that infects and replicates within bacteria.

Basalt– dense, volcanic rock (type of igneous rock) which is usually dark gray or black and fine-grained. It comprises most of the oceanic crust and was extruded by volcanoes or through fissures. Basalt commonly comprises the minerals feldspar and pyroxene.

Batholith– a very large volume of igneous intrusive (also called plutonic) rock that forms from cooled magma. Batholiths (from Greek *bathos*, depth, and *lithos*, rock) are almost always granite, quartz monzonite, or diorite.

Bedding plane– a surface that separates layers of sedimentary rocks. Bedding is layering or strata that develops as sediments are deposited. Cross-bedding refers to beds of sandstone that are at an angle to the main horizontal bed or layer.

Biblical creationism– the belief that humanity, life, the earth, and the universe were created within six solar days by the God of the Bible about 6,000 years ago.

Biodegradation– the process by which organic materials are broken down by enzymes (biochemical catalysts) produced by microorganisms. Organic material can be degraded aerobically, with oxygen, or anaerobically, without oxygen.

Biogenesis, Law of– states that life can only arise from pre-existing life, not from nonliving matter. This law has become the cornerstone of biological sciences. No one has ever seen or demonstrated "spontaneous generation" of life from nonlife.

Biosphere– that part of the earth, including air, land, surface rocks, and water, where life occurs.

Black hole– a theoretical region of space in which the gravitational field is so powerful that nothing, not even electromagnetic radiation or visible light, can escape its pull.

Bloom– a rapid increase or accumulation in the population of planktonic organisms in an aquatic system. Blooms may occur in freshwater as well as salt water environments. Cause is related to nutrients, especially nitrogen, phosphorus, and iron, warming temperature, and turbulence.

Breccia– a rock composed of angular clastic fragments that are cemented in a matrix. (Cementing materials may include dissolved calcium carbonate, silica, iron, and moisture.) Variety of different origins may include sedimentary breccia and igneous breccia.

Butte– a flat-topped mountain with several steep cliff faces. A butte is smaller and more "tower-like" than a *mesa*.

Calcite– a carbonate mineral and the most stable polymorph of calcium carbonate ($CaCO_3$). The other polymorph is *aragonite*. Aragonite will change to calcite at 470°C.

Carbonate minerals– minerals containing the carbonate ion CO_3^{2-}.

Catastrophism– maintains that normal geological and physical processes of the earth have been interrupted by a cataclysmic worldwide flood as described in Genesis 6-8. Does a worldwide flood covering the mountains seem far-fetched from a geologic and hydraulic standpoint? See Phases Diagram of the Worldwide Flood in chapter 4. There is a logical explanation and an impressive array of scientific evidence for a global flood and recent creation of the earth as presented in chapters 4 through 9.

Causality, Law of– see Law of Causality.

Cavitation– a general term used to describe the behavior of bubbles in a liquid. *Inertial* cavitation is the process where a bubble rapidly collapses, producing a shock wave. Such cavitation often occurs in pumps, propellers, and impellers.

Cementation– the process of deposition of dissolved mineral components in the pore spaces of sediments. It is an important factor in the lithification process—that is, converting unconsolidated sediments into solid rock such as sandstones, conglomerates, or breccias. Cementing materials may include dissolved calcium carbonate, silica, iron, and moisture.

Chalk– see limestone.

Chromosome– cell structure that carries DNA which consists of thousands of genes.

Circular reasoning– assumes as fact what one is trying to prove...***x*** is used to prove ***y*** and ***y*** is used to prove ***x***. For example, secular scientists may assert "evolution is true ***because*** living organisms created themselves, organisms tend to move from simple to more complex forms, and natural processes have remained constant during the world's history." The reasons offered after the word ***because*** provide ***no evidence***—these reasons are assumptions and offer no proof of the claim. This is called blind trust, or circular reasoning.

Clastic fragments– weathered or broken down rocks that have been transported from their place of origin. Clastic rocks are composed of fragments of preexisting rock. The term is commonly applied to sedimentary rocks.

Coal (also known as a fossil fuel)– a readily combustible black or brownish-black rock. The softer bituminous coal is regarded as a sedimentary rock, and the harder forms, such as anthracite coal, are regarded as metamorphic rocks because of exposure to elevated temperature and pressure.

Cogenetic– a term that implies that a rock material was created at the same time and place from a homogeneous pool of magma.

Combustion or burning– an exothermic chemical reaction (reaction that releases heat) between a fuel and an oxidant, accompanied by the production of heat, or heat and light in the form of a flame. For example: $CH_4 + 2O_2 \rightarrow CO_2 + 2H_2O$ + heat. A simpler example can be seen in the combustion of hydrogen and oxygen, which is a commonly used reaction in rocket engines: $2H_2 + O_2 \rightarrow 2H_2O$ + heat. The result is simply water vapor.

Condensation– process by which water is converted from a gaseous form (water vapor) to a liquid. See evaporation.

Conformity– the continual deposition of sediments with no interruption or erosional break in the depositional process.

Constant– something that does not change over time; that is, it is invariable. The term is an antonym of "variable."

Continental shelf– a shallow, nearly level area that is submerged below sea level at the edge of a continent (between the shoreline and the continental slope).

Continental shield– large stable area of low relief at or near the earth's surface comprising Precambrian igneous and metamorphic rock, e.g., Canadian, Amazonian, Baltic, West African, Indian, China-Korean, Australian, and Antarctic shields. Continental shields are found in all continents.

Continuum– anything that goes through a gradual transition from one condition to a different condition without any abrupt changes.

Convection current– a circular current in semi-fluid material (e.g., asthenosphere in the upper mantle of the earth) formed when heated materials rise and cooler materials sink.

Coulees– dry, braided canyons. The term is sometimes applied to a kind of valley or drainageway.

Crinoids– marine animals, also known as sea lilies or feather-stars, that make up the class Crinoidea of the echinoderms (phylum Echinodermata). They live in shallow water and at great depths (6,000 meters). Crinoids are characterized by a mouth on the top surrounded by feeding arms.

Cross-bedding– term referring to beds of sandstone that are at an angle to the main horizontal bed or layer.

Crust– the earth's outermost layer (about 4 to 45 miles thick, or 6.4 km to 72.4 km) composed of silicate rocks. (All rocks are composed principally of silicate minerals, SiO_2, silicate dioxide which includes the mineral quartz.) See diagram in Appendix D.

Deltaic deposit– a deposit of sediment (usually sand and silt) formed where a river enters a lake or an ocean.

Denudation– the removal of surface rocks and sediment by erosion and weathering—this geologic process leads to a reduction of elevation and relief in landforms and landscapes.

Deoxyribonucleic acid (DNA)– a nucleic acid that contains the genetic instructions used in the development and functioning of all known living organisms. The main role of DNA molecules is the storage of genetic information within genes. Each gene serves as a recipe on how to build a protein molecule and other components of a cell. Proteins comprise amino acids that perform important tasks for cell functions or serve as building blocks. Chemically, DNA consists of two long polymers of simple units called *nucleotides*.

Discordant (disagreeing)– a term that describes data (age of rocks) from different radioisotope techniques that do not agree with known historical dates, or with each other, and data using the same technique on different specimens of the same rock that do not agree with each other.

Discrepant– a term implying that data (age of rocks) from radioisotope dating techniques do not agree with fossil or stratigraphic studies.

Dolomite– a sedimentary carbonate rock composed of calcium magnesium carbonate, $CaMg(CO_3)_2$.

Drumlin– an elongated hill formed when a glacier flows over and reshapes a mound of till (unsorted glacial drift).

Dry braided canyon or coulee– see coulees.

Dry lake basin– a major topographic depression that was filled with water thousands of years ago.

Earthquake– the sudden release of energy in the earth's crust that creates seismic waves. See Appendix D.

Element– see radioisotope dating.

Endorheic basin (internal, terminal, or enclosed basin)– a closed drainage basin that retains water and allows no outflow to other bodies of water such as rivers or oceans.

Empirical analysis– verification through repeated measurement and testing. This is the basis for what is known as the "scientific method." Common steps include observation, prediction (hypothesis), data collection, experimentation to test the hypothesis under controlled conditions, and conclusions. Experimentation is testing that is carried out under controlled conditions to validate a hypothesis or perhaps to verify a known law.

Energy– the ability to do work or to move something. **Types of energy**– generally classified as potential energy (stored energy) and kinetic energy (energy of movement or released energy); types include heat energy (thermal energy), fossil fuels (coal, oil, and natural gas), chemical energy, electrical energy, solar energy, biomass energy (energy from plants), wind energy, hydropower, geothermal energy, ocean energy, and atomic energy. Matter is a form of energy.

> Where does energy come from? Most of earth's energy comes from the sun as solar energy. Through solar energy we are able to produce different kinds of energy. How is energy measured? One of the basic measuring units is called a BTU (British thermal unit). Also, energy can be measured in joules (approximately 1055 joules = 1 BTU).

Thermodynamics (and Energy Transformation)– is the study of heat power—heat is "energy in transit." It is the branch of physics that examines kinetic energy, or the efficiency of energy transfer, or energy transformation; for example, hot to cold; high pressure to low pressure; solar energy to fossil fuel to mechanical energy to electrical energy to light energy. When we eat, our bodies transform energy stored in food (potential energy) into energy to do work (kinetic energy). When we run or walk, think, read or write, we "burn" food energy in our bodies. Cars, planes, boats, light bulbs, and machinery also transform energy into work. Work means moving, lifting, warming, or lighting something.

The entire universe owes its continued existence to energy transformations—without such transformations there would be nothingness. Regardless of how energy is transformed, the total amount of energy after the transformation is the same as it was before the transformation. Energy can be neither created nor destroyed (First Law of Thermodynamics).

Entropy– a measure of disorder or unusable energy. It represents energy that is no longer available for doing work. As usable energy decreases and unusable energy increases, entropy increases. Entropy is increasing to a maximum as energy in the universe degrades to an eventual state of inert uniformity or randomness—that is, a state of no usable energy. Every energy transformation results in a reduction in the usable or free energy of the system. While the *quantity* of matter/energy remains unchanged, the *quality* of matter/energy has the tendency to decline or deteriorate over time. This is commonly known as **The Law of Increasing Entropy.**

Enzyme– type of protein molecule; specifically arranged amino acids that accelerate chemical reactions.

Eolian (aeolian)– a term referring to processes pertaining to the activity of winds and their ability to shape the surface of the earth. Winds may erode, transport, and deposit materials, and are effective erosional agents in regions with sparse vegetation such as deserts. Water is a much more (significantly more) powerful erosional agent than wind.

Erosion– the movement of rocks and minerals by agents such as water, ice, wind, and gravity.

Erratic boulder– a boulder that was transported to its present location by a glacier. (Such a boulder is usually different from the bedrock in the immediate area.)

Esker– a snake-like ridge formed by deposition in a stream that flowed on, within, or beneath a glacier.

Evaporation– process by which water is converted from a liquid form to a gas (water vapor). See condensation.

Evaporites— a type of sedimentary rock that forms when water evaporates leaving behind minerals such as halite (NaCl) and gypsum ($CaSO_4$-$2H_2O$).

Evolution— the belief that all living organisms created themselves by their own natural processes with no supernatural input. It is a doctrine of increasing organization and complexity in the universe, or a doctrine of continual design without a designer. Evolutionists believe that man descended from the apes, all vertebrates descended from fish, all fish descended from invertebrates, and all life descended from single-celled organisms which arose spontaneously from nonliving chemicals (dust or rock). Evolution has *never been observed* or known to occur within living populations. See abiogenesis.

Extinction— the cessation of existence of a species or group of animals (or plants). Extinction is usually brought about by adaptation to a specific environment following a drastic change in that environment. Some believe there may be other causes such as disease and meteorological events.

Feldspar— the name of a group of rock-forming minerals that make up 60 percent or more of the earth's crust. Feldspars crystallize from magma and are found in igneous, metamorphic, and sedimentary rocks. The mineralogical composition of most feldspars can be expressed in terms of orthoclase ($KAlSi_3O_8$), albite ($NaAlSi_3O_8$), and anorthite ($CaAl_2Si_2O_8$). Chemically, feldspars are silicates of aluminum comprising sodium, potassium, iron, calcium, or barium, or combinations of these elements.

First Law of Thermodynamics— see thermodynamics.

Fission tracks— see chapter 8.

Flocculation— a process where a dissolved mineral (solute) precipitates out of solution in the form of *floc* or "flakes." It also refers to the process by which fine particulates clump together into floc.

Formation— a distinct body of sedimentary, igneous, or metamorphic rock with similar characteristics. A formation usually comprises a group of similar sediment layers. Examples of sedimentary rock formations are Tapeats Sandstone, Bright Angle Shale, Redwall Limestone, Hermit Shale, Coconino Sandstone, Toroweap Limestone, and Kaibab Limestone of the Grand Canyon in Arizona.

Fossilization (petrification)— is the process by which organic material is converted into mineralized fossils. It is essentially the replacement of organic material with silica (silicon dioxide, SiO_2) and other minerals such as calcite and pyrite. See petrification and cementation.

> **Silicification** is the process in which organic matter is replaced with silica. Volcanic material is a common source of silica. Silicification most often occurs when a specimen is quickly buried in sediments of deltas and floodplains or organisms are buried in volcanic ash. Water must be present because it reduces the amount of oxygen and reduces deterioration of the organism by fungi and bacteria.

Fumarole (from Latin *fumus*, smoke)— an opening in the earth's crust, usually in the vicinity of volcanoes. It typically emits steam and gases such as carbon dioxide, sulfur dioxide, hydrochloric acid, and hydrogen sulfide.

Galaxy— a massive, gravitationally bound system consisting of stars, interstellar gas and dust, and dark matter orbiting around a common center of mass. Typical galaxies range from dwarfs with as few as 10 million stars up to giants with one trillion stars, all orbiting a common center of mass. Galaxies also contain multiple star systems, star clusters, and interstellar clouds. The Sun is one of the stars in the **Milky Way** galaxy; the **Solar System** includes the earth and all the other objects that orbit the Sun.

Gamete— a cell (e.g., egg) that fuses with another gamete (e.g., sperm) during fertilization (conception) in organisms that reproduce sexually. A *zygote* is a fertilized egg.

Gene— a segment of DNA located on the chromosome. Each gene is one set of instructions or blueprint to build one particular component of the organism. Genes consist of nucleotides which carry instructions to build a protein molecule. Proteins are biological molecules consisting of one or more chains of amino acids; the sequence of amino acids in a protein is defined by the gene. See Appendix B.

Gene pool— genetic information *present* in the population for a "kind" of animal or plant. The original gene pool comprised a large or wide variety of genetic information *present* in the original population.

General relativity theory— the geometric theory of gravitation published by Albert Einstein in 1916. It is the description of gravity in modern physics. It unifies special relativity and Newton's law of universal gravitation (see chapter 9, Gravitational Time Dilation and Distant Starlight), and describes gravity as a property of the geometry of space and time, or spacetime.

Genetic burden— see mutational load.

Genetic code— the set of rules by which information encoded in genetic material (DNA or RNA sequences) is translated into proteins by living cells.

Genome— the entire hereditary information of an organism which is encoded in the DNA. More precisely, the genome of an organism is a complete genetic sequence on one set of chromosomes.

Geochronometers— techniques used to date the earth and universe.

Geologic column, time scale, and fossil record— a chronology of 'standard' geologic time subdivided into units: eons, eras, periods, and epochs, comprising many millions of years as defined by uniformity theory, and portraying evolution from single celled organisms to humans. As stated in chapter 7, the standard geological column is ***fictitious***—it does not exist anywhere on earth.

Geomorphology— the study of the earth's erosional and depositional landforms and the processes that may have shaped them.

Glacial striations— gouges cut into bedrock by glacial abrasion. Striations usually occur as straight, parallel grooves that represent the movement of the glacier.

Gneiss— a common type of metamorphic rock with a banded appearance formed from pre-existing formations of igneous or sedimentary rocks.

Granite— an igneous rock that solidified from hot magma and comprises minerals such as quartz (white to glassy) and feldspar (gray-pink), and some biotite (dark mica flakes) and hornblende (black-dark green). Massive volumes of granite comprise large portions of the continents and the central core of most mountain ranges. Various types of granite are called rhyolite, obsidian, and pumice.

Gravitational time delay— one of the four classic Solar System tests of general relativity. Radar signals passing near a massive object take longer to travel to a target and longer to return (as measured by the observer) than if the mass of the object were not present.

Gravitational time dilation— the effect of time passing at different rates in regions of different gravitational pull; the greater the pull, the slower time passes and the lesser the pull (near the outer fringes of the universe), the faster time passes. Einstein originally predicted this effect in his theory of relativity and it has since been confirmed by empirical science (i.e., precise measurements). Thus, billions of years would be available for light to reach the earth while less than an ordinary day is passing on earth. In other words, light from the extremities of the universe has the potential of reaching earth in a relatively short period of time (from earth's perspective).

Gumbotil— mature weathered clay soil comprising small stones.

Half-life— see radioisotope dating.

Heat death— a state of maximum entropy; it is a possible final state of the universe in which it has "run down" to a state of no usable energy to sustain motion or life.

Helium (symbol, He)— is an element with an atomic number of 2; it is colorless, odorless, tasteless, non-toxic and inert, and it is generated during radioactive decay of uranium-238 to lead-206. During the decay process, each emitted alpha particle consists of two protons and two neutrons which is equivalent to the helium atom (1 uranium atom produces 8 alpha particles, or 8 helium atoms, during the decay process). Helium is the second lightest element and is the second most abundant element in the observable universe.

Historical geology– branch of geology that deals with the history of earth's physical changes—events of the past which are not reproducible and cannot be tested. Secular geologists make certain assumptions within the realm of "naturalistic" or "known science" as their basis for interpretation. These assumptions include:

> **Uniformitarianism (also known as uniformity theory)**– maintains that geological, meteorological, and other physical processes have remained constant ('slow and gradual') during earth's history. It is the belief that existing natural processes are sufficient to account for all past changes in our world and the universe. Essentially, the doctrine assumes that conditions on earth have never been interrupted by worldwide cataclysmic events.
>
> **Evolution**– the belief that all living organisms created themselves by their own natural processes with no supernatural input—it is a doctrine of increasing organization and complexity in the universe; or a doctrine of continual design without a designer. Evolutionists believe that man descended from the apes, all vertebrates descended from fish, all fish descended from invertebrates, and all life descended from single-celled organisms which arose spontaneously from non-living chemicals (abiogenesis). Evolution has never been observed or known to occur within living populations.
>
> **Biostatic fossil correlation**– the belief that if an index fossil is found in a rock formation, that rock formation is the same approximate age as the index fossil. Ages of index fossils are hypothetically assigned based on the assumption of evolution and uniformity theory. Isn't this "circular reasoning"? See circular reasoning and chapter 7, The Fossil Record, Uniformitarian Approach and Evolution.

Hoodoos– pinnacle-like erosional formations with variable thickness from top to bottom, whereas a pinnacle or spire has a smoother profile that tapers from the ground upward.

Humanism– is a belief that human beings are able to determine right and wrong through universal qualities such as reason—that is, prime importance is placed on human ability and self-reliance rather than the divine or supernatural. The basic philosophy of *humanism* is that man created himself, has control over his own destiny, and can rely on his own superior "intellect"—essentially, that man is God.

Hypothesis– an educated guess based upon observation but not yet proved. Most hypotheses can be supported or refuted by experimentation or continued observation.

Igneous– a type of rock that solidified from magma.

Index fossils– used to define and identify geologic periods (or faunal stages) as defined by uniformity theory. The term refers to mineralized animal or plant fossils preserved in rock strata that are characteristic of a particular span of standard geologic time or environment. This is an enterprise of old-earth advocates and evolutionists and is based on the assumption of evolution and uniformity doctrine. See chapter 7, The Fossil Record.

Inert– a term used to describe something that is not chemically active. For example, the noble gases were described as being inert because they did not react with other elements or themselves. The noble gases are the nonmetal, inert elements in group 18 of the periodic table. The five noble gases that have stable isotopes are helium (He), neon (Ne), argon (Ar), krypton (Kr), and xenon (Xe)—for example, these atoms do not combine with other atoms to form molecules.

Inner core of the earth– the deepest core, approximately 3,958 miles (6,370 kilometers) below the crust; so hot and under so much pressure that it is believed to be a solid. See diagram in Appendix D.

Invertebrate (also known as *Protochordata*)– an animal lacking a backbone or spinal column. The group includes 97 percent of all animal species—all animals except those in the Chordate phylum (vertebrate fish, reptiles, amphibians, birds, and mammals).

Irreducible complexity– see Appendix B for more information.

Isochron– means equal time; implies that the rock or mineral samples were formed at the same time.

Isotopes– different types of atoms of the same chemical element, each having a different atomic mass. Isotopes of the same element have the same number of protons (same atomic number which uniquely identifies the element) but different numbers of neutrons. Therefore, isotopes have different mass numbers (neutrons plus protons).

Kame– a small mound of layered sediment deposited by a stream that flowed on, within, or beneath a glacier. See diagram in chapter 4.

Kettle lake– a depression created by the melting of a very large piece of ice left buried by a receding glacier. See diagram in chapter 4.

Kind (group)– a term referring to distinct groups of animals (i.e., mankind, horse kind, bear kind, dog kind, cat kind, etc.) and plants. The classification of living things is by life, domain, kingdom, phylum, class, order, family, genus, and species, whereby order or family may have represented the biblical kinds or groups.

Kinetic energy– energy in motion or released energy. It is defined as the work needed to accelerate a given mass at rest to a given velocity.

Kolking– an erosional process whereby rapidly moving water creates a rotating spiral or underwater tornado that can remove large pieces of bedrock along the bottom.

Lagerstätten (German; literally, *places of storage, resting place*)– sedimentary deposits that exhibit extraordinary fossil richness or completeness; that is, fossils are extremely well-preserved and diverse.

Law of Causality– the relationship between one event (called a "cause") and another event (called an "effect"). The "effect" is the direct result of the "cause." The First Law of Thermodynamics implies that creation occurred at sometime in the past—the universe had a beginning! "Everything which has a beginning has a cause—the universe had a beginning—therefore it had a cause...God, unlike the universe, had no beginning, so does not need a cause." See chapter 9, The Big Bang and Its Problems.

Laws of Motion– see Newton's Three Laws of Motion.

Laws of Science– facts of nature. Laws are generally accepted to be true and absolute throughout the known universe. They have been subjected to extensive measurements and experimentation and have repeatedly proved to be unwavering or inflexible (e.g., the law of gravity, the laws of motion, and laws of thermodynamics). A "law" differs from hypotheses, theories, postulates, and principles in that a law can be expressed by a single mathematical equation with an empirically determined constant.

Light year (symbol: ly)– a unit of length, equal to just under 10 trillion kilometers. As defined by the International Astronomical Union (the organization that has jurisdictional authority to promulgate the definition), a light-year is the **distance** that light travels in a vacuum in one Julian year. (Julian year (symbol: a)– a unit of measurement of time defined as exactly 365.25 days of 86,400 seconds per day, totaling 31,557,600 seconds.) See speed of light.

Limestone– a sedimentary rock composed largely of the mineral calcite (calcium carbonate: $CaCO_3$). Like most other sedimentary rocks, limestone is composed of grains; however, most grains in limestone are a mixture of skeletal fragments of marine organisms and sand and silt. Some limestone is fine-textured and is formed by the chemical precipitation of calcite or aragonite (e.g., travertine).

Travertine– a form of limestone deposited by mineral springs, especially hot springs. See Appendix F.

Chalk– a form of limestone composed of the mineral calcite, a fine grain calcium carbonate. It is a soft, compact white rock that is a very pure type of limestone. Unlike other forms of limestone, chalk consists of tiny shells called tests, which comprise billions of planktonic microorganisms called foraminifera and calcareous algae known as cocoliths and rhabdoliths.

Planktonic– are drifting microscopic plants and animals that float in great numbers at or near the surface in fresh and salt water and serve as food for fish and other larger organisms.

Foraminifera (forams)– are amoeboid (irregular shaped), single-celled microorganisms called protists; they are testate (possessing a shell), eukaryotic (having their own nucleus), and diverse. Foraminifera are typically included in the kingdom Protozoa, although some put them in the equivalent Protista. Many have chloroplasts with which they carry on photosynthesis. These organisms either live on the sea bottom (benthic) or float in the upper water column (planktonic).

Calcareous algae– are simple plants lacking roots, leaves, and a vascular system. Calcareous algae are capable of forming calcium carbonate ($CaCO_3$) within or on their bodies. These algae are abundant and they are major components of planktonic communities in fresh and salt water environments. They principally belong to red and green algae with only two calcified genera of brown algae.

Lithification (from the Greek word *lithos* meaning "rock")– the process in which sediments compact under heavy overburden (pressure), expel interstitial fluids, and gradually become solid rock. It is a process of porosity destruction through compaction and cementation—it includes all the processes which convert unconsolidated sediments into sedimentary rocks. See petrification and cementation.

Lithosphere– the relatively cool, rigid outer layer of the earth, about 62 miles (100 kilometers) thick, that includes the crust and part of the mantle. See diagram in Appendix D.

Macroevolution– another name for evolution. It is theoretical changes in an animal or plant because of new genetic information introduced into the gene pool which, in turn, produces a new kind (or category) of organism. Such changes (i.e, genetic uphill drift) have *never been observed* within living populations.

Magma– molten rock (igneous rock) produced within the earth.

Magnetic field (earth's magnetic field)– a magnetic dipole with one pole near the North Pole and the other near the South Pole. The earth's magnetic field is created by electric currents in the earth's core which extend tens of thousands of kilometers into space, an area called the magnetosphere.

Geomagnetic reversal is a change in the orientation of earth's magnetic field such that the positions of magnetic north and magnetic south become interchanged.

Magnetism– a phenomenon by which materials exert attractive or repulsive forces on other materials. Well-known materials that exhibit detectable magnetic properties (called magnets) are nickel, iron, cobalt, and their alloys. All materials are influenced to a greater or lesser degree by the presence of a magnetic field.

Mantle– the mostly solid layer of the earth, lying beneath the crust and above the outer core. The mantle extends from the base of the crust to a depth of about 1,802 miles (2,900 kilometers). See diagram in App D.

Meiosis– is a type of cell division necessary for sexual reproduction; a process in which living things are programmed genetically to pass on their information by making copies of themselves. Genetically, it is cell division in which two diploid cells (2n) produce four haploid cells (*n*) called gametes. A diploid cell (2n) is a cell with two copies of each type of chromosome and a haploid cell (n) is a cell that has half the number of chromosomes, one of each type of chromosome that makes up the genotype. The genetic code (DNA) of a father is passed on by the sperm gamete, and the genetic code (DNA) of a mother is passed on by the egg gamete. See Appendix B.

Melanin– a substance that exists in the plant and animal kingdoms as a pigment. In humans, melanin is found in skin and hair, and in the pigmented tissue of the iris, and is the primary determinant of human skin color.

Mesa– flat-topped mountain that is smaller than a plateau and larger than a butte.

Metamorphic rock– forms when rock (e.g., igneous, sedimentary, and metamorphic rock) changes or recrystallizes because of increased temperature and pressure. Examples are slate (metamorphosed shale), schist (metamorphosed slate), gneiss (metamorphosed igneous or sedimentary rock), quartzite (metamorphosed quartz sandstone), and marble (metamorphosed limestone). Although metamorphism occurs in recent times on a limited scale when hot magma is intruded into sedimentary rocks, *enormous areas of metamorphic rock found throughout the world are because of forces more intense than those observed today.*

Microevolution– a "natural selection" process that is not evolution—that is, there is No creation of new genetic information to form a new species. Rather, it is a term commonly used by scientists to describe reshuffling of genes within an "existing gene pool" of the original kind of animal (or plant) in response to a specific environment. It is a genetic process when organisms with favorable genes suitable for a specific environment are able to survive, reproduce, and pass on these genes to the next generation through adaptation. This is sometimes referred to as "survival of the fittest." Most people are unaware that natural selection (or survival of the fittest) is a thinning out process that *leads to a loss of genetic information.*

Milky Way– the galaxy in which our Solar System is located.

Minerals– inorganic solids in rocks with a specific chemical composition and crystalline structure. The various kinds of minerals in rocks are dependent on the temperature and pressure and concentration of various elements during rock formation. Common minerals are quartz, feldspar, mica, hornblende, and pyroxene. Minerals range in composition from pure elements to very complex silicates with thousands of known forms. The study of minerals is called mineralogy. Common elements in minerals are oxygen and silicon; other common elements include aluminum, iron, calcium, magnesium, potassium, and sodium. A rock, by comparison, is an aggregate of minerals and need not have a specific chemical composition.

Mollusca– a large phylum of invertebrate (without a backbone) animals.

Monadnock or inselberg– an isolated small mountain that rises abruptly from a gently sloping or virtually level surrounding peneplain. See monolith.

Monocline– a bend in a rock formation with only one limb.

Monolith– a single massive mountain or rock feature. Such geological formations are most often very hard metamorphic or sedimentary rock, or igneous rock. See monadnock.

Moraine– a ridge of till (sediment deposited directly by glacial ice).

Mutations– random errors or defects in DNA. Such defects may affect the whole chromosome or just one gene.

Mutational load (genetic burden)– is the accumulation of defective genes over many generations. Genetic mistakes are inherited—the next generation makes a copy of the defective DNA, so the defect is passed on. Somewhere down the line another mistake happens, and the mutational defects accumulate. This is known as the problem of increasing mutational load or genetic burden which is consistent with the Second Law.

Natural selection (also known as speciation, variation, adaptation, or "survival of the fittest")– a mechanism for change in animal population when organisms with favorable genes (or variations) to a specific environment are able to survive, reproduce, and pass these genes on to the next generation through *adaptation*. This is most commonly known as "survival of the fittest." Natural selection is *NOT evolution*. It is the genetic process within populations which selects favorable gene traits from an "existing gene pool" best suited for a given environment. The gene pool for a particular "kind" of animal is genetic information *already present* in the population for a "kind" of animal. Most people are unaware that natural selection is a thinning out process that *leads to a loss of genetic information.* See chapter 3.

Nautiloids– marine animals called cephalpods (kingdom Animalia; phyllum Mollusca; class Cephalopoda; subclass Nautiloidea) that possess an external shell. The best-known examples are the modern nautiluses.

Newton's Three Laws of Motion– 1) law of inertia states that an object will persist in its state of rest or uniform motion unless acted on by another force; all objects resist change in their state of rest or motion, 2) law describing the relationship between force (f), mass (m), and acceleration (a) states that a force applied to a mass undergoes acceleration, described by f = ma. For example, 96 newtons (force) will accelerate an object weighing 1,200 kg (mass) about 0.08 m/s/s (acceleration)—heavier objects require more force than lighter objects to move the same distance, and 3) for every action, there is an equal and opposite reaction.

Nuclear fission– a nuclear reaction in which the nucleus of an atom splits into smaller parts that release energy in the form of heat.

Nuclear fusion– the process by which multiple atomic nuclei join together to form a single heavier nucleus. It is accompanied by the release or absorption of large quantities of energy. Nuclear fusion occurs naturally in stars.

Nucleotides– the structural units of DNA and RNA. A nucleic acid is a chain of nucleotides—these molecules carry genetic information.

Orogeny– all the tectonic (structural) processes associated with mountain building.

Outer core of the earth– the portion that lies about 1,802 miles (2,900 kilometers) below the crust and is believed to be molten liquid comprising mostly nickel and iron. See diagram in Appendix D.

Overfit valley– a valley much too large for modern day rivers or river meanders.

Oxidation– the *loss* of electrons (hydrogen atom) or *gain* of oxygen (*increase* in oxidation state by a molecule). **Reduction** describes the *gain* of electrons (hydrogen atom) or *loss* of oxygen (*decrease* in oxidation state by a molecule).

Paleomagnetism– the study of the remnant magnetization in rocks; the earth's magnetic field as it existed in the past. Paleomagnetic dating involves collecting rock samples to determine how the earth's magnetic poles were oriented when the rocks were formed.

Paleontologist– a person who studies prehistoric life forms on earth by examining plant and animal fossils.

Pangaea– the primeval supercontinent composed of all the major landmasses that were primarily centralized in the northern hemisphere. According to many historical geological textbooks and evolutionary time scale, Pangaea I existed 1.8 billion years ago; Pangaea II existed 700 million to 1 billion years ago; and Pangaea III existed 200 to 300 million years ago. From a Biblical and young earth perspective, the primeval supercontinent was created by God about 6,000 years ago according to Genesis 1:9.

Paraconformity– a sedimentary layer that has remained dormant for supposedly millions of years according to geologic time scale—it is a surface of non-deposition and non-erosion. It is a *hypothetical concept* that has never been observed to occur in nature, and is an attempt by secular geologists to overcome the problems associated with conformable layers that are separated by vast periods of time while displaying no evidence of erosion. In actuality, there is no stagnant or dormant surface on land today and a paraconformity is actually a conformity; that is, there is no missing layer or formation. See diagram in Appendix G.

Parasite– an organism that lives on or in a host and gets its food from or at the expense of its host.

Partial pressure– the pressure of a gas in a mixture whereas the total pressure of a gas mixture is the sum of the partial pressures of each individual gas in the mixture.

Parting lineation– sedimentary features comprising parallel ridges and grooves formed on the bedding plane parallel to the current. Also known as current lineation.

Pathogenic organism– any organism which causes any infection or disease. The word comes from the Greek pathos ("disease") and genesis ("creation").

Peat– an accumulation of partially decayed vegetative matter. Peat forms in wetlands called bogs, moors, pocosins, mires, and peat swamp forests.

Pelletoids– an aggregation (flocculation) of fine-grain crystals of aragonite or calcite in the formation of marine lime muds.

Peneplain– a flat, nearly featureless plain formed by sheetlike erosion. Common erosional features of peneplains are buttes, monadnocks or monoliths.

Petrification (fossilization)– the replacement of organic material with silica (silicon dioxide, SiO_2) in the formation of mineral fossils. See fossilization and cementation.

Phanerozoic– current Eon in the geologic time scale with abundant animal life. According to standard time scale, it covers 545 million years since the Precambrian Eon. Phanerozoic Eon comprises three eras: Paleozoic, Mesozoic, and Cenozoic Eras (oldest to youngest) and 12 periods: Cambrian, Ordovician, Silurian, Devonian, Mississippian and Pennsylvanian, Permian, Triassic, Jurassic, Cretaceous, Tertiary, and Quaternary (oldest to youngest). See diagram in chapter 7, Standard Geologic Column and Time Scale.

Photosynthesis– the conversion of light energy into chemical energy by plants. Light energy is converted to chemical energy and then stored. The chemical formula for photosynthesis is $6CO_2 + 6H_2O$ + light energy (chlorophyll) $\rightarrow C_6H_{12}O_6 + 6O_2$ whereby carbon dioxide is converted (reduced) to simple sugars.

Pigment– a material that changes the color of light it reflects as the result of selective color absorption.

Pinnacle or spire– an erosional formation that uniformly tapers from the ground upward.

Plants– belong to the kingdom Plantae. They are the producers—that is, they are able to make their own food by extracting carbon from carbon dioxide and obtain most of their energy from sunlight via photosynthesis using chlorophyll contained in chloroplasts, which gives them their green color.

Plateau– a highland area of relatively flat terrain. It is much larger than a mesa.

Plate tectonics– a theory of global crustal movement. The lithosphere (earth's crust) is segregated into crustal plates that move relative to one another by floating on a semi-liquid asthenosphere. Seismic and tectonic activity (i.e., volcanoes and earthquakes) occurs primarily along plate boundaries. See asthenosphere.

Playa– a desert basin with no outlet which periodically fills with water to form a temporary lake.

Plucking– an erosional process whereby rapidly moving water and suspended debris are able to dislodge loose chunks of bedrock along the bottom.

Plutonic– a very large volume of igneous intrusive rock that forms from cooled magma. See batholith.

Polystrate– a term that describes fossils (e.g., polystrate trees and animals) that cut across several sedimentary layers.

Postulate (axiom)– something that is not proved but is self-evident. For example, a finite whole is greater than any of its parts or it is impossible for something to be and not be at the same time.

Potential energy– energy stored within a physical system; it is free energy. It is called *potential* energy because it has the potential to be converted into other forms of energy to do work. For example, the conversion of light energy to chemical energy, or conversion of wind energy to electrical energy to light energy. The standard unit (SU) of measure for potential energy is the *joule*—the same unit for work or energy in general.

Principles– a set of rules or standards that forms the basis for ethical or moral behavior. For example, the Ten Commandments of the Bible.

Progressive creation (sometimes called the Gap Theory or Day-Age Theory)– an idea that maintains the days of Genesis equal standard geologic ages (that is, millions of years), and each kind of animal life was supernaturally created at various times throughout earth's history.

Proteins (also known as polypeptides)– biological molecules consisting of one or more chains of amino acids; the sequence of amino acids in a protein is defined by the gene. See amino acids.

Protozoa– a class of unicellular invertebrates. Amoeba is an organism within the class Protozoa.

Quartz– a silicate rock with the chemical formula SiO_2. It is widespread and abundant in continental rocks but is rare in the oceanic crust.

Quasar– an extremely powerful and distant active galactic nucleus (compact region at the center of a galaxy) that has a much higher than normal luminosity.

Quaternary– a standard geologic time period comprising two geologic epochs: Pleistocene and Holocene. See diagram in chapter 7, Geologic Column and Time Scale.

Radiohalos– see chapter 8.

Radioisotope– a version of a chemical element that has an unstable nucleus and emits radiation during its decay to a stable form. Radioisotopes have important uses in medical diagnosis, treatment, and research.

Radioisotope dating– an attempt by secular geologists to determine a rock's age by measuring the ratio of radioactive isotopes in the rock sample (parent and daughter isotopes) and the rate at which isotopes decay while making various assumptions: 1) constant rate of decay, 2) no loss or gain of parent or daughter isotopes (closed system), and 3) known amounts of daughter isotopes at the start of cooling. These assumptions are considered to be 'fatally flawed' by RATE research scientists. See chapter 8 and Appendix I.

Atomic number (Z) or proton number– the number of protons found in the nucleus of an atom.

Mass number (A)– sometimes referred to as the atomic weight, is basically the number of protons and neutrons in an atomic nucleus with just a little extra added by the orbiting electrons.

Element– a type of atom that is distinguished by its atomic number; that is, by the number of protons in its nucleus. In an atom of neutral charge, atomic number is equal to the number of electrons. For example, carbon has an atomic number of 6 and atomic weight of 12 (6 protons, 6 neutrons, and 6 electrons). Elements with atomic numbers greater than 82 are unstable (radioactive) and undergo radioactive decay into smaller, more stable elements (atoms). Uranium has an atomic number of 92.

Half-life– the time required for the mass of an atom to decay to half of its initial value.

Isotopes– different types of atoms of the same chemical element, each having a different atomic mass. Isotopes of the same element have the same number of protons (same atomic number which uniquely identifies the element) but different numbers of neutrons. Therefore, isotopes have different mass numbers (neutrons plus protons).

Radiocarbon dating– a radiometric dating method that uses the naturally occurring isotope carbon-14 to determine the age of carbonaceous materials. This method cannot be used to date igneous or sedimentary rocks or most fossils (mineralized plants and animals; organic structures replaced by silica), but it can be used to date fossils that contain carbon—for example, flesh, charcoal, wood, non-mineralized bone, carbonate deposits including marine and freshwater shells unlike most rocks. Assumptions are shown to be "fatally flawed." This method is valid only for "recent" times—that is, within 5,000 years—and when correlated with known historical dates.

Radioactive decay– the process in which an unstable element loses energy by emitting radiation in the form of particles (alpha and beta particles) or electromagnetic waves. This decay, or loss of energy, results in an atom of one type, called the *parent*, transforming to an atom of a different type, called the *daughter*. For example, uranium (parent) decays into lead (daughter), potassium (parent) decays into argon (daughter), and rubidium (parent) decays into strontium (daughter).

Regression; Least Squared Regression— is a statistical technique in which a known value x (for example, the parent isotope) is used to predict the corresponding value y (for example, the daughter isotope). Least squared is the minimum sum of the squared values of the vertical differences.

Relict landforms— geologic landforms that remain after surrounding rock formations have disappeared. Such features have survived decay and disintegration. Plateaus, mesas, buttes, spires, hoodoos, and coulees are examples of relict landforms.

Respiration— the conversion of oxygen and simple sugars to carbon dioxide and water. It can be loosely thought of as the opposite of photosynthesis. Animal respiration is the transport of oxygen from air to the cells within animal tissues and the transport of carbon dioxide in the opposite direction. It is the process of oxidizing food molecules, like glucose, to carbon dioxide and water. (*Anaerobic* respiration refers to the oxidation of molecules in the absence of oxygen to produce energy, in opposition to *aerobic* respiration which uses oxygen.)

Ribonucleic acid (RNA)— very similar to DNA, but differs in a few important structural details. In the cell, RNA is usually single stranded, while DNA is usually a double helix. RNA nucleotides contain ribose while DNA contains deoxyribose (a type of ribose that lacks one oxygen atom), and RNA has the nucleotide uracil rather than thymine, which is present in DNA. RNA is transcribed from DNA by enzymes called RNA polymerases.

Rickets— the softening of bones in children which commonly lead to fractures and deformity. The predominant cause is vitamin D deficiency.

Rifting— the tectonic process in which crustal plates separate or diverge along the plate boundary. Rift is the zone of separation, or divergence.

Rhythmites— successional sedimentary layers, sometimes referred to as turbidites.

Runaway subduction— rapid plunging of the crustal plate beneath the continents during the worldwide flood. See subduction.

Salt— a crystalline solid, composed primarily of sodium chloride (NaCl); white, pale pink, or light gray in color; and normally obtained from sea water or rock deposits.

Salt Lake— a hypersaline water body within an enclosed basin.

Salt Layers— sedimentary salt layers, primarily sodium chloride (halite or rock salt), hundreds of feet thick found throughout the world.

Sapping— a process of erosion when soft sediment at the base of cliffs disintegrates by the leaching of water causing the collapse of large overhead rocks. This is one of the processes that forms gullies or alcoves.

Schist— a metamorphic rock containing abundant mica (sheet-like or platy silicate mineral). When shale is subjected to heat and pressure, it is metamorphosed into slate, and if slate continues to undergo heat and pressure, it will become schist. Schist contains more than 50% platy minerals, often finely interleaved with quartz and feldspar. See metamorphic rock.

Science— an organized body of knowledge in the form of ***testable*** explanations and predictions about all matter and energy. By the turn of the 19th century, the word "science" became associated with the scientific method, a systematic way to study the natural world, including physics, chemistry, geology, and biology.

Scientific creationism— maintains basic kinds or groups of life (e.g., mankind, bear kind, horse kind, dog kind, cat kind, and so forth) appeared abruptly, without arising from a different kind. Much variation within a kind of animal (or plant) is expected because of natural selection and, over time, there is a genetic loss or thinning out of information in response to a specific environment. See chapter 3, Natural Selection and Extinction.

Scientific method— common steps that scientists use to gather information to solve problems. These common steps include observation, prediction (hypothesis), experimentation (to test the hypothesis under controlled conditions), data collection, and conclusions.

Seamount– a marine mountain, usually of volcanic origin, that rises one kilometer or more above the sea floor.

Seawater (salt water)– water from the ocean or sea. On average, seawater in the world's oceans has a salinity of ~3.5 percent, or 35 parts per thousand. This means that every 1 kg of seawater has approximately 35 grams of dissolved salt. Dissolved salts include the ions of chloride (Cl^-), sodium (Na^+), magnesium (Mg^{2+}), sulfate (SO_4^{2-}), calcium (Ca^+), potassium (K^+), and other minor salts.

Second Law of Thermodynamics– see thermodynamics.

Secular scientists (as defined within the context of this book)– base their opinions of the origin of man, the earth, and the universe on evolutionary doctrine without a supernatural Creator. The term secular implies naturalistic, earthly, worldly, materialistic, and non-religious.

Sedimentary rocks– rocks formed by the deposition and consolidation of loose, unconsolidated clastic fragments from preexisting rocks, or from the precipitation of minerals from solution (calcium carbonate forming limestone). In just about every case, sediments are deposited underwater as long, flat layers by oceans, rivers, or lakes, or by flood events, and then exposed as dry land. Layers are cemented together (lithified) by dissolved silica, calcium carbonate, or iron as sedimentary rock layers. Sediment layers can be seen in mountains and canyons throughout the world. Clastic sedimentary rocks are classified as sandstone (cemented quartz sand), shale (cemented clay and silt), and conglomerate (cemented fragments larger than sand). Chemical sedimentary rocks include limestone (calcium carbonate) and dolomite (magnesium calcium carbonate).

Shale (also called mudstone)– a fine-grained sedimentary rock comprising clay minerals. Metamorphosed shale is slate.

Silica– silicon dioxide, SiO_2, found in quartz, opal, chert, and other mineral varieties.

Soil– the naturally occurring, unconsolidated or loose covering of clastic rock fragments (primarily clay, silt, and sand) and decaying organic matter (humus) on the surface of the earth, capable of supporting life.

Solar System– the system consisting of the Sun and those celestial objects bound to it by gravity. These objects are the eight planets (Mercury, Venus, Earth, Mars, Jupiter, Saturn, Uranus, and Neptune) and two dwarf planets (Pluto and Eris), the planets' 166 known moons, five other dwarf planets, and billions of small bodies.

Spacetime– any mathematical model that combines space and time into a single construct called the spacetime continuum. Spacetime is usually interpreted with space being three-dimensional and time playing the role of the fourth dimension. By combining space and time into a single construct, physicists have simplified a large number of physical theories, as well as described in a more uniform way the workings of the universe at both the supergalactic and subatomic levels.

Specific complexity– describes something with high informational content or intelligent design. Examples are an encyclopedia, an automobile, and a supercomputer. It is also a commonly used term for preprogrammed genetic systems such as a DNA molecule consisting of precisely arranged amino acids—basic building blocks of protein molecules. On the other hand, repetitive sequences such as XYZXYZXYZ or space pulsars or the geometric arrangement of crystals are not specific complexity or proof of intelligent design. See chapter 2, Preprogrammed Design; also see chapter 3 and Appendix B.

Speed of light– the speed of all electromagnetic radiation, including visible light, in the vacuum of free space. The speed of light (denoted by the letter *c*) is about 670,616,629.4 miles per hour or 983,571,056.4 feet per second (roughly one foot per nanosecond), which is about 186,282.397 miles per second.

Spire (or pinnacle)– see pinnacle.

Spontaneous generation– a hypothetical and unproven process whereby life arose spontaneously from nonlife (also known as abiogenesis). No one has ever observed or demonstrated spontaneous generation. As proved by the French scientist Louis Pasteur in the mid-1800s, life cannot be produced from nonliving matter. Biogenesis states that life can only arise from pre-existing life, not from nonliving matter. This biological law (biogenesis) has become the cornerstone of biological sciences. See biogenesis and chapter 3.

Stratigraphic rules– several principles of stratigraphy.

Principle of superposition– In any undisturbed geologic sequence, older beds lie below younger beds.

Principle of original horizontality– Whether rock beds are deformed or not, they were originally deposited in a horizontal position.

Principle of lateral continuity– A water-laid stratum, at the time it was formed, must continue laterally in all directions until it thins out as the result of nondeposition.

Principle of inclusion– If a body of rock contains fragments of another type of rock, the host rock must be younger than the rock fragment.

Principle of cross-cutting– Any geologic feature that intersects a series of beds (such as a fault) must be younger than the beds it cuts across. Such intrusions may be clastic or igneous dikes, sills, polystrate fossils, and igneous (magma) intrusions.

Stratigraphy– the branch of geology that studies rock layers and layering of sedimentary and volcanic rocks.

Subduction– the tectonic process in which a crustal plate descends beneath an adjacent plate and dives into the asthenosphere. See runaway subduction.

Submerged canyon– a deep, V-shaped, gorge.

Submerged terrace– a prehistoric coastline that was above sea level but has been inundated due to land subsidence or rising sea level.

Superluminal speed– speed that is faster than the speed of light.

Superposition (Principle of)– see stratigraphic rules.

Super volcano– a volcano that produces the largest and most voluminous kind of eruption on earth (for example, super volcano at Yellowstone National Park). See chapter 4, Beginnings of Catastrophic Volcanoes.

Syncline– a bend or fold in a rock formation that is concaved upward (U-shaped). Opposite of anticline.

Tectonic– a branch of geology dealing with geologic structures within the earth's lithosphere and the forces and movements that created these structures.

Teleonomy– the quality of apparent purposefulness of structures and functions in living organisms; implies intention, purpose, and foresight.

Theistic evolution– essentially the same as evolution, except that God created life and started the evolutionary process. See progressive creation.

Theory– an explanation of a set of related observations based upon hypotheses and verified multiple times over an extended period of time by unrelated groups of researchers. A theory is not scientific law and *remains unproven*. Theories are most often disproven over time and replaced with other theories. (Note: Evolution has never been observed in nature or proven by empirical science and does not meet the definition of "theory.")

Thermodynamics– is the study of heat power—heat is "energy in transit." It is the branch of physics that examines kinetic energy, or the efficiency of energy transfer, or energy transformation; for example, hot to cold; high pressure to low pressure; solar energy to fossil fuel to mechanical energy to electrical energy to light energy. When we eat, our bodies transform energy stored in food (potential energy) into energy to do work (kinetic energy). When we run or walk, think, read or write, we "burn" food energy in our bodies. Cars, planes, boats, light bulbs, and machinery also transform energy into work. Work is moving, lifting, warming, or lighting something.

The First Law of Thermodynamics (Law of Conservation of Matter) states that matter/energy cannot be created or destroyed. Although the amount of matter/energy remains the same, energy can be transferred from one form to another. For example, it can change from solar energy to fossil fuel to mechanical energy to electrical energy to light energy; or from hot to cold; high pressure to low pressure, etc. This law confirms that creation is no longer occurring which, in turn, implies that creation existed at sometime in the past. In today's world, there is no creation of new matter/energy rising to higher levels of organized complexity!

The Second Law of Thermodynamics states that matter/energy in the universe available for work is decaying or running down. While the *quantity* remains unchanged, the *quality* of matter/energy has the tendency to decline or deteriorate over time. This is commonly known as **The Law of Increasing Entropy.** While usable energy is used for growth and repair, it is "irretrievably lost in the form of unusable energy."

Thrust fault— a special type of reverse fault that is nearly horizontal. A **reverse fault** is when the hanging wall has been moved up relative to the foot wall. See chapter 5, Mountain Overthrust, with diagram of a thrust fault.

Time— a continuum lacking spatial dimensions and in which events proceed from past through present to future. Time has been a major subject of religion, philosophy, and science, but defining it in a non-controversial manner applicable to all fields of study has consistently eluded the greatest scholars. (Crouch, Will (2006-2008). Is there a defensible argument for the non-existence of time? *On Philosophy*. http://www.onphilosophy.co.uk/time_-_a_dialogue.html. Retrieved August 2009, from Wikipedia.)

Transform fault— horizontal slipping (instead of vertical slipping) along a fault line such as the San Andreas Fault of California.

Trilobites— extinct arthropods. Arthropods are the largest phylum of animals and include the insects, arachnids, crustaceans, and others.

Tsunami— a massive wave created when a body of water is rapidly displaced, usually in the open ocean. A tsunami has a much smaller amplitude (wave height) and a very long wavelength (often hundreds of kilometers long) as compared with a "**wind wave**" which has a much larger amplitude but a much smaller wavelength. Tsunamis are much more dangerous because of the exceedingly greater volume of water. See chapter 4, Beginnings of Catastrophic Earthquakes and Tsunamis, showing diagram of these waves.

Turbidites— pancake-like sediment layers that originate in catastrophic underwater avalanches or mudslides. In some cases turbidite deposits were formed by massive slope failures where rivers deposit large deltas. These slopes fail in response to earthquake shaking or excessive sedimentation load. See Sandstone Turbidites - Rapid Formation of Layered Sediments in chapter 4, showing diagram formation; Sandstone Turbidites in chapter 5; Presence of Sharp Contact Bedding Planes in chapter 6; and Appendix D, Earthquakes and Tsunamis.

Type (form)— refers to so-called transitional animals or plants that have supposedly evolved when new genetic information is introduced into the gene pool. Genetic "uphill" drift is utterly fictitious and transitional forms have *never been observed* within living populations.

Uinkaret Plateau— a volcanic area that covers about 500 to 600 square miles (1,500 square km) north of the Colorado River. Volcanic features (i.e., basaltic cinder cones and lava flows) lie within the canyon and on the plateau. See chapter 5, The Colorado Plateau and Grand Canyon and chapter 8, Unreliable Results.

Unconformity— an erosional break between sedimentary strata. See diagram in Appendix G.

Underfit river (see overfit valley)— a river that is much too small to have eroded the valley. The valley is a relict landform, much too large for modern day rivers or river meanders.

Uniformitarianism (also known as Uniformity Theory)— a theory maintaining that geological, meteorological, and other natural processes have remained constant during the world's history. It is the belief that existing physical processes are sufficient to account for all past changes in our world and the universe. The doctrine assumes that conditions on earth have never been interrupted by worldwide cataclysmic events. It essentially states that scientific laws and processes operating today also operated in the past.

Universe— everything that physically exists: the entirety of space and time, all forms of matter and energy.

Unreliable— a term that implies the data is inconsistent, or the dating technique provides inconsistent data.

Variation— see adaptation or natural selection.

Vertebrates— animals with backbones or spinal columns. Vertebrates comprise fish, amphibians, reptiles, birds, and mammals (including humans). See animal phyla.

Virus— a non-living entity or protein that cannot sustain itself or reproduce until it enters a living cell or bacteria where it will then replicate. Viruses are the smallest infectious agents known and are found everywhere in abundance.

Volcano— an opening or rupture in the earth's crust that allows molten rock, ash, and gases to discharge into the atmosphere. Volcanoes tend to exist along the boundaries of tectonic plates. Volcanoes are generally found where tectonic plates are pulled apart or come together. As with earthquakes, about 80 percent of all volcanoes exist within the "Ring of Fire" around the Pacific Rim. The Mid-Atlantic Ridge is also known for volcanoes caused by "divergent tectonic plates" pulling apart whereas the Pacific "Ring of Fire" is known for volcanoes caused by "convergent tectonic plates" coming together. See chapter 4, Beginnings of Catastrophic Volcanoes.

Weathering— the breakdown of rocks, soils, and minerals by heat, water, ice, and wind, or biologically produced chemicals.

White hole— the theoretical time reversal of a black hole. While a black hole acts as a vacuum, sucking up any matter that crosses the event horizon, a white hole acts as a source that ejects matter on the other side of a black hole.

Wormhole— a hypothetical topological feature of spacetime that is basically a "shortcut" through space and time. Spacetime can be viewed as a two-dimensional surface, and when "folded" over, a wormhole bridge (tube or throat) can be formed. A wormhole has at least two mouths which are connected to a single tube. If the wormhole is traversable, matter can "travel" from one mouth to the other by passing through the tube or throat. While there is no observational evidence for wormholes, spacetimes-containing wormholes are known to be valid solutions in general relativity.

Zircon ($ZrSiO_4$)— a crystalline mineral found in igneous rocks throughout the earth's crust—it is also found in metamorphic and sedimentary rocks. Zirconium is a chemical element with the symbol Zr.

Zygote— a fertilized egg.

notes:* *Appendix A - Glossary

Some definitions were obtained from Wikipedia, the free encyclopedia, under the Creative Commons Attribution Share-Alike License and GNU Free Documentation License. Other sources include scientific creation books, geology and biology textbooks, science periodicals, and personal sources.

Appendix B
Genetic Variations

Living things are programmed genetically to pass on their information by making copies of themselves through a process known as meiosis. The genetic code (DNA) of a father is passed on by the sperm gamete, and the genetic code (DNA) of a mother is passed on by the egg gamete. Simplified example: an egg gamete or sperm gamete comprises two pairs of chromosomes (n = 2), or two sets of parallel "ropes" of information (diploid cell), one set from the mother and one from the father. In this example, four kinds of gametes are possible, 2^2 (haploid cells for each gamete). As shown in the table below, the union of these gametes in forming a zygote (fertilized egg), $2^2 \times 2^2$, produces 16 possible combinations or 16 variations.

Meiosis and Genetic Variation

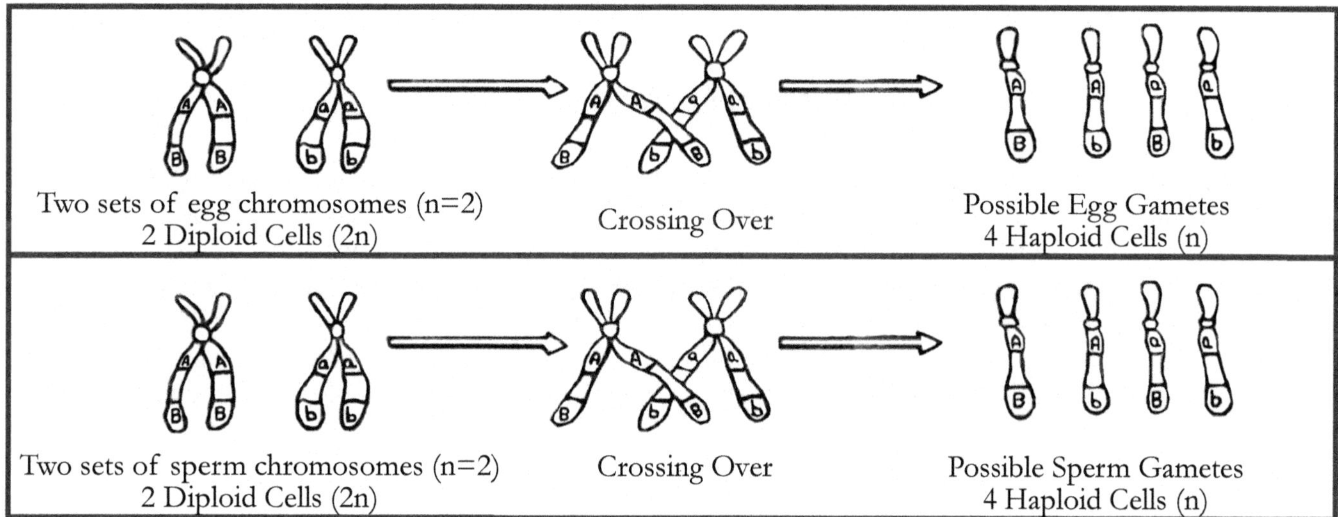

Possible Combinations of Chromosomes in Zygotes

		Sperm Gamete 2^2 = 4 variations			
		AB	Ab	aB	ab
Egg Gamete 2^2 = 4 variations	AB	AABB	AABb	AaBB	AaBb
	Ab	AABb	AAbb	AaBb	Aabb
	aB	AaBB	AaBb	aaBB	aaBb
	ab	AaBb	Aabb	aaBb	aabb

Unlike the example above, humans have 23 pairs of chromosomes (n = 23), or a total of 46 chromosomes (2n = 46). Half of each chromosomal pair is inherited from the mother and half from the father. Statistically, the number of different genetic variations of eggs or sperm a person can produce is more than 8 million (2^{23} haploid cells for each gamete). During the initial phase of fertilization (a process known as meiosis), each parent copies half of their genetic information (or half the number of chromosomes), which allows the offspring to have the same number of chromosomes as their parents (46 chromosomes). When fertilization occurs, $2^{23} \times 2^{23}$ different zygotes (fertilized eggs) are possible, or 70 trillion possible variations!

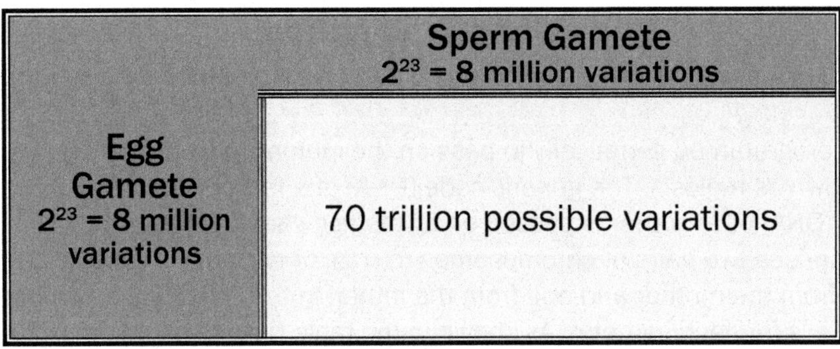

One can see why each individual is unique and in the animal kingdom there are many possibilities for variation (speciation). Gene shuffling (reshuffling or genetic recombination) of the same information in many different ways will result in much variation in any "kind" population—plants or animals.

Chromosomes, DNA, and Genes

Each cell in the human body comprises 23 pairs of chromosomes located within the nucleus of the cell (which contains most of the cell's genetic material). Each chromosome consists of strands of DNA (deoxyribonucleic acid) which comprise thousands of genes. The human cell contains about 100,000 genes. Chemically, DNA consists of two long polymers or intertwining strands called *nucleotides* (sugar-phosphate backbone). The horizontal rungs consist of chemical bases: adenine (A), cytosine (C), guanine (G), and thymine (T). DNA bases pair up with each other, A with T and C with G—thus, there are four possible rungs: A-T, T-A, C-G, and G-C. Each gene consists of nucleotides (portions of these DNA strands) which carry the genetic blueprint to build a protein molecule or one particular component of the organism.

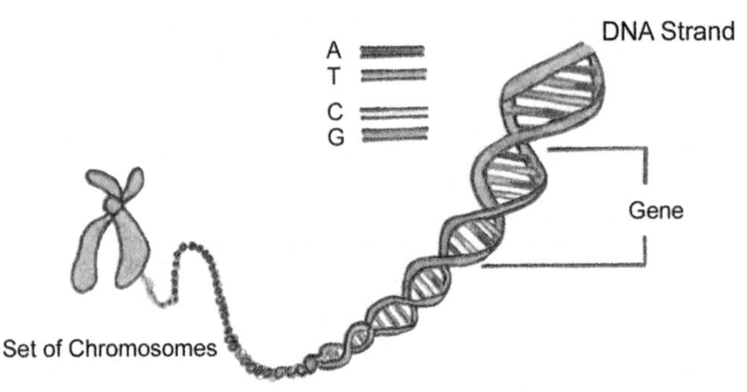

> If the DNA strand is split down the center through each horizontal rung, each half-ladder of the strand will contain all the genetic information to rebuild a copy of the original strand.

Human DNA is similar to a computer software program located within the nucleus of a cell. Similar to the ones and zeros in computer programming (binary code, base-2 number system of two symbols, 0 and 1), three billion of these chemical bases (letters A, T, C, G) are arranged in a similar fashion—a coding language of four letters with two pairs ("or two bits, in computer terms").[1] The precise sequence of these letters (similar to a building and maintenance manual) directs the cell to produce a protein (also known as a polypeptide) made up of amino acids that accomplish specific tasks. The letter sequence in the DNA specifically identifies you as an individual.

> To visualize the complexity of a DNA molecule, "a live reading of that code at a rate of three letters per second would take thirty-one years, even if reading continued day and night."[2]

Statistically, a DNA molecule, consisting of billions of precisely arranged chemical bases (A,T,C,G), and a protein molecule, consisting of thousands of precisely arranged left-handed amino acids (complex genetic machinery) functioning interdependently, could never arise by random chance. And these cellular parts must be performing at the same time and from the very beginning. The probability of just getting an average-size protein of left-handed amino acids (300 amino acids) occurring naturally is zero. (See chapter 3, Evolution - A Statistical Impossibility.) How did we get a brain, heart, lungs, stomach, teeth, tongue, eyes, ears, skin, muscles, bones, a nervous and vascular system, and the other organs, all working together as male and female?

Imagine the odds against evolution—and remembering that evolution is in opposition to the First and Second Laws of Thermodynamics and Law of Biogenesis. If all the parts of a Porsche sports car (including the individual nuts and bolts) were thrown into a small pond, how long would it take for the parts to assemble themselves into a car? The answer is *NEVER*. According to evolutionary doctrine, the only difference between the car and the origin of the first living cell is that parts of the car were created by intelligent beings, whereas the cell with unparalleled complexity and design was supposedly made and assembled by random interaction of lifeless molecules.[4] See chapter 3, Law of Biogenesis - Origin of Life.

> **Irreducible complexity** is another reason why evolution is impossible. A practical example is a common household mousetrap—its individual parts have no value or function, yet the machine could not function if any one of its parts were missing. Microscopic examples are the human eye, the flagellum of bacteria, and blood clotting mechanisms.[3]

The bottom line is the genetic code is so enormously complex and precise in its design that such information could not have arisen by random chance. So the question is, who wrote the genetic software program? Who placed this functioning code inside the nucleus of the cell?

> Dr. Francis Collins, director of the Human Genome Project, stated, "It is humbling for me, and awe inspiring, to realize that we have caught the first glimpse of our own [genetic biochemical code] instruction book, previously known only to God."[5]

Genetic Mutations

Mutations are random errors or defects in the cell's DNA chemical structure—and such errors may affect whole chromosomes or just one gene. If a gene is damaged, incomplete, missing or duplicated, the altered information for building and maintenance may lead to disease. Genetic mistakes are inherited—the next generation makes a copy of the defective DNA, so the defect is passed on. Somewhere down the line another mistake happens, and so the mutational defects accumulate. This is known as the problem of *increasing mutational load* or "genetic burden," which is consistent with the Second Law of Thermodynamics. See chapter 2, Biblical Explanation, and chapter 3, Biblical Account - The Edenic Curse.

By pinpointing the location of each gene in the human body and understanding the gene's function, "scientists are discovering promising new ways to approach the prediction, prevention, diagnosis, and treatment of disease."[6]

---

notes: Appendix B

1. Collins, Francis S. (2006). *The Language of God*. New York: Free Press, 1, 102, 110. (Dr. Francis S. Collins, director of the Human Genome Project, is one of the world's leading scientists in the study of DNA and the human genome.)

2. Ibid., 1. Also see EveryStudent.com, Is God Real? Why the DNA Structure Points to God. Retrieved April 2010, from http://www.everystudent.com/wires/Godreal.html.

3. Sarfati, J. (2011). *Refuting Evolution 2.* Atlanta, GA: Creation Book Publishers, 167-179; and Sarfati, J. (2007). *Refuting Evolution.* Atlanta, GA: Creation Book Publishers, 122.

4. Bartholomew, Stephen (2010). *Scopes Retried.* New York: S. Bartholomew, 165; and Denton, Michael (1986). *Evolution: A Theory In Crisis*. Bethesda, MD: Adler & Adler, 328.

5. Collins (2006), op. cit., 3.

6. Larson, David E. (Ed.-in-Chief, 1996). *Mayo Clinic Family Health Book*. New York: William Morrow and Company, Inc., B-8.

Appendix C
Equivocal and Unequivocal Beneficial Mutations

In an article titled "Mutations: The Raw Material for Evolution?" Dr. Barney Maddox, M.D., explains.[1]

> Virtually all the 'beneficial mutations' known are only equivocally beneficial, not unequivocally beneficial. In bacteria, several mutations in cell wall proteins may deform the proteins enough so that antibiotics cannot bind to the mutant bacteria. This creates bacterial resistance to that antibiotic. Does this support evolutionary genetic theory? No, since the mutant bacteria do not survive as well in the wild as the native (nonmutant) bacteria. That is, the resistant (mutant) bacteria will only do well in an artificial situation, where it is placed in a culture medium with the antibiotic. Only then can it overgrow at the expense of the native bacteria. In the wild, the native bacteria are always more vigorous than the mutant bacteria. (p. 12)
>
> In humans there is one equivocally beneficial mutation, out of 4,000 devastating mutations: sickle cell anemia... Sickle crisis occurs when red cells sickle and clog the arteries to parts of organs. Organs then undergo infarction (death from lack of blood supply). Without medical support the homozygotes are likely to die in young to middle age. (p. 12)
>
> But there is one positive. Heterozygotes in Africa, where malaria is endemic, are more resistant to malaria than people with normal hemoglobin, and the heterozygote genotype may have a survival advantage, but only in those areas. Could this be a limited example of evolutionary progress? Not really. When the mutant sickle gene is latent (i.e., sickling isn't occurring), there is a survival advantage in areas with malaria. But whenever sickling occurs, in the heterozygote or the homozygote, it obstructs blood vessels and causes pain and death to organs. (p. 12)
>
> However, the biology textbooks, when discussing mutation in evolution, only discuss the very rare 'positive' mutation, like sickle cell anemia. The fact of some 4,000 devastating genetic diseases is suppressed from publication. (p. 12)
>
> Sickling is always *negative* when it occurs, so it remains a very poor example of evolution, and in fact refutes it. ***Evolution theorists have yet to demonstrate the unequivocally positive nature of a single mutation.*** [Italics and bold added] (p. 12)
>
> Evolutionary science teaches that all the wonderful organs and enzymes in humans and animals—eyes, hemoglobin, lungs, hearts, and kidneys, all coded with DNA—arose totally by random chance through mutations in DNA. Consider the construction and operation of a machine. If random changes are made to a machine or the blueprint that codes for the construction of the machine, will that help its function? Absolutely not. Random changes occur every day that destroy the manufacture and function of machines. Likewise, random changes to information destroy the function and outcome of that information. (p. 13)
>
> Biology textbooks in theory present positive and negative mutations to students as though these were commonplace and roughly equal in number. However, these books fail to inform students that ***unequivocally positive mutations are unknown to genetics, since they have never been observed*** (or are so rare as to be irrelevant). [Italics and bold added] (p. 11)
>
> While instructing students that harmful mutations were more numerous than 'beneficial' mutations, this textbook failed to disclose that even equivocally beneficial mutations (which still have a downside) are extremely rare (about one in 10,000), and that ***unequivocally beneficial mutations are nonexistent in nature.*** [Italics and bold added] (p. 13)

Assuming an "unequivocal" mutation—a mutation that is nonexistent in nature—at the very least "the next 10,000 mutations...would each be fatal or crippling, and ... **would bring the evolution process to a halt.**" (p. 12) [Bold added]

---

notes: Appendix C

1. Maddox, B. (September 2007). Mutations: The raw material for evolution? *Acts & Facts*, 36 (9), Dallas, TX: Institute for Creation Research, 10-13. Copyright © 2007 Institute for Creation Research, used by permission.

Appendix D
Basic Geology of Planet Earth

The earth has a radius of about 3,958 miles (6,370 km) and is divided into four zones: *inner core, outer core, mantle*, and *crust.* The crust overlies the continents to a depth of about 12 to 45 miles (19 to 73 km) and ocean basins to a depth of about 4 to 12 miles (6 to 19 km). Although no one has drilled to these depths, the consistency of the zones can be estimated by analyzing seismic waves. The continental crust is composed of granite rock overlain by sedimentary rock, and the oceanic crust is mainly basalt (dense volcanic, igneous rock).

The solid portion of the earth comprises the lithosphere which includes the crust and part of the upper mantle. This is a relatively cool, rigid, outer layer of the earth that extends to a depth of about 62 miles (100 km). Earthquakes originate in the lithosphere.

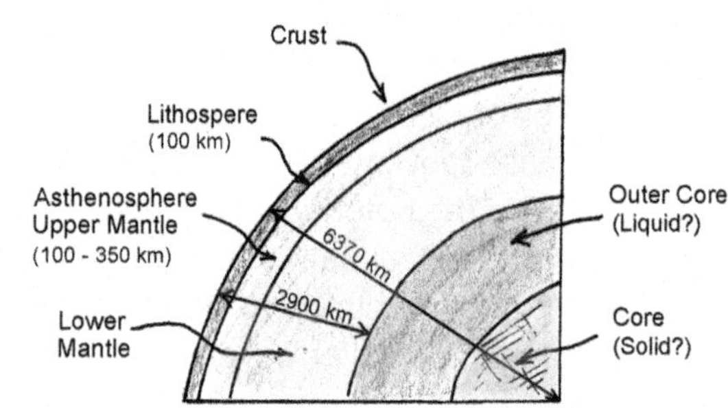

Most of the earth is made up of the mantle, which comprises solid rock and semi-liquid magma that extends to a depth of about 1,802 miles (2,900 km). The asthenosphere is that portion of the upper mantle beneath the lithosphere which is semi-liquid rock or magma. It is believed that the lithosphere "floats" on the asthenosphere.

The earth's inner and outer core is believed to consist of primarily nickel and iron. The outer core is believed to be molten liquid and the inner core is so hot and under so much pressure that it is believed to be a solid.

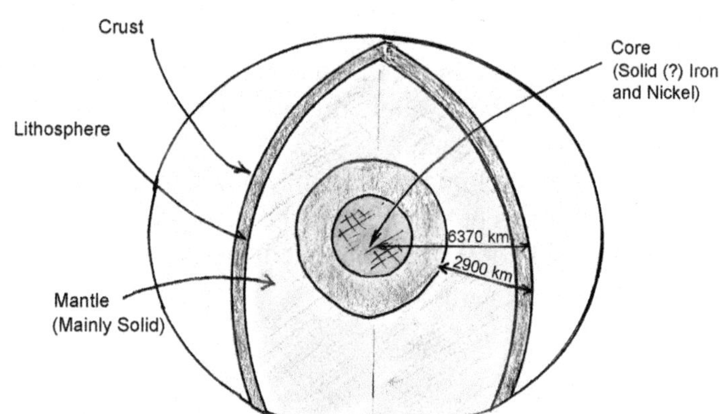

Source: Thompson, G. R., & Turk, J. (1991). Modern Physical Geology

Granite is an igneous rock that solidified from hot magma and comprises large portions of the continents and the central core of most mountain ranges. Mineral components of granite include quartz (white to glassy), feldspar (gray-pink), and some biotite (dark mica flakes), and hornblende (black-dark green).

Sedimentary rock comprises clastic fragments of other rocks that were transported and deposited primarily by water (or sometimes wind) as soft sediment layers in an unconsolidated form but eventually cemented together. (Cementing materials include dissolved calcium carbonate, silica, iron, and moisture.) In just about every case, sediments are deposited underwater by oceans, rivers, lakes, or flood events, often as long, flat layers known as turbidites. Such layers can be seen in mountains and canyons throughout the world.

Metamorphic rock forms when rock (e.g., igneous, sedimentary, or other metamorphic rock) changes or recrystallizes in response to increased temperature and pressure. Like igneous rock, metamorphic rock occurs deep within the core of many mountain ranges and within continental shields exposed at or near the earth's surface. Although metamorphism occurs in recent times on a very limited scale when hot magma intrudes into sedimentary rocks, enormous areas of metamorphic rock found throughout the world are the result of far greater forces than those observed today.

Earthquakes and Tsunamis

An earthquake is an abrupt motion of the earth caused by the release of clastic stress within the lithosphere. About 80 percent of all the planet's earthquakes occur along the edge of the Pacific Ocean basin, commonly known as the "Ring of Fire," because of the frequency of volcanic activity along crustal plate boundaries. See diagram, Beginnings of Catastrophic Earthquakes and Tsunamis in chapter 4.

Most earthquakes occur within these fault zones because of the movement of tectonic plates—gigantic rock slabs in the earth's lithosphere. These movements are usually gradual and go unnoticed on the surface, but enormous stress can build up between colliding crustal plates. When stress is released, it sends massive vibrations, called seismic waves, hundreds of miles through the rock.

Scientists assign a magnitude rating to earthquakes based on the strength and duration of the seismic waves. A quake measuring 3 to 5 is considered mild; 5 to 7 is moderate; 7 to 8 is strong; and 8 or more is great. On average, a magnitude 8 quake strikes somewhere each year and tens of thousands of people die annually worldwide. Although collapsing structures claim the majority of lives, destruction and loss of life is often compounded by mudslides, fires, or tsunamis.

Earthquakes, mass movements (i.e., mudslides, landslides, avalanches) above or below water, and volcanic eruptions have the potential to generate a tsunami—a series of massive waves created when a body of water is rapidly displaced, usually in the open ocean. A tsunami has a smaller amplitude (wave height) and a very long wavelength (often hundreds of kilometers long) as compared with a "wind wave" (usually with a larger amplitude but much smaller wavelength). See diagram, Beginnings of Catastrophic Earthquakes and Tsunamis, in chapter 4.

Typically, tsunamis are produced when the sea floor abruptly collapses because of an earthquake and vertically displaces the overlying water mass. Water fills the low spot and overcompensates, creating the tsunami. Because tsunamis have long wavelengths, even the most devastating tsunamis may pass unnoticed at sea, often forming only a passing "hump," but as the tsunami comes ashore, it often produces massive floodwaters rather than the breaking surf usually seen at the beach.

> In open oceanic waters, the crest of a tsunami may be 1–5 meters high but 62 to 186 miles long (100–300 km) and moving 200 to 300 miles per hour (322–483 km/hr). As it comes ashore, it is a wall of water ten to hundreds of feet high. In Noah's Flood, gigantic tsunamis were likely hundreds of feet high.

The Sri Lanka-Sumatra tsunami on December 26, 2004, and the Japanese tsunami on March 11, 2011, resulted in the loss of hundreds of thousands of lives. These tsunamis were caused by undersea earthquakes with a magnitude of 9.0, or perhaps higher.

Appendix E
Evidence of Rapid Limestone Formations

In today's marine environment, calcareous muds form by the "mechanical breakdown of carbonate-containing sea creatures" and by precipitation of calcium carbonate ($CaCO_3$).[1] Marine organisms such as corals, clams, and mussels are able to extract $CaCO_3$ from seawater (that is, biochemically fix bicarbonate, HCO_3^-, and calcium, Ca^{+2}) to build their skeletal structure. When these marine organisms die, their shells disintegrate and accumulate as a calcareous ($CaCO_3$) mud.

According to most geology textbooks, accumulation rates of this bottom sediment (lime mud) are about one foot per 1,000 years. Because evolutionists believe that "the present is the key to the past"—that is, geological and meteorological processes have remained constant throughout earth's history—it is not possible to explain massive limestone formations (e.g., the Redwall, Kaibab, and Toroweap limestone of the Grand Canyon) and chalk beds found throughout the world. However, visual and empirical evidence offers *compelling support for rapid formation of lime muds.* Geochemical mechanisms for rapid accumulation of lime muds are presented in Appendix F.

Redwall Limestone (vertical formation in foreground) and Supai Group (intermix limestone and sandstone), South Rim of Grand Canyon, Arizona. *Photos by Roger Gallop*

Kaibab Limestone, Grand Canyon, Arizona

Lagerstätten

There are many examples found worldwide of extraordinary fossilization of animals that were rapidly covered with lime mud of fine texture. Such fossil assemblies are found worldwide—they are diverse and extremely well-preserved and earn the label of Lagerstätten (meaning extraordinary fossil richness). For example, some of "the 'world's most perfect fossils'... come from the fine-grained limestones of the Santana Formation of northeast Brazil."[2] While describing fossil fish of the Santana Formation, it was concluded that except "for a few notable exceptions lithification was instantaneous and fossilization may even have been the cause of death."[3] According to Dr. Steven A. Austin, "These limestones appeared to have formed as animals were smothered in lime mud."[4]

Some of the world's major Lagerstätten are found in limestone (lime muds), volcanic ash, shale, slate, chert, and chalk, and a mixture of limestone, sandstone, and shale. Such formations are found in the United States (Montana, Illinois, Kansas, New York, Utah, New Mexico, Nebraska, Kansas, Alabama, Idaho, California), Brazil, Argentina, France, Kazakhstan, China, Australia, Canada, Greenland, Sweden, South Africa, England, Scotland, Germany, Dominican Republic, and Italy.[5]

> Lagerstätten (German; literally *places of storage*) are sedimentary deposits that exhibit extraordinary fossil richness or completeness.

Fossils and Cross-bedding in Redwall Limestone

In his book, *Grand Canyon: Monument to Catastrophe*, Dr. Steven Austin discusses fossil evidence and cross-bedding that support flood deposition of the Redwall Limestone.[6] These evidences are summarized below.

In 1966, a geologist discovered numerous large, cigar-shaped fossils known as nautiloids in the lowest part of the Redwall Formation. They are a class of animals called cephalopods, which include octopuses, squids, and cuttlefishes. These fossil animals resemble the modern-day nautilus—a coiled, free-swimming, deep-sea mollusk found in the Pacific and Indian oceans.

Nautiloid fossils in the Redwall Limestone differ from the modern nautilus by being straight rather than coiled. The shape of this fossilized animal allowed it to swim with great speed and agility throughout the ocean. So how did so many of these fast-swimming deep-sea animals become entombed in lime mud of fine texture and fossilized in one location? Studies of these fossils suggest that they were "aligned" with the current and smothered by extreme rapid burial. Rapid accumulation of lime mud is also confirmed by fossils of "crinoid heads," an animal that lives in a cup-like structure that would quickly disintegrate after death.

Cross-bedding has been reported at several locations in the Grand Canyon. Cross-beds in the Redwall Limestone at Kanab Creek (with a vertical thickness of 30 feet) are "remnants of large sand waves (underwater dunes) composed of coarser lime sediment" that were formed by steady ocean currents.[7] (Refer to chapter 5, Coconino Sandstone and "sand waves," and chapter 6, Presence of Cross-Bedding and Fragile Surface Features.)

Also, quartz sand grains are intermixed with ancient limestones such as Kaibab and the Supai Group. The presence of coarse sand grains provides additional evidence that these limestones derived from sediment that was transported by moving water, not a steady drizzle of lime mud in a tranquil sea over millions of years. These limestones of the Grand Canyon "accumulated from sediment which was being transported by moving water, not simply deposited from a slow, steady rain of carbonate mud on the floor of a calm and placid sea."[8]

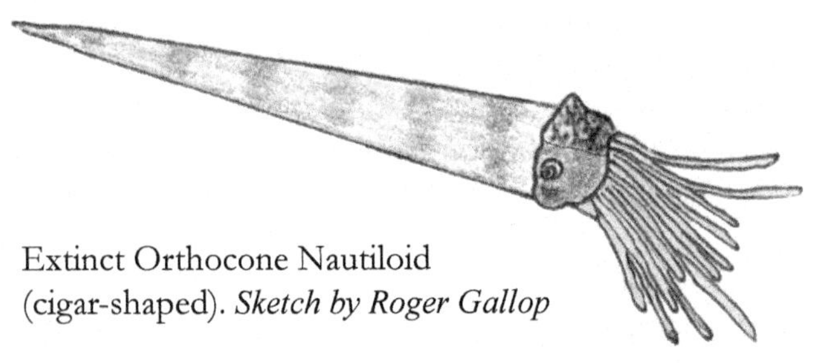

Extinct Orthocone Nautiloid (cigar-shaped). *Sketch by Roger Gallop*

Modern-Day Nautilus
Nautilus pompilus

Kingdom: Animalia
Phylum: Mollusca
Class: Cephalopoda
Subclass: Nautiloidea

In order to swim, the modern-day nautilus uses jet propulsion as it draws water into and out of the chamber through a small, flexible propulsion tube (hyponome). Nautiluses inhabit depths of about 300 meters (980 feet), rising to about 100 meters (330 feet) at night to feed.

Lack of Coral and Sponge Reefs

Some evolutionists maintain that fossiliferous limestones are ancient coral or sponge "reefs" that slowly accumulated for millions of years within the shallows of ancient seas. If this were the case, one would expect to find large, organic reef formations buried in the limestone formation. Do such formations occur within the Grand Canyon limestones? Are sponges, corals, and algae, which comprise the fabric of modern reefs, found in the canyon?

Perhaps the most wide-ranging study of the Grand Canyon limestone was conducted by McKee and Gutschick, who state, "Coral reefs are not known from the Redwall Limestone."[9] Further a report on the Kaibab Formation by Hopkins indicates, "Discrete organic build-ups, such as sponge patch reefs, have not been documented."[10] Suffice it to say that coral and sponge reef formations have not been shown to characterize any of the limestones of the Grand Canyon.

---

notes: Appendix E

1. Austin, S.A. (Ed.). (1994). *Grand Canyon: Monument to Catastrophe*. Institute for Creation Research, El Cajon, CA, 24.
2. Ibid., 26.
3. Martill, D.M. (1989). The medusa effect: Instantaneous fossilization. *Geology Today*, 5: 201. Also cited in Austin (1994), 26. Also, see Martill, D.M. (1989). A new 'solenhofen' in Mexico. *Geology Today*, 5: 25-28.
4. Austin (1994), op. cit., 26. Also see Brett, C.E., and Seilacher, A. (1991). *Fossil Lagerstatten: A Taphonomic Consequence of Event Sedimentation*. (G. Einsele, W. Ricken, and S. Seilacher, Eds.). *Cycles and Events in Stratigraphy.* New York: Springer-Verlag, 296.
5. Lagerstätte (May 2009). In Wikipedia, the free encyclopedia. Last Modified, May 2009. Retrieved May 2009, from http://en.wikipedia.org/wiki/Lagerst%C3%A4tten.
6. Austin, op. cit., 25-28.
7. Ibid., 28.
8. Ibid.
9. McKee, E.D. and Gutschick, R.G. (1969). *History of the Redwall Limestone in Northern Arizona.* Boulder, CO: Geological Society of America, Memoir 114, p. 557; as cited in Austin, 26.
10. Hopkins, R.L. (1990). Kaibab Formation. In: *Grand Canyon Geology.* (S.S. Beus and M. Morales, Eds.). New York: Oxford University Press and Museum of Northern Arizona Press, chapter 12, 243. Also cited in Austin, 26.

Appendix F
Mechanisms for Rapid Accumulation of Lime Muds

As stated in previous chapters, evolutionary doctrine does not take into account a worldwide flood but rather, maintains that natural geological and meteorological processes have remained constant during earth's history. It is the belief that existing physical processes are sufficient to account for all past changes in our world and the universe. *There is no basis for this belief other than the assumption of evolution as the only "scientific" mechanism for the existence of man and all living things*—the *"modern scientific establishment has bound itself to a single system of interpretation...and alternatives must be rejected out of hand."*[1]

Creationist doctrine maintains that God directly interceded in the normal physical processes of the world, causing significant geologic cataclysmic events at certain points in time (creation and Noah's Flood). As presented throughout this book, there is overwhelming visual and empirical evidence for a worldwide catastrophic flood and a young earth.

> There are three primary geochemical mechanisms for rapid limestone formation: 1) aggregation or flocculation of fine-grain crystals of calcite or aragonite in a high energy environment, 2) warming of oceanic waters and direct precipitation of dissolved calcium carbonate, and 3) precipitation of dissolved calcium carbonate due to the depletion of carbon dioxide in the atmosphere, all as a result of a worldwide flood.

Effects of a Worldwide Flood

A worldwide flood would have greatly altered the carbon balance of the world. The flood would have buried and sealed massive amounts of carbon in the form of plants that became coal and oil (the vast coal and oil reserves found today), thus depleting the total carbon (^{14}C and ^{12}C) available in the biosphere—that is, total carbon within plant and animal communities and the atmosphere.

During the flood event, the normal production of CO_2 by means of animal respiration and decaying plants would have been substantially curtailed because of rapid burial by millions of tons of water-deposited sediment. Also, new plants regrowing after the flood would have absorbed the remaining or residual CO_2 (by means of photosynthesis) that was not replaced by decay of vegetation, thereby continuing to reduce CO_2 in the atmosphere.

This dramatic change of *partial pressure* of gases across the sea-atmosphere interface would have caused a major shift in carbon dioxide equilibrium (between the oceans and the atmosphere)—resulting in massive precipitation of calcium carbonate and immense limestone formations we observe today. With this in mind, let's take a look at the basic mechanisms for rapid accumulation of lime muds.

> Plants absorb carbon dioxide and use it to grow, but when plants decay or burn, carbon dioxide is released back into the atmosphere.

Carbon Dioxide Equilibrium

The solubility of carbon dioxide in water is great compared to oxygen because CO_2 reacts chemically with water. Seawater in contact with air will gain or lose CO_2 depending on whether the partial pressure of the gas in the water is lower or higher than that in the atmosphere. Seawater can absorb considerably more CO_2 from the atmosphere if the partial pressure of the atmospheric CO_2 increases; and if it decreases, as would have happened during a catastrophic worldwide flood, seawater would liberate considerable quantities of CO_2. Thus, the oceans act as a CO_2 regulator for the atmosphere.

A visual, practical example of partial pressure is a sealed bottle that is half full of water. Gaseous CO_2 from the air dissolves in the water and dissolved CO_2 in the water escapes to the air: CO_2 in the air = CO_2 dissolved in water. The reaction continues until a balance or equilibrium is reached. Although an equilibrium may be reached (at a constant temperature and pressure), the chemical equilibrium remains dynamic.

If we upset this equilibrium by **increasing** the amount of CO_2 in the air, the reaction adjusts for the disturbance—that is, **carbonate reaction moves to the right,** removing CO_2 from the air and dissolving excess CO_2 in the water in order to establish equilibrium. Forward direction is favored and pH decreases (acidic environment; increases hydrogen ion concentration, H^+).

Any process that increases the amount of CO_2 in the air

a) causes the CO_2 equilibrium reaction [1] to shift to the right to use up increased CO_2 in the air

$$CO_2 + H_2O \rightarrow H_2CO_3 \rightarrow HCO_3^- + H^+ \rightarrow CO_3^{-2} + 2H^+ \quad [1]$$

weak carbonic acid → bicarbonate → carbonate

b) in turn, it causes reaction [2] to shift to the right, which means that $CaCO_3$ (calcium carbonate, or calcite) dissolves or goes into solution.

$$CaCO_3 \text{ (base)} + H_2CO_3 \rightarrow Ca^{+2} + 2HCO_3^- \text{ (acidic)} \quad [2]$$

solid calcite → dissolved calcite - CO_2 in solution

Catastrophic Worldwide Flood - Depletion of CO_2 in the Biosphere

If, on the other hand, the equilibrium is disturbed by **decreasing** the amount of CO_2 in the air, such as the depletion of CO_2 production during a catastrophic worldwide flood (see chapter 4), the reaction adjusts for the disturbance—that is, carbonate reaction shifts to the left, releasing CO_2 into the air and precipitating $CaCO_3$ in order to establish equilibrium. The **carbonate reaction moves to the left**—a backward direction is favored and pH increases (basic or alkalizing environment; decreases hydrogen ion concentration, H^+).

Any process that decreases the amount of CO_2 in the air

a) causes the CO_2 equilibrium reaction [1] to shift to the left to offset the loss of CO_2 in the air

$$CO_2 + H_2O \leftarrow H_2CO_3 \leftarrow HCO_3^- + H^+ \leftarrow CO_3^{-2} + 2H^+ \quad [1]$$

weak carbonic acid ← bicarbonate ← carbonate

b) in turn, it causes reaction [2] to shift to the left, which means that $CaCO_3$ (calcite or calcium carbonate) precipitates or comes out of solution.

$$CaCO_3 \text{ (base)} + H_2CO_3 \leftarrow Ca^{+2} + 2HCO_3^- \text{(acidic)} \quad [2]$$

solid calcite ← dissolved calcite- CO_2 in solution

A worldwide flood would have resulted in *rapid* depletion of CO_2 in the air and deposition of massive quantities of lime mud deposits (to maintain CO_2 equilibrium between air and water)—all within a period of just one year! Following the flood, increasing amounts of CO_2 were *gradually* released into the atmosphere as regrowing plants began to decay. The reaction to the left would slowly decrease until equilibrium was reached (CO_2 in the air = CO_2 dissolved in water). Although limestone rock below sea level would eventually be subject to gradual dissolution, *most lime mud deposits during the flood were immediately exposed to the atmosphere because of the much lowered post-flood sea level elevations*. See chapter 4, The Great Ice Age. Additionally, pre-flood plants remained buried and sealed, and became today's vast coal and oil reserves throughout the world.

Increase in Temperature Due to Volcanism and Superheated Geysers

During the flood, great tectonic events occurred throughout the world. These included catastrophic rifting and subduction of continental plates, uplift of oceanic basins, flooding of low-lying continents, sediment deposition, mountain and continent uplift, and torrential erosional drainage of floodwaters, all processes resulting in land features we observe today. (See chapter 4 for more information.)

These events included volcanisms, earthquakes, and superheated geysers on a massive scale—consequently, floodwaters would have become very warm. In turn, warming of seawaters would have precipitated massive quantities of lime mud deposits throughout the world. Today, $CaCO_3$ is often found precipitating near hot springs. In contrast, $CaCO_3$ is not found in sediments in the colder, deep marine environment.

> Gases like CO_2 and O_2 are less soluble in warm water and more soluble in cold water. As seawater is heated, carbonate (CO_3^{-2}) comes out of solution, combines with calcium (Ca^{+2}), flocculates or clumps together, and quickly precipitates.

Aggregation of Fine-Grain Crystals of Calcite in a High Energy Environment

In recent times, there have been numerous observations of rapid accumulation of lime muds in a high-energy water environment. It has been suggested that these lime mud layers were formed by "direct precipitation" of calcium carbonate during storm events. Microscopic examination reveals that modern lime mud particles aggregate into "pelletoids [which] exhibit hydraulic characteristics of sand."[2] In tidal channels or during storms, fine grain aragonite particulates clump together as a "creamy white lime mud" and quickly settle to the bottom.

> During the worldwide flood, as seawater warmed in a high energy environment and CO_2 was depleted in the atmosphere, pelletoids would have rapidly accumulated on the ocean bottom—and in response to earthquakes, massive slope failures formed turbidites (pancake-like strata) comprising lime mud, sand, and shell fragments. Bulging, pancake-like lime mud layers of Kaibab Limestone (see photos in chapter 6, Limestone Formations) indicate huge mudflow and turbidite formation.

***notes:* Appendix F**

1. Evolution's evangelists. (May 2008). *Acts & Facts*, 37 (5), Dallas, TX: Institute for Creation Research, 10. Copyright © 2008 Institute for Creation Research, used by permission. 2008. Evolution's Evangelists. *Acts & Facts*, 37 (5): 10.

2. Austin, S.A. (Ed.). (1994). *Grand Canyon: Monument to Catastrophe*. Institute for Creation Research, El Cajon, CA, 26.

Appendix G
Unconformities and Paraconformities

Geology textbooks generally use the term unconformities and paraconformities (pseudo-conformity) to describe eroded or missing layers. An *unconformity* represents an erosional break between sedimentary strata, and *paraconformity* is an assumed or hypothetical erosional event between sedimentary strata over time although there is *no actual indication* of erosion. *Conformity* represents continual deposition of sedimentary layers with no erosional break.

> As discussed in chapters 4 and 5, sediment is deposited underwater and eroded by wind, rain, frost, and water while exposed as dry land, or eroded as a result of rapid drainage and scouring processes in a high energy water environment (i.e., Phase 3 of the worldwide flood). Today, sediments exist as either horizontal deposits (turbidites) or horizontal deposits that have been tilted and folded by tectonic uplift.

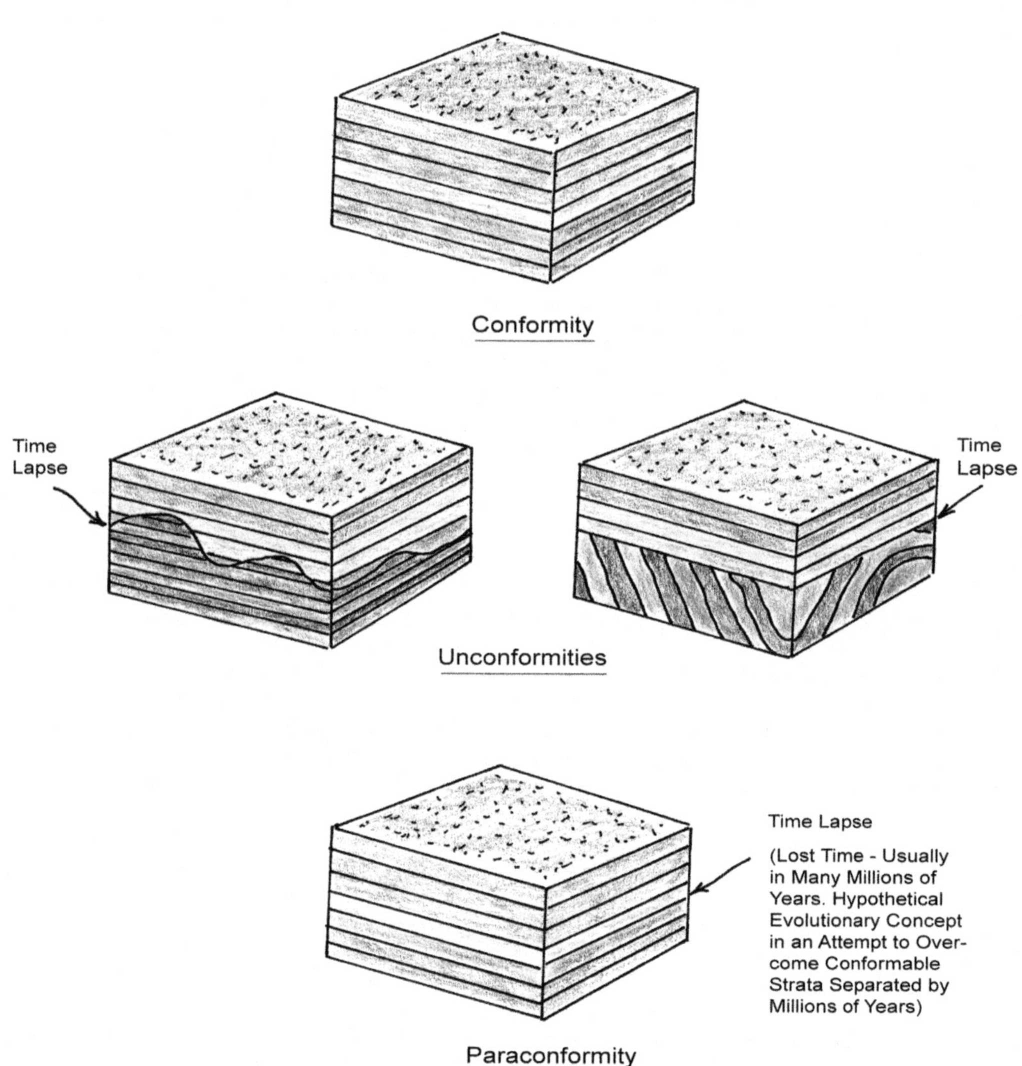

Types of Erosional Episodes in Sedimentary Rock

Unconformity – A Regional and Rapid Erosional Event

The most prominent unconformity (an angular unconformity) of the Grand Canyon, sometimes referred to as the Great Unconformity, is the Precambrian rocks which underlie the Tapeats Sandstone. Most creation geologists believe this feature was formed during worldwide tectonic events at the beginning of the Flood event.

Although regional unconformities (layers or formations with observable erosional breaks) do exist, such as those between Zion and Bryce Canyons and the Grand Canyon, evidence suggests that these unconformities are the result of rapid erosion while underwater—not erosion over millions of years while the land is exposed to dry conditions. For example, it appears the upper layers of plateaus north and south of the Grand Canyon—upper layers such as Moenkopi and Chinle formations found in erosional remnants such as Red Butte of the Coconino Plateau just north of Flagstaff—were rapidly swept away by "broad, sheetlike erosion" toward the end of the Flood.[1] (See chapter 5, Great Peneplains and Monoliths, and chapter 7, Contradictions in Uniformity Theory.)

Regional and Rapid Erosional Event

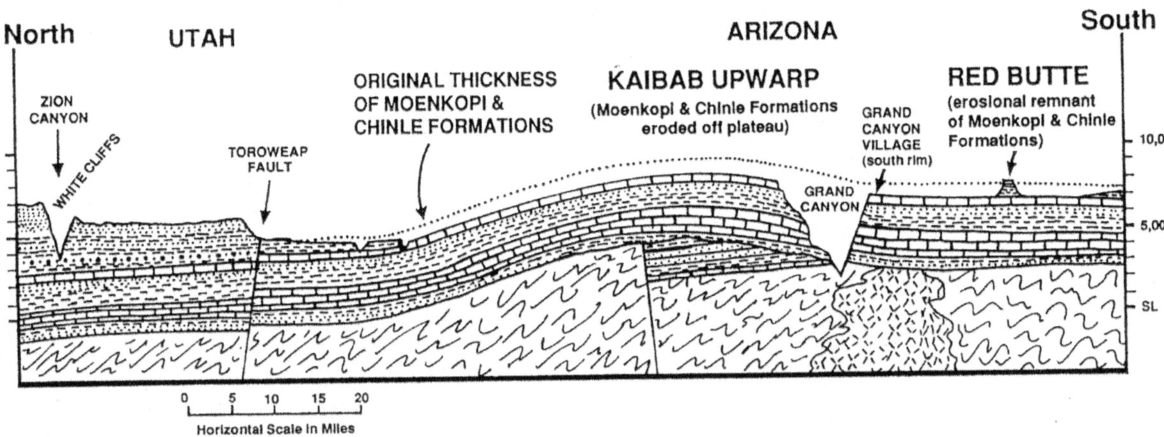

Copyright © 1994 Institute for Creation Research, used by permission. Figure 5.3 in Austin, S. (1994). *Grand Canyon: Monument to Catastrophe.* Santee, CA: Institute for Creation Research, 84.

Paraconformity – A Hypothetical Erosional Event

According to secular geologists, a *paraconformity* is a sedimentary layer that has remained dormant for millions of years; that is, a surface of non-deposition and non-erosion. Essentially, it is a *missing time gap* between two sediment formations.

For example, the Chinle formation (sedimentary rocks comprising mudstones, siltstones, sandstones, and clay soils) was supposedly deposited more than 200 million years ago at the end of the Triassic Period. This formation makes up most of the landforms of the Petrified Forest National Park in Arizona and is spread across northern Arizona, Nevada, Utah, western New Mexico, and western Colorado. The Bidahochi Formation, which directly overlies the Chinle Formation in the national park, was supposedly deposited 3 to 6 million years ago. The Chinle/Bidahochi paraconformity represents 200 million years of missing geologic history.

Another example is the contact between Hermit Shale and the overlying Coconino Sandstone. As noted in the photo, the sharp contact between these formations exhibits no evidence of erosion—although, from an evolutionist standpoint, the time break between Hermit and Coconino supposedly represents more than 10 million years. Other formations of the canyon supposedly representing millions of years grade into each other such as Tapeats sandstone, Bright Angel shale, and Muav limestone (collectively called the Tonto Group). These sharp contacts between formations indicate there were no long periods of time separating the formations.

The term, paraconformity, is an attempt by secular geologists to overcome the problems associated with conformable layers that are separated by vast periods of time while displaying no evidence of erosion. In actuality, there is no stagnant or dormant surface on land today and a paraconformity is actually a conformity; that is, there is no missing layer or formation.

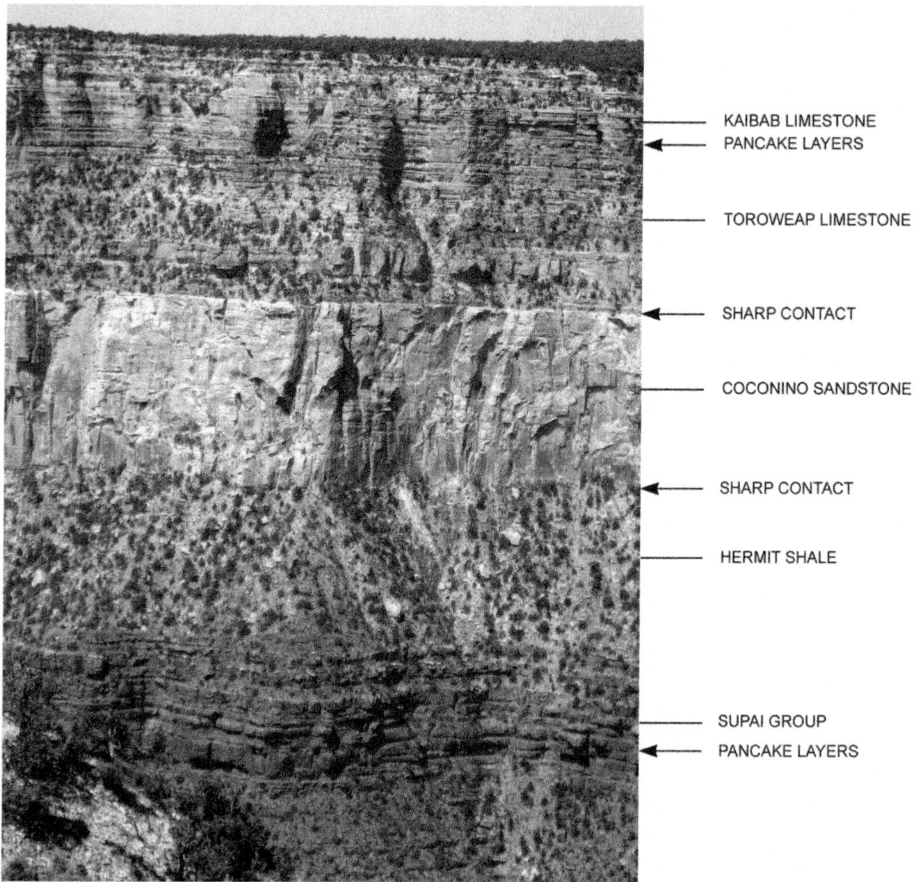

Grand Canyon, Arizona. The lack of erosion along bedding planes of rock formations (often referred to as paraconformities) supposedly separated by many millions of years provides strong evidence of rapid deposition of sediment strata. *Photo by Roger Gallop*

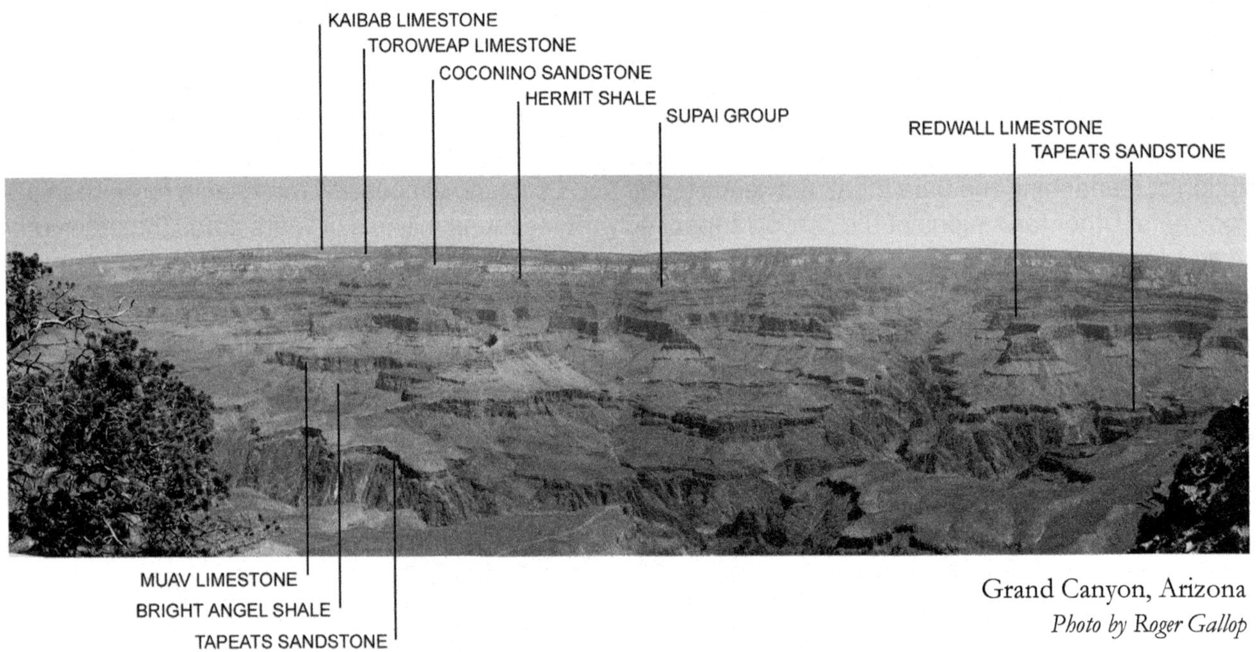

Grand Canyon, Arizona
Photo by Roger Gallop

---

notes: Appendix G

1. Austin, S.A. (Ed.). (1994). *Grand Canyon: Monument to Catastrophe*. Institute for Creation Research, El Cajon, CA, 84. Also see Austin, S.A. (November 1984). Ten misconceptions about the geologic column. El Cajon, CA: Institute for Creation Research, ICR *Impact,* Article 137 (Item #3). Retrieved from http://www.cnt.ru/users/chas/imp-137.htm; Austin, S.A. (1984). Ten misconceptions about the geologic column. *Acts & Facts,* 13 (11); Woodmorappe, J. (1981). The essential nonexistence of the evolutionary-uniformitarian geologic column: a quantitative assessment. *Creation Research Society Quarterly,* 18: 46-71; and Woodmorappe, J. (1999). The geologic column: Does it exist? *Journal of Creation,* 13 (2): 77-82. Retrieved from http://creation.com/journal-of-creation-archive-index.

Appendix H
Carbon-14 Dating and Effects of a Worldwide Flood

A cataclysmic global flood would have greatly altered the carbon balance of the world with massive tectonic upheaval. The flood would have buried and sealed massive amounts of carbon in the form of plants that became coal and oil (see chapters 4 and 6), thus exhausting the total carbon (C-14 and C-12) available in the biosphere—that is, the total carbon in earth's plant and animal communities and atmosphere. The normal production of CO_2 by *decaying plants would have been substantially curtailed* because of burial under millions of tons of water-laid sediment. Also, new plants regrowing after the flood would have absorbed the remaining CO_2 (by photosynthesis) that was not replaced by decay of vegetation.

Although C-14 and C-12 were depleted in the biosphere because of burial by the flood, C-14 was continually being produced in the atmosphere from nitrogen-14. Thus, C-14 was increasing at a faster rate than C-12. Although the amount of C-14 and C-12 was substantially depleted, the ratio of C-14 to C-12 would have continued to increase after the flood—so one would reason the ratio of C-14 to C-12 in plants and animals during and after the flood was lower than it is today.

As stated by Dr. John D. Morris, "It is now admitted by all investigators that equilibrium does not exist—that the C-14 concentration is constantly increasing."[1] This *implies less carbon-14 in the past—so animals and plants that supposedly died hundreds or thousands of years ago would date much older than their true age.* See end note 2.

> How is carbon-14 formed? When cosmic radiation strikes nitrogen-14 in the atmosphere, N-14 is converted into C-14—in turn, C-14 decays back into N-14 with a half- life of 5,730 years.

> "The rate of carbon-14 production today is 18% higher [18 to 25% higher] than the rate of decay. This means that today we are experiencing a net *increase* in the proportion of C-14 in the atmosphere. It is impossible to determine whether it has *always* been increasing... But one thing is certain: there is no reason to believe that the carbon-14 to carbon-12 proportion has been constant throughout time, and *good* reason to believe it has been...lower in the past than it is today."[2] See end note 2 for additional information.

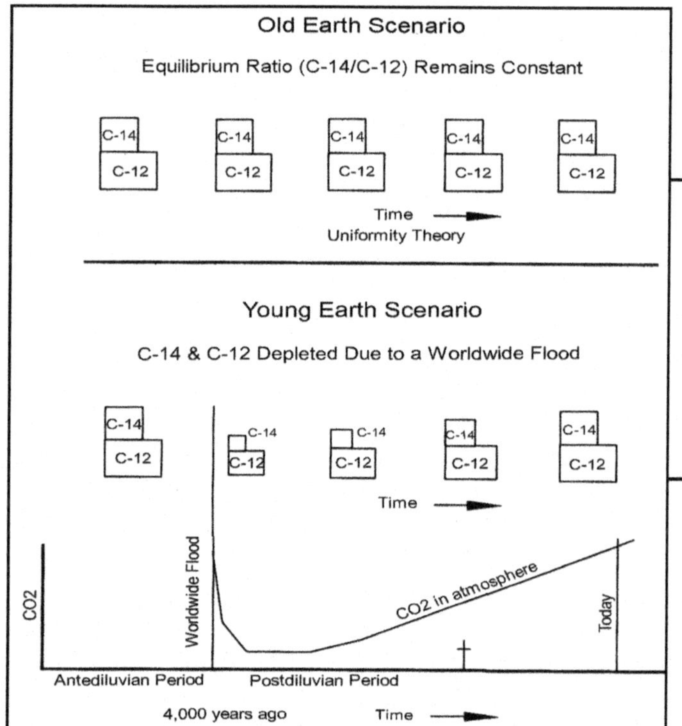

Note: The ratio of C-14 to C-12 is extremely small (~1:1 x 10^{12}) and is not proportionally represented in the graphics. *Sketch by Roger Gallop*

> About 99% of all natural carbon is C-12, about 1% is C-13, and trace amounts are found as C-14. The atmospheric ratio of C-14 to C-12 is approximately one to one trillion (1:1 x 10^{12}).
>
> **Old Earth Scenario** assumes a constant ratio of C-14/C-12 and a constant production/decay rate of C-14 throughout history. For example, if the sampling of a dead animal yielded 50 atoms of C-14 and 100 x 10^{12} atoms of C-12, then one would conclude the animal at the time of death had 100 atoms of C-14. So decay of C-14 from 100 to 50 atoms would yield a half-life of 5,730 years.
>
> **Young Earth Scenario** maintains that total carbon in the biosphere was depleted during and following the flood—and C-14 has been increasing at a faster rate relative to C-12. This implies less C-14 in the past relative to C-12—so animals and plants would date much older than their true age.

Carbon was depleted from the atmosphere in several ways:

- Plants, including ocean plankton, were smothered and sealed by hot volcanic ash and sediment deposits—the normal production of CO_2 by decaying plants would have been abruptly curtailed because of burial under millions of tons of water-laid sediment.
- Plants regrowing after the flood absorbed remaining CO_2 through photosynthesis.
- Atmospheric diffusion, precipitation from rainfall, and volcanic particulates.

Carbon was eventually released back into the atmosphere in several ways:

- As the carbonate reaction shifted left during the flood, CO_2 was released back into the atmosphere from the oceans–the reaction would slowly decrease until equilibrium was reached.
- Volcanic eruptions would have released gases into the atmosphere. (Volcanic gases are primarily water vapor, carbon dioxide, and sulfur dioxide.)
- Respiration by plants and animals; also decay of animal and plant matter, deforestation, and the industrial age (that is, burning of coal, oil, and gas).

In summary, plants and animals that lived and died during and following the flood would have absorbed *lower* carbon-14 levels—and their fossils would yield ages much older than their true ages (lower $^{14}C/^{12}C$ ratio yields an older age). Measurement of C-14 in historically dated objects (for example, seeds in the graves of dated tombs) enables the level of C-14 in the atmosphere at that time to be estimated, and so partial calibration of the C-14 "clock" is possible. Therefore, carbon dating *carefully applied to items from historical times* can be useful.

Radiocarbon dating has proven accurate to about 5,000 years, when artifacts could be calibrated to known dates, and *older dates could not be calibrated since there is no known historical material beyond that time.*[3] However, even with such historical calibration, archaeologists do not regard C-14 dates as reliable because of frequent differences, and they rely more on dating methods that link the date to historical records.

***notes:* Appendix H**

1. Morris, J.D. (1994). *The Young Earth*. Green Forest, AR: Master Books, 65; and Morris, J.D. (2007), 64. Also see Catchpoole, D., Sarfati, J., and Wieland, C. (2008). *The Creation Answers Book*. (D. Batten, Ed.). Atlanta, GA: Creation Book Publishers, 69-71; DeYoung, D. (May 2006). *Thousands ... Not Billions*. Green Forest, AR: Master Books, 59; and Wieland, C. (April 1979). Carbon-14 dating – explained in everyday terms. *Creation*, 2 (2): 14-18.

2. Carbon-14 dating (November 2008, last modified). In Encyclopedia of Creation Science. Creation Wikipedia. Retrieved May 2009, from http://creationwiki.org/Carbon-14_dating. See Whitelaw, R.L. (1993). A review and critique of pertinent creationist writing. 1950-1990. *Creation Research Society Quarterly*, 29 (4): 170-183; and Cook, M.A. (1986). Nonequilibrium radiocarbon dating substantiated. *Proceedings of the First International Conference on Creationism*, vol. 2, Pittsburg, PA: Creation Science Fellowship, 59-68.

Carbon-14 production rates are affected by the amount of cosmic rays penetrating the earth's atmosphere which, in turn, are affected by the strength of the earth's magnetic field which deflects cosmic rays. Precise measurements during this past century have shown a steady decay in the earth's magnetic field (see chapter 9, Rapid Magnetic Field Reversals and Decay) resulting in a steady increase in carbon-14 production.

3. Livingston, David (2003). The date of Noah's flood. Retrieved April 2008, from http://www.ancientdays.net/flooddate.htm. Also see Libby, W.F. (1958). Chemistry and the atomic nucleus. *American Journal of Physics*, 26: 528–541. (Dr. Willard Libby was the inventor of the radiocarbon technique.)

Appendix I
Isochron Dating

Isochron techniques were developed when scientists first recognized difficulties with the assumptions of radiometric dating: 1) constant rate of decay, 2) no loss or gain of parent or daughter isotopes (closed system), and 3) known amounts of daughter isotopes at the start of cooling. Scientists have attempted to address assumptions 2 and 3 with the use of a technique known as "least squares regression" on multiple specimens of rocks and minerals, all from the same geologic unit. (Isochron means equal time; assume sampled rocks and minerals formed at the same time.) A graphical representation and brief explanation of this technique are provided in the text box.

Cogenetic means they formed at the same time and place from the parent homogeneous pool of magma and each rock sample comprised the same amount of daughter isotopes at the start of cooling.

Assumption 1 is **not** addressed by the isochron method. The slope of the regression curve *indicates the total amount of decay, not the rate of decay. Regardless of the past rate of decay, the isochron graph would appear the same.*[1]

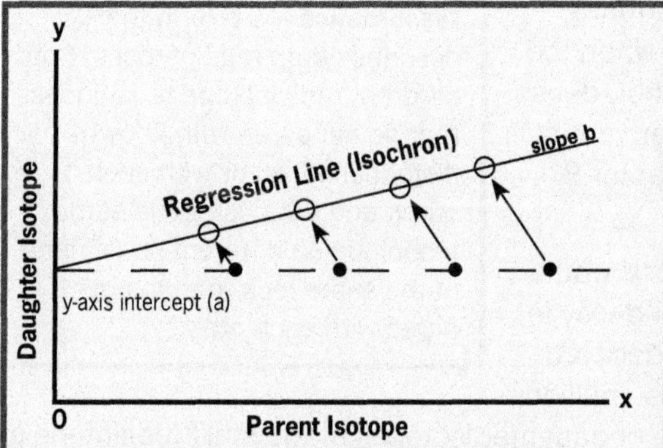

Graph of Ideal Whole Rock Isochron

The horizontal axis (x) represents the amount of parent atoms present in the four mineral samples, and the vertical axis (y) represents the amount of daughter atoms present in the four samples. The small circles (o) represent rock samples that were actually analyzed and measured for daughter and parent isotopes. The four rock samples are assumed to be cogenetic; that is, they solidified at the same time and place from a homogeneous pool of magma with the same daughter content. The solid circles (●) represent assumed daughter and parent isotopes at the time of cooling.

A "least squares regression" procedure is utilized to find the slope (b) and y-axis intercept (a). The slope of the line (called the regression coefficient) provides the isochron age of four mineral or rock samples (statistical average, thereby allegedly addressing the second assumption). The intersection of the slope with the vertical axis indicates the amount of "inherited" daughter isotopes at the time of cooling (thereby allegedly addressing the third assumption). By knowing the half-life of the isotopes and how much the daughter isotope changes per unit change of the parent isotope, the scientist can supposedly date the rock.

Assumption 2 is addressed by sampling multiple rocks presumably from the same magma—rocks that are considered *cogenetic*. Supposedly, if mixing or contamination has occurred, data points should not fall on a straight line.

As explained by Dr. Steven A. Austin, "incomplete mixing of two magmas [contamination] having different strontium isotope ratios produces a mixing diagram where all mixtures lie on a straight line...*with the slope of the line having no identifiable time significance!*"[2] These straight line isochrons may be the result of mixing or contamination, and are commonly referred to as "false isochrons."[3]

Critics have also pointed out that if the system has remained closed and contamination has not occurred, then the whole rock data should lie on a "single point"—not straight, multiple data points as most often depicted on a graph. According to many competent scientists, the second assumption is invalid because it assumes no tectonic activity (i.e., no uplift, folding, and thrusting of the earth's crust), leaching, or migration over the supposed millions of years of earth's history—consequently, the technique is "fatally flawed."[4]

Assumption 3 is addressed by plotting the intersection of the slope with the vertical axis which indicates the amount of "inherited" daughter isotopes at the time of cooling. The isochron technique appears to address this assumption, but when creation geologists sample basaltic rocks known to have flowed in *recent times* (see chapter 8, Unreliable Results), and after being analyzed by reputable and independent laboratories, they almost *always date in the millions of years*.[5] This indicates that assumptions of radioisotope dating are simply ***wrong***.

The RATE project results clearly show that discordance (results do not agree with known historical dates or with other techniques) exists among the various radioisotope dating methods. In fact, discordance for rock samples from the same geologic location is commonplace.

Nevertheless, when secular geologists are asked to document the most reliable radiometric age dates, they usually turn to the whole rock isochron-diagram technique. The precision of measurements is not in question, but rather the assumptions of the procedure.

The following are primary reasons for discordance—and reasons that ultimately invalidate the isochron (radioisotope) dating technique:[6]

- ***Assumption 1*** (Constant rate of decay) – The isochron technique assumes a constant rate of decay based on uniformity theory—*it does not test for accelerated nuclear decay.* The slope of the regression curve *indicates the total amount of decay, not the rate of decay* that may have occurred in the past. *Regardless of the past rate of decay, the isochron graph would appear the same.*

  The RATE research team has found substantial physical evidence that decay rates were, in fact, much higher in the past—that accelerated nuclear decay would have been associated primarily with creation week and to a lesser extent, the worldwide flood.

  The larger portion of accelerated decay would have occurred during creation week as original rocks were being formed—a period when the universe was expanding and accelerating from an extremely hot, dense phase when matter and energy were concentrated—in secular vernacular, during the Big Bang. Such evidences are discussed in chapter 8, Accelerated Nuclear Decay.

  Accelerated decay during creation and during a worldwide flood would *completely invalidate the isochron technique.* A higher rate of decay in the past would yield a greater amount of daughter isotopes (decayed end products) in a short period of time and much higher ages—millions of years of age rather than thousands of years. The presence of daughter isotopes at the start (beginning of creation or start of the worldwide flood) represents millions or billions of years instead of zero age—this discrepancy cannot be detected or accounted for by the isochron technique. See Assumption 3.

> **Discordance**— is a term that describes data (age of rocks) from different radioisotope techniques that do not agree with known historical dates, or with each other, and data using the same technique on different specimens of the same rock that do not agree with each other.

- ***Assumption 2*** (No loss or gain of parent or daughter isotopes, assumes a closed system) – As stated in the first assumption, the isochron technique assumes neither the parent nor daughter isotopes (including the reactive and mobile intermediate isotopes, radium-226 and radon-222) were gained or lost through migration or contamination (mixing of magma) over the supposed many millions of years of earth's history. This assumes no tectonic activity (that is, no uplift, folding, and thrusting of the earth's crust in forming mountain ranges). With such mixing, rocks are "open systems" (not cogenetic) which invalidates the radioisotope technique. And, as previously stated, if mixing or contamination has not occurred, then the whole rock data should lie on a "single point." This assumption is invalid and the technique is "fatally flawed."

- ***Assumption 3*** (Known amounts of daughter isotopes present at the start) – In order for the isochron technique to work, secular geologists must assume no daughter isotopes at the start (that is, assume that melting resets the clock to zero) or be able to account for inherited daughter isotopes. If daughter material (for instance, lead-206) was present at the start, the rock would have the appearance of age. The isochron technique assumes it can test for inherited daughter isotopes, but it consistently finds millions of years of "age" from recently solidified magma. Evidence indicates that melting does not reset the clock to zero and isochron dating is simply unable to differentiate between inherited daughter isotopes and non-inherited daughter isotopes.

> The parent/daughter ratio (P/D) is used to determine or calculate the age of rock. If parent or daughter isotopes were added or removed as a result of leaching by groundwater, or if daughter isotopes were present at the start of solidification of magma, the P/D ratio would have nothing to do with age—the ratio would be *meaningless.*
>
> ---
>
> Parent isotopes (i.e., uranium, potassium, and rubidium) are more soluble in water in their oxidized form than their daughter isotopes (i.e., lead, argon, and strontium); thus, parents are more subject to groundwater leaching, especially in a flood environment.[7] If parent isotopes (P) were removed by leaching, the rock would date older than the actual age.

Rock samples formed from recently solidified magma should indicate no age—that is, the slope of the isochron graph should be flat and the y-axis intercept should indicate inherited daughter isotopes. Instead, the isochron graph indicates increasing daughter and parent isotopes along a straight line. But reality is not quite as simple as plotting present day isotope measurements and finding a regression line (equation). Mixing of magmas and accelerated nuclear decay during creation and the worldwide flood would result in "false isochrons" indicating millions of years instead of zero age.

The question is: Why should anyone believe any radioisotope date if dates from recently solidified rocks indicate vast ages? Daughter isotopes are present at the start contrary to assumption 3, and isochron dating is simply unable to differentiate between inherited daughter isotopes and non-inherited daughter isotopes.

Let's take a hypothetical look at the original basement (Precambrian) igneous rocks formed at the time of creation and at the beginning of the worldwide flood. When the universe was expanding and the original rocks were being formed, daughter isotopes (such as lead or argon) would have been present—created or derived from a variety of nuclear processes. To assume otherwise is wishful thinking. The presence of daughter isotopes at the start (day one of earth's formation) and mixing in "open systems" in the upper mantle would erroneously represent millions or billions of years of age, and would not be accounted for (as "inherited" daughter isotopes) in the isochron technique.

According to RATE project scientists, "It appears that all three of these dating essentials commonly fail at some level."[8] No analytical equipment or technique can 1) test the constancy of the decay rate, 2) determine if the system was open or closed, or 3) determine the initial condition in the rock. Isotope concentrations or ratios can be measured accurately, but isotope concentrations are not dates, but rather *interpretations of data based on assumptions that are plagued with multiple flaws*—**yet scientists contend as fact what they cannot prove.**

The bottom line is this: Ancient and recent lava flows have the same chemistry derived from the earth's upper mantle. P/D ratios have nothing to do with age of the rock—they are just ratios.

Scientists are simply unable to look at a rock specimen and tell you if the amount of daughter isotopes was the result of accelerated nuclear decay, or if there was mixing or inheritance of daughter isotopes from crustal rock at the time of cooling, or the amount of daughter isotopes at the start of creation and earth's formation. The bottom line is that these dates are *meaningless*.

---

notes: Appendix I

1. DeYoung, D. (May 2006). *Thousands ... Not Billions.* Green Forest, AR: Master Books, 120.
2. Austin, S.A. (April 1988). Grand Canyon lava flows: A survey of isotope dating methods. Institute for Creation Research, ICR *Impact*, Article 178, 4. Retrieved September 2009, from http://www.icr.org/article/grand-canyon-lava-flows-survey-isotope-dating-meth/; and Faure, G. (1986). *Principles of Isotope Geology.* (2nd edition). New York: John Wiley & Sons, 144-147.
3. DeYoung (2006), op. cit., 136, 177. Also see Morris (1994), op. cit., 59-60; Austin, Steven A., Ten misconceptions about the geologic column. ICR *Impact,* Article 137 (Item #8), November 1984. Retrieved from http://www.cnt.ru/users/chas/imp-137.htm; and Faure, G. and Mensing, T.M. (2005). *Isotopes: Principles and applications*. Hoboken, NJ: John Wiley and Sons.
4. Overn, W. (March 15, 2008, last modified). Isochron rock dating is fatally flawed. Twin Cities Creation Science Association. Retrieved December 2007, from http://www.tccsa.tc/articles/isochrons2.html.
5. Wieland, C. (2001). *Stones and Bones*. Green Forest, AR: Master Books, 37. Also see Snelling, A.A. (2000). Radioactive dating failure: Recent New Zealand lava flows yield ages of millions of years. *Creation,* 22 (1): 18-21; DeYoung (2006), 125-126, 178; Morris (1994), 54-55; and Morris (2003), 135.
6. DeYoung (2006), op. cit., 42, 119-120, 138-139, 177-178.
7. Sharp, D. (1986). *The Revolution Against Evolution*, Douglas Bruce Sharp, 17-18.
8. DeYoung (2006), op. cit., 139.

Subject Index

Note: Glossary pages are not listed in the subject index.

Name Index

Note: Glossary pages are not listed in the subject index.